Kawasaki 750
Air-cooled Fours
Owners Workshop Manual

by Pete Shoemark

Models covered

KZ/Z750E. UK 1980, US 1980 to 1982

KZ/Z750 H (Ltd) UK 1980 to 1982, US 1980 to 1983

KZ/Z750L. UK 1981 to 1987, US 1983

KZ/Z750R (GPz). UK and US 1982

ZX750A (GPz). UK 1983 to 1988, US 1983 to 1985

Z750P (GT750). UK 1982 to 1991

KZ750N (Spectre). US 1982 to 1983

KZ750F (Ltd Shaft). US 1983

All models have a 738 cc air-cooled engine

(574-3AG10)

© Haynes Publishing 2007

A book in the **Haynes Service and Repair Manual Series**

ISBN 978 1 85010 828 3

US Library of Congress Control Number 91-74069

British Library Cataloguing in Publication Data
A catalogue record for this book is available from the British Library.

Printed in the UK

Haynes Publishing
Sparkford, Yeovil, Somerset BA22 7JJ, England

Haynes North America, Inc
859 Lawrence Drive, Newbury Park, California 91320, USA

ABCDE
FGHIJ
KLMNO
PQ
3

Acknowledgements

Our thanks are due to Kawasaki Motors (UK) Limited who provided the Z750 R1 model featured in Chapters 1 to 6 and who supplied much of the technical data and line drawings. Thanks are also due to Eric Roberts of Bristol, who supplied the ZX750 A3 model featured in the supplementary Chapter, and to CW Motorcycles of Dorchester who supplied the Z750 P5 shown on the front cover. The Avon Rubber Company provided illustrations and advice about tyre fitting, and NGK Spark Plugs (UK) Ltd furnished information on spark plug grades and conditions.

About this manual

The purpose of this manual is to present the owner with a concise and graphic guide which will enable him to tackle any operation from basic routine maintenance to a major overhaul. It has been assumed that any work would be undertaken without the luxury of a well-equipped workshop and a range of manufacturer's service tools.

To this end, the machine featured in the manual was stripped and rebuilt in our own workshop, by a team comprising a mechanic, a photographer and the author. The resulting photographic sequence depicts events as they took place, the hands shown being those of the author and the mechanic.

The use of specialised, and expensive, service tools was avoided unless their use was considered to be essential due to risk of breakage or injury. There is usually some way of improvising a method of removing a stubborn component, providing that a suitable degree of care is exercised.

The author learnt his motorcycle mechanics over a number of years, faced with the same difficulties and using similar facilities to those encountered by most owners. It is hoped that this practical experience can be passed on through the pages of this manual.

Where possible, a well-used example of the machine is chosen for the workshop project, as this highlights any areas which might be particularly prone to giving rise to problems. In this way, any such difficulties are encountered and resolved before the text is written, and the techniques used to deal with them can be incorporated in the relevant section. Armed with a working knowledge of the machine, the author undertakes a considerable amount of research in order that the maximum amount of data can be included in the manual.

A comprehensive section, preceding the main part of the manual, describes procedures for carrying out the routine maintenance of the machine at intervals of time and mileage. This section is included particularly for those owners who wish to ensure the efficient day-to-day running of their motorcycle, but who choose not to undertake overhaul or renovation work.

Each Chapter is divided into numbered sections. Within these sections are numbered paragraphs. Cross reference throughout the manual is quite straightforward and logical. When reference is made 'See Section 6.10' it means Section 6, paragraph 10 in the same Chapter. If another Chapter were intended, the reference would read, for example, 'See Chapter 2, Section 6.10'. All the photographs are captioned with a section/paragraph number to which they refer and are relevant to the Chapter text adjacent.

Figures (usually line illustrations) appear in a logical but numerical order, within a given Chapter. Fig. 1.1 therefore refers to the first figure in Chapter 1.

Left-hand and right-hand descriptions of the machines and their components refer to the left and right of a given machine when the rider is seated normally.

Motorcycle manufacturers continually make changes to specifications and recommendations, and these, when notified, are incorporated into our manuals at the earliest opportunity.

Contents

1982 Z750 R1 (GPz750) model

Engine / transmission unit

Introduction to the Kawasaki 750 Fours

In 1972 Kawasaki launched their first four-cylinder four-stroke motorcycle, the 900cc Z1 model. It soon established itself as a fast and powerful machine, its dohc engine gaining the reputation of being unburstable. The Z1 evolved during the next decade, appearing in 1000cc and 1100cc guises, but the basic layout of the machine remained largely unchanged, an indication of how well the original formula worked.

When a 650cc dohc four appeared in 1976 it was of no surprise that it bore a close resemblance to the larger machines. Though lacking the sheer power of the Z1 it was nonetheless a creditable performer, and reflected the demands imposed by increasing oil prices during the latter part of the 1970s. The 650 model has appeared in a number of guises since 1976, and remains in production at the time of writing.

The KZ750 range (known as the Z750 range in the UK) was introduced in 1980 and it too has appeared in various configurations designed to appeal to a broad market. The various models reflect the growing sophistication in suspension technology and the tentative application of microprocessors in the machine's instrumentation.

The model prefixes and suffixes may appear somewhat confusing at first sight, especially where they do not coincide with the model badge on the side panel. To avoid complications a brief summary of the models covered by this manual appears below. Throughout this manual the single letter and number suffix (e.g E1, R1 etc) is used to identify the models in question.

Kawasaki KZ750 (Z750, UK) models 1980-82

1980 KZ750 E1 (Z750 E1, UK). Frame No KZ750 E-000001 onwards. The original KZ750, called a 'light compact sports model' by Kawasaki. It featured CV carburettors, air-assisted front forks, rear suspension units with adjustable damping, cast wheels, twin front and single rear disc brakes. Finished in 'Luminous Dark Red'. Chrome front mudguard.

1980 KZ750 H1 (Z750 H1, UK). Frame No KZ750 H-000001 onwards. LTD version of the above model distinguished by high pullback handlebars and more pronounced stepped seat with separate front and rear cushion sections. Finished in 'Ebony'. Sidepanel marked 'LTD 750'. Chrome front mudguard.

1981 KZ750 E2 (E1 continued for UK). Frame No KZ750 E-012001 or JKAKZDE1 – BA013722 onwards: As E1 model except colour and graphics changes and introduction of low fuel level warning system. Finished in 'Candy Persimmon Red' or 'Moondust Silver'. Painted front mudguard.

1981 KZ750 H2 (Z750 H2, UK). Frame No KZ750 H-013501 or JKAKZDH1 – BA017121 onwards: As H1 model except colour and graphics changes and introduction of self-cancelling direction indicators. Finished in 'Ebony' or 'Luminous Midnight Red'. Chrome front mudguard.

1981 Z750 L1 (Not sold in US). Frame No KZ750 E-012001 onwards. 'European' styled version of the basic 'E' model. Painted parts finished in 'Luminous gun blue' or 'Moondust silver'.

1982 KZ750 E3 (Not sold in UK). Frame No JKAKZDE1 – CA032401 onwards: Similar to E2 model, no detail change information available.

1982 KZ750 H3 (Z750 H3, UK). Frame No KZ750 H-031601 or JKAKZDH1 – CA031601 onwards: Similar to H2 with minor modifications and additions.

1982 Z750 L2 (Not sold in US). Frame No KZ750 E-032401 onwards. Similar to L1 model.

1982 KZ750 R1 (Z750 R1, UK). Frame No KZ750 R-000001 or JKAKZDR1 – CA000001. The GPz 750 model.

1982 KZ750 N1 (Not sold in UK). Frame No JKAKZDN1 – CA000001. Shaft drive Spectre model.

1982 Z750 P1 (Not sold in US). Frame No KZ750 P-000001. Shaft drive GT750 model.

Later models, from 1983 onwards, are covered in the supplementary Chapter at the end of this manual.

Model dimensions and weights

	E1,E2,E3,L1,L2	H1,H2,H3	R1	N1	P1
Overall length	2192 mm (86.3 in)	2195 mm (86.4 in)	2170 mm (85.4 in)	2205 mm (86.8 in)	2255 mm (88.8 in)
Overall width	780 mm (30.7 in)	810 mm (31.9 in)	780 mm (30.7 in)	860 mm (33.9 in)	760 mm (29.9 in)
Overall height	1135 mm (44.7 in)	1235 mm (48.6 in)	1220 mm (48.0 in)	1180 mm (46.5 in)	1105 mm (43.5 in)
Wheelbase	1420 mm (55.9 in)	1450 mm (57.1 in)	1460 mm (57.5 in)	1484 mm (58.4 in)	1480 mm (58.3 in)
Ground clearance	150 mm (5.9 in)	155 mm (6.1 in)	140 mm (5.5 in)	150 mm (5.9 in)	150 mm (5.9 in)
Dry weight	210 kg (463 lb)	211 kg (465 lb)	217 kg (478 lb)	219 kg (482 lb)	220 kg (485 lb)

Note: slight variations in dimensions may be found depending on the country or state of original importation

Ordering spare parts

When ordering spare parts for the KZ750 models it is recommended that the owner should deal with an official Kawasaki dealer. Most parts should be available from stock, though a number of less frequently needed items will have to be ordered. Be wary of using 'pattern' parts. Whilst in many cases these will will be of acceptable quality, there is always some risk of failure, perhaps causing extensive damage to the machine or even the rider. Note also that the use of non-official spares will invalidate the warranty.

Retain any worn or broken parts until the correct replacements have been obtained; they may prove essential to identify the correct replacement where detail design changes have taken place. When ordering, always quote the engine and frame numbers in full. These are located on the crankcase and steering head respectively.

Some of the more expendable parts such as spark plugs, bulbs, tyres, oils and greases etc., can be obtained from accessory shops and motor factors, who have convenient opening hours, charge lower prices and can often be found not far from home. It is possible to obtain parts on Mail Order basis from a number of specialists who advertise regularly in the motor cycle magazines.

Engine number is stamped on raised boss near oil filler cap

Frame number is stamped on the right-hand side of the steering head

Safety first!

Professional motor mechanics are trained in safe working procedures. However enthusiastic you may be about getting on with the job in hand, do take the time to ensure that your safety is not put at risk. A moment's lack of attention can result in an accident, as can failure to observe certain elementary precautions.

There will always be new ways of having accidents, and the following points do not pretend to be a comprehensive list of all dangers; they are intended rather to make you aware of the risks and to encourage a safety-conscious approach to all work you carry out on your vehicle.

Essential DOs and DON'Ts

DON'T start the engine without first ascertaining that the transmission is in neutral.

DON'T suddenly remove the filler cap from a hot cooling system – cover it with a cloth and release the pressure gradually first, or you may get scalded by escaping coolant.

DON'T attempt to drain oil until you are sure it has cooled sufficiently to avoid scalding you.

DON'T grasp any part of the engine, exhaust or silencer without first ascertaining that it is sufficiently cool to avoid burning you.

DON'T allow brake fluid or antifreeze to contact the machine's paintwork or plastic components.

DON'T syphon toxic liquids such as fuel, brake fluid or antifreeze by mouth, or allow them to remain on your skin.

DON'T inhale dust – it may be injurious to health (see *Asbestos* heading).

DON'T allow any spilt oil or grease to remain on the floor – wipe it up straight away, before someone slips on it.

DON'T use ill-fitting spanners or other tools which may slip and cause injury.

DON'T attempt to lift a heavy component which may be beyond your capability – get assistance.

DON'T rush to finish a job, or take unverified short cuts.

DON'T allow children or animals in or around an unattended vehicle.

DON'T inflate a tyre to a pressure above the recommended maximum. Apart from overstressing the carcase and wheel rim, in extreme cases the tyre may blow off forcibly.

DO ensure that the machine is supported securely at all times. This is especially important when the machine is blocked up to aid wheel or fork removal.

DO take care when attempting to slacken a stubborn nut or bolt. It is generally better to pull on a spanner, rather than push, so that if slippage occurs you fall away from the machine rather than on to it.

DO wear eye protection when using power tools such as drill, sander, bench grinder etc.

DO use a barrier cream on your hands prior to undertaking dirty jobs – it will protect your skin from infection as well as making the dirt easier to remove afterwards; but make sure your hands aren't left slippery. Note that long-term contact with used engine oil can be a health hazard.

DO keep loose clothing (cuffs, tie etc) and long hair well out of the way of moving mechanical parts.

DO remove rings, wristwatch etc, before working on the vehicle – especially the electrical system.

DO keep your work area tidy – it is only too easy to fall over articles left lying around.

DO exercise caution when compressing springs for removal or installation. Ensure that the tension is applied and released in a controlled manner, using suitable tools which preclude the possibility of the spring escaping violently.

DO ensure that any lifting tackle used has a safe working load rating adequate for the job.

DO get someone to check periodically that all is well, when working alone on the vehicle.

DO carry out work in a logical sequence and check that everything is correctly assembled and tightened afterwards.

DO remember that your vehicle's safety affects that of yourself and others. If in doubt on any point, get specialist advice.

IF, in spite of following these precautions, you are unfortunate enough to injure yourself, seek medical attention as soon as possible.

Asbestos

Certain friction, insulating, sealing, and other products – such as brake linings, clutch linings, gaskets, etc – contain asbestos. *Extreme care must be taken to avoid inhalation of dust from such products since it is hazardous to health.* If in doubt, assume that they *do* contain asbestos.

Fire

Remember at all times that petrol (gasoline) is highly flammable. Never smoke, or have any kind of naked flame around, when working on the vehicle. But the risk does not end there – a spark caused by an electrical short-circuit, by two metal surfaces contacting each other, by careless use of tools, or even by static electricity built up in your body under certain conditions, can ignite petrol vapour, which in a confined space is highly explosive.

Always disconnect the battery earth (ground) terminal before working on any part of the fuel or electrical system, and never risk spilling fuel on to a hot engine or exhaust.

It is recommended that a fire extinguisher of a type suitable for fuel and electrical fires is kept handy in the garage or workplace at all times. Never try to extinguish a fuel or electrical fire with water.

Note: *Any reference to a 'torch' appearing in this manual should always be taken to mean a hand-held battery-operated electric lamp or flashlight. It does* **not** *mean a welding/gas torch or blowlamp.*

Fumes

Certain fumes are highly toxic and can quickly cause unconsciousness and even death if inhaled to any extent. Petrol (gasoline) vapour comes into this category, as do the vapours from certain solvents such as trichloroethylene. Any draining or pouring of such volatile fluids should be done in a well ventilated area.

When using cleaning fluids and solvents, read the instructions carefully. Never use materials from unmarked containers – they may give off poisonous vapours.

Never run the engine of a motor vehicle in an enclosed space such as a garage. Exhaust fumes contain carbon monoxide which is extremely poisonous; if you need to run the engine, always do so in the open air or at least have the rear of the vehicle outside the workplace.

The battery

Never cause a spark, or allow a naked light, near the vehicle's battery. It will normally be giving off a certain amount of hydrogen gas, which is highly explosive.

Always disconnect the battery earth (ground) terminal before working on the fuel or electrical systems.

If possible, loosen the filler plugs or cover when charging the battery from an external source. Do not charge at an excessive rate or the battery may burst.

Take care when topping up and when carrying the battery. The acid electrolyte, even when diluted, is very corrosive and should not be allowed to contact the eyes or skin.

If you ever need to prepare electrolyte yourself, always add the acid slowly to the water, and never the other way round. Protect against splashes by wearing rubber gloves and goggles.

Mains electricity and electrical equipment

When using an electric power tool, inspection light etc, always ensure that the appliance is correctly connected to its plug and that, where necessary, it is properly earthed (grounded). Do not use such appliances in damp conditions and, again, beware of creating a spark or applying excessive heat in the vicinity of fuel or fuel vapour. Also ensure that the appliances meet the relevant national safety standards.

Ignition HT voltage

A severe electric shock can result from touching certain parts of the ignition system, such as the HT leads, when the engine is running or being cranked, particularly if components are damp or the insulation is defective. Where an electronic ignition system is fitted, the HT voltage is much higher and could prove fatal.

Tools and working facilities

The first priority when undertaking maintenance or repair work of any sort on a motorcycle is to have a clean, dry, well-lit working area. Work carried out in peace and quiet in the well-ordered atmosphere of a good workshop will give more satisfaction and much better results than can usually be achieved in poor working conditions. A good workshop must have a clean flat workbench or a solidly constructed table of convenient working height. The workbench or table should be equipped with a vice which has a jaw opening of at least 4 in (100 mm). A set of jaw covers should be made from soft metal such as aluminium alloy or copper, or from wood. These covers will minimise the marking or damaging of soft or delicate components which may be clamped in the vice. Some clean, dry, storage space will be required for tools, lubricants and dismantled components. It will be necessary during a major overhaul to lay out engine/gearbox components for examination and to keep them where they will remain undisturbed for as long as is necessary. To this end it is recommended that a supply of metal or plastic containers of suitable size is collected. A supply of clean, lint-free, rags for cleaning purposes and some newspapers, other rags, or paper towels for mopping up spillages should also be kept. If working on a hard concrete floor note that both the floor and one's knees can be protected from oil spillages and wear by cutting open a large cardboard box and spreading it flat on the floor under the machine or workbench. This also helps to provide some warmth in winter and to prevent the loss of nuts, washers, and other tiny components which have a tendency to disappear when dropped on anything other than a perfectly clean, flat, surface.

Unfortunately, such working conditions are not always available to the home mechanic. When working in poor conditions it is essential to take extra time and care to ensure that the components being worked on are kept scrupulously clean and to ensure that no components or tools are lost or damaged.

A selection of good tools is a fundamental requirement for anyone contemplating the maintenance and repair of a motor vehicle. For the owner who does not possess any, their purchase will prove a considerable expense, offsetting some of the savings made by doing-it-yourself. However, provided that the tools purchased meet the relevant national safety standards and are of good quality, they will last for many years and prove an extremely worthwhile investment.

To help the average owner to decide which tools are needed to carry out the various tasks detailed in this manual, we have compiled three lists of tools under the following headings: *Maintenance and minor repair, Repair and overhaul,* and *Specialized.* The newcomer to practical mechanics should start off with the simpler jobs around the vehicle. Then, as his confidence and experience grow, he can undertake more difficult tasks, buying extra tools as and when they are needed. In this way, a *Maintenance and minor repair* tool kit can be built-up into a *Repair and overhaul* tool kit over a considerable period of time without any major cash outlays. The experienced home mechanic will have a tool kit good enough for most repair and overhaul procedures and will add tools from the specialized category when he feels the expense is justified by the amount of use these tools will be put to.

It is obviously not possible to cover the subject of tools fully here. For those who wish to learn more about tools and their use there is a book entitled *Motorcycle Workshop Practice Manual* available from the publishers of this manual. It also provides an introduction to basic workshop practice which will be of interest to a home mechanic working on any type of motor vehicle.

As a general rule, it is better to buy the more expensive, good quality tools. Given reasonable use, such tools will last for a very long time, whereas the cheaper, poor quality, item will wear out faster and need to be renewed more often, thus nullifying the original saving. There is also the risk of a poor quality tool breaking while in use, causing personal injury or expensive damage to the component being worked on.

For practically all tools, a tool factor is the best source since he will have a very comprehensive range compared with the average garage or accessory shop. Having said that, accessory shops often offer excellent quality tools at discount prices, so it pays to shop around. There are plenty of tools around at reasonable prices, but always aim to purchase items which meet the relevant national safety standards. If in doubt, seek the advice of the shop propietor or manager before making a purchase.

The basis of any toolkit is a set of spanners. While open-ended spanners with their slim jaws, are useful for working on awkwardly-positioned nuts, ring spanners have advantages in that they grip the nut far more positively. There is less risk of the spanner slipping off the nut and damaging it, for this reason alone ring spanners are to be preferred. Ideally, the home mechanic should acquire a set of each, but if expense rules this out a set of combination spanners (open-ended at one end and with a ring of the same size at the other) will provide a good compromise. Another item which is so useful it should be considered an essential requirement for any home mechanic is a set of socket spanners. These are available in a variety of drive sizes. It is recommended that the ½-inch drive type is purchased to begin with as although bulkier and more expensive than the ⅜-inch type, the larger size is far more common and will accept a greater variety of torque wrenches, extension pieces and socket sizes. The socket set should comprise sockets of sizes between 8 and 24 mm, a reversible ratchet drive, an extension bar of about 10 inches in length, a spark plug socket with a rubber insert, and a universal joint. Other attachments can be added to the set at a later date.

Maintenance and minor repair tool kit

> Set of spanners 8 – 24 mm
> Set of sockets and attachments
> Spark plug spanner with rubber insert – 10, 12, or 14 mm
> as appropriate
> Adjustable spanner
> C-spanner/pin spanner
> Torque wrench (same size drive as sockets)
> Set of screwdrivers (flat blade)
> Set of screwdrivers (cross-head)
> Set of Allen keys 4 – 10 mm
> Impact screwdriver and bits
> Ball pein hammer – 2 lb
> Hacksaw (junior)
> Self-locking pliers – Mole grips or vice grips
> Pliers – combination
> Pliers – needle nose
> Wire brush (small)
> Soft-bristled brush
> Tyre pump
> Tyre pressure gauge
> Tyre tread depth gauge
> Oil can
> Fine emery cloth
> Funnel (medium size)
> Drip tray
> Grease gun
> Set of feeler gauges
> Brake bleeding kit
> Strobe timing light
> Continuity tester (dry battery and bulb)
> Soldering iron and solder
> Wire stripper or craft knife
> PVC insulating tape
> Assortment of split pins, nuts, bolts, and washers

Repair and overhaul toolkit

The tools in this list are virtually essential for anyone undertaking major repairs to a motorcycle and are additional to the tools listed above. Concerning Torx driver bits, Torx screws are encountered on some of the more modern machines where their use is restricted to fastening certain components inside the engine/gearbox unit. It is therefore recommended that if Torx bits cannot be borrowed from a local dealer, they are purchased individually as the need arises. They are not in regular use in the motor trade and will therefore only be available in specialist tool shops.

> Plastic or rubber soft-faced mallet
> Torx driver bits
> Pliers – electrician's side cutters
> Circlip pliers – internal (straight or right-angled tips are
> available)
> Circlip pliers – external
> Cold chisel
> Centre punch
> Pin punch
> Scriber
> Scraper (made from soft metal such as aluminium
> or copper)
> Soft metal drift
> Steel rule/straight edge
> Assortment of files
> Electric drill and bits

> Wire brush (large)
> Soft wire brush (similar to those used for cleaning suede
> shoes)
> Sheet of plate glass
> Hacksaw (large)
> Valve grinding tool
> Valve grinding compound (coarse and fine)
> Stud extractor set (E-Z out)

Specialized tools

This is not a list of the tools made by the machine's manufacturer to carry out a specific task on a limited range of models. Occasional references are made to such tools in the text of this manual and, in general, an alternative method of carrying out the task without the manufacturer's tool is given where possible. The tools mentioned in this list are those which are not used regularly and are expensive to buy in view of their infrequent use. Where this is the case it may be possible to hire or borrow the tools against a deposit from a local dealer or tool hire shop. An alternative is for a group of friends or a motorcycle club to join in the purchase.

> Valve spring compressor
> Piston ring compressor
> Universal bearing puller
> Cylinder bore honing attachment (for electric drill)
> Micrometer set
> Vernier calipers
> Dial gauge set
> Cylinder compression gauge
> Vacuum gauge set
> Multimeter
> Dwell meter/tachometer

Care and maintenance of tools

Whatever the quality of the tools purchased, they will last much longer if cared for. This means in practice ensuring that a tool is used for its intended purpose; for example screwdrivers should not be used as a substitute for a centre punch, or as chisels. Always remove dirt or grease and any metal particles but remember that a light film of oil will prevent rusting if the tools are infrequently used. The common tools can be kept together in a large box or tray but the more delicate, and more expensive, items should be stored separately where they cannot be damaged. When a tool is damaged or worn out, be sure to renew it immediately. It is false economy to continue to use a worn spanner or screwdriver which may slip and cause expensive damage to the component being worked on.

Fastening systems

Fasteners, basically, are nuts, bolts and screws used to hold two or more parts together. There are a few things to keep in mind when working with fasteners. Almost all of them use a locking device of some type; either a lock washer, lock nut, locking tab or thread adhesive. All threaded fasteners should be clean, straight, have undamaged threads and undamaged corners on the hexagon head where the spanner fits. Develop the habit of replacing all damaged nuts and bolts with new ones.

Rusted nuts and bolts should be treated with a rust penetrating fluid to ease removal and prevent breakage. After applying the rust penetrant, let it 'work' for a few minutes before trying to loosen the nut or bolt. Badly rusted fasteners may have to be chiseled off or removed with a special nut breaker, available at tool shops.

Flat washers and lock washers, when removed from an assembly should always be replaced exactly as removed. Replace any damaged washers with new ones. Always use a flat washer between a lock washer and any soft metal surface (such as aluminium), thin sheet metal or plastic. Special lock nuts can only be used once or twice before they lose their locking ability and must be renewed.

If a bolt or stud breaks off in an assembly, it can be drilled out and removed with a special tool called an E-Z out. Most dealer service departments and motorcycle repair shops can perform this task, as well as others (such as the repair of threaded holes that have been stripped out).

Spanner size comparison

Jaw gap (in)	Spanner size
0.250	$\frac{1}{4}$ in AF
0.276	7 mm
0.313	$\frac{5}{16}$ in AF
0.315	8 mm
0.344	$\frac{11}{32}$ in AF; $\frac{1}{8}$ in Whitworth
0.354	9 mm
0.375	$\frac{3}{8}$ in AF
0.394	10 mm
0.433	11 mm
0.438	$\frac{7}{16}$ in AF
0.445	$\frac{3}{16}$ in Whitworth; $\frac{1}{4}$ in BSF
0.472	12 mm
0.500	$\frac{1}{2}$ in AF
0.512	13 mm
0.525	$\frac{1}{4}$ in Whitworth; $\frac{5}{16}$ in BSF
0.551	14 mm
0.563	$\frac{9}{16}$ in AF
0.591	15 mm
0.600	$\frac{5}{16}$ in Whitworth; $\frac{3}{8}$ in BSF
0.625	$\frac{5}{8}$ in AF
0.630	16 mm
0.669	17 mm
0.686	$\frac{11}{16}$ in AF
0.709	18 mm
0.710	$\frac{3}{8}$ in Whitworth; $\frac{7}{16}$ in BSF
0.748	19 mm
0.750	$\frac{3}{4}$ in AF
0.813	$\frac{13}{16}$ in AF
0.820	$\frac{7}{16}$ in Whitworth; $\frac{1}{2}$ in BSF
0.866	22 mm
0.875	$\frac{7}{8}$ in AF
0.920	$\frac{1}{2}$ in Whitworth; $\frac{9}{16}$ in BSF
0.938	$\frac{15}{16}$ in AF
0.945	24 mm
1.000	1 in AF
1.010	$\frac{9}{16}$ in Whitworth; $\frac{5}{8}$ in BSF
1.024	26 mm
1.063	$1\frac{1}{16}$ in AF; 27 mm
1.100	$\frac{5}{8}$ in Whitworth; $\frac{11}{16}$ in BSF
1.125	$1\frac{1}{8}$ in AF
1.181	30 mm
1.200	$\frac{11}{16}$ in Whitworth; $\frac{3}{4}$ in BSF
1.250	$1\frac{1}{4}$ in AF
1.260	32 mm
1.300	$\frac{3}{4}$ in Whitworth; $\frac{7}{8}$ in BSF
1.313	$1\frac{5}{16}$ in AF
1.390	$\frac{13}{16}$ in Whitworth; $\frac{15}{16}$ in BSF
1.417	36 mm
1.438	$1\frac{7}{16}$ in AF
1.480	$\frac{7}{8}$ in Whitworth; 1 in BSF
1.500	$1\frac{1}{2}$ in AF
1.575	40 mm; $\frac{15}{16}$ in Whitworth
1.614	41 mm
1.625	$1\frac{5}{8}$ in AF
1.670	1 in Whitworth; $1\frac{1}{8}$ in BSF
1.688	$1\frac{11}{16}$ in AF
1.811	46 mm
1.813	$1\frac{13}{16}$ in AF
1.860	$1\frac{1}{8}$ in Whitworth; $1\frac{1}{4}$ in BSF
1.875	$1\frac{7}{8}$ in AF
1.969	50 mm
2.000	2 in AF
2.050	$1\frac{1}{4}$ in Whitworth; $1\frac{3}{8}$ in BSF
2.165	55 mm
2.362	60 mm

Standard torque settings

Specific torque settings will be found at the end of the specifications section of each chapter. Where no figure is given, bolts should be secured according to the table below.

Fastener type (thread diameter)	kgf m	lbf ft
5mm bolt or nut	0.45 – 0.6	3.5 – 4.5
6 mm bolt or nut	0.8 – 1.2	6 – 9
8 mm bolt or nut	1.8 – 2.5	13 – 18
10 mm bolt or nut	3.0 – 4.0	22 – 29
12 mm bolt or nut	5.0 – 6.0	36 – 43
5 mm screw	0.35 – 0.5	2.5 – 3.6
6 mm screw	0.7 – 1.1	5 – 8
6 mm flange bolt	1.0 – 1.4	7 – 10
8 mm flange bolt	2.4 – 3.0	17 – 22
10 mm flange bolt	3.0 – 4.0	22 – 29

Choosing and fitting accessories

The range of accessories available to the modern motorcyclist is almost as varied and bewildering as the range of motorcycles. This Section is intended to help the owner in choosing the correct equipment for his needs and to avoid some of the mistakes made by many riders when adding accessories to their machines. It will be evident that the Section can only cover the subject in the most general terms and so it is recommended that the owner, having decided that he wants to fit, for example, a luggage rack or carrier, seeks the advice of several local dealers and the owners of similar machines. This will give a good idea of what makes of carrier are easily available, and at what price. Talking to other owners will give some insight into the drawbacks or good points of any one make. A walk round the motorcycles in car parks or outside a dealer will often reveal the same sort of information.

The first priority when choosing accessories is to assess exactly what one needs. It is, for example, pointless to buy a large heavy-duty carrier which is designed to take the weight of fully laden panniers and topbox when all you need is a place to strap on a set of waterproofs and a lunchbox when going to work. Many accessory manufacturers have ranges of equipment to cater for the individual needs of different riders and this point should be borne in mind when looking through a dealer's catalogues. Having decided exactly what is required and the use to which the accessories are going to be put, the owner will need a few hints on what to look for when making the final choice. To this end the Section is now sub-divided to cover the more popular accessories fitted. Note that it is in no way a customizing guide, but merely seeks to outline the practical considerations to be taken into account when adding aftermarket equipment to a motorcycle.

Fairings and windscreens

A fairing is possibly the single, most expensive, aftermarket item to be fitted to any motorcycle and, therefore, requires the most thought before purchase. Fairings can be divided into two main groups: front fork mounted handlebar fairings and windscreens, and frame mounted fairings.

The first group, the front fork mounted fairings, are becoming far more popular than was once the case, as they offer several advantages over the second group. Front fork mounted fairings generally are much easier and quicker to fit, involve less modification to the motorcycle, do not as a rule restrict the steering lock, permit a wider selection of handlebar styles to be used, and offer adequate protection for much less money than the frame mounted type. They are also lighter, can be swapped easily between different motorcycles, and are available in a much greater variety of styles. Their main disadvantages are that they do not offer as much weather protection as the frame mounted types, rarely offer any storage space, and, if poorly fitted or naturally incompatible, can have an adverse effect on the stability of the motorcycle.

The second group, the frame mounted fairings, are secured so rigidly to the main frame of the motorcycle that they can offer a substantial amount of protection to motorcycle and rider in the event of a crash. They offer almost complete protection from the weather and, if double-skinned in construction, can provide a great deal of useful storage space. The feeling of peace, quiet and complete relaxation encountered when riding behind a good full fairing has to be experienced to be believed. For this reason full fairings are considered essential by most touring motorcyclists and by many people who ride all year round. The main disadvantages of this type are that fitting can take a long time, often involving removal or modification of standard motorcycle components, they restrict the steering lock and they can add up to about 40 lb to the weight of the machine. They do not usually affect the stability of the machine to any great extent once the front tyre pressure and suspension have been adjusted to compensate for the extra weight, but can be affected by sidewinds.

The first thing to look for when purchasing a fairing is the quality of the fittings. A good fairing will have strong, substantial brackets constructed from heavy-gauge tubing; the brackets must be shaped to fit the frame or forks evenly so that the minimum of stress is imposed on the assembly when it is bolted down. The brackets should be properly painted or finished — a nylon coating being the favourite of the better manufacturers — the nuts and bolts provided should be of the same thread and size standard as is used on the motorcycle and be properly plated. Look also for shakeproof locking nuts or locking washers to ensure that everything remains securely tightened down. The fairing shell is generally made from one of two materials: fibreglass or ABS plastic. Both have their advantages and disadvantages, but the main consideration for the owner is that fibreglass is much easier to repair in the event of damage occurring to the fairing. Whichever material is used, check that it is properly finished inside as well as out, that the edges are protected by beading and that the fairing shell is insulated from vibration by the use of rubber grommets at all mounting points. Also be careful to check that the windscreen is retained by plastic bolts which will snap on impact so that the windscreen will break away and not cause personal injury in the event of an accident.

Having purchased your fairing or windscreen, read the manufacturer's fitting instructions very carefully and check that you have all the necessary brackets and fittings. Ensure that the mounting brackets are located correctly and bolted down securely. Note that some manufacturers use hose clamps to retain the mounting brackets; these should be discarded as they are convenient to use but not strong enough for the task. Stronger clamps should be substituted; car exhaust pipe clamps of suitable size would be a good alternative. Ensure that the front forks can turn through the full steering lock available without fouling the fairing. With many types of frame-mounted fairing the handlebars will have to be altered or a different type fitted and the steering lock will be restricted by stops provided with the fittings. Also check that the fairing does not foul the front wheel or mudguard, in any steering position, under full fork compression. Re-route any cables, brake pipes or electrical wiring which may snag on the fairing and take great care to protect all electrical connections, using insulating tape. If the manufacturer's instructions are followed carefully at every stage no serious problems should be encountered. Remember that hydraulic pipes that have been disconnected must be carefully re-tightened and the hydraulic system purged of air bubbles by bleeding.

Two things will become immediately apparent when taking a motorcycle on the road for the first time with a fairing – the first is the tendency to underestimate the road speed because of the lack of wind pressure on the body. This must be very carefully watched until one has grown accustomed to riding behind the fairing. The second thing is the alarming increase in engine noise which is an unfortunate but inevitable by-product of fitting any type of fairing or windscreen, and is caused by normal engine noise being reflected, and in some cases amplified, by the flat surface of the fairing.

Luggage racks or carriers

Carriers are possibly the commonest item to be fitted to modern motorcycles. They vary enormously in size, carrying capacity, and durability. When selecting a carrier, always look for one which is made specifically for your machine and which is bolted on with as few separate brackets as possible. The universal-type carrier, with its mass of brackets and adaptor pieces, will generally prove too weak to be of any real use. A good carrier should bolt to the main frame, generally using the two suspension unit top mountings and a mudguard mounting bolt as attachment points, and have its luggage platform as low and as far forward as possible to minimise the effect of any load on the machine's stability. Look for good quality, heavy gauge tubing, good welding and good finish. Also ensure that the carrier does not prevent opening of the seat, sidepanels or tail compartment, as appropriate. When using a carrier, be very careful not to overload it. Excessive weight placed so high and so far to the rear of any motorcycle will have an adverse effect on the machine's steering and stability.

Luggage

Motorcycle luggage can be grouped under two headings: soft and hard. Both types are available in many sizes and styles and have advantages and disadvantages in use.

Soft luggage is now becoming very popular because of its lower cost and its versatility. Whether in the form of tankbags, panniers, or strap-on bags, soft luggage requires in general no brackets and no modification to the motorcycle. Equipment can be swapped easily from one motorcycle to another and can be fitted and removed in seconds. Awkwardly shaped loads can easily be carried. The disadvantages of soft luggage are that the contents cannot be secure against the casual thief, very little protection is afforded in the event of a crash, and waterproofing is generally poor. Also, in the case of panniers, carrying capacity is restricted to approximately 10 lb, although this amount will vary considerably depending on the manufacturer's recommendation. When purchasing soft luggage, look for good quality material, generally vinyl or nylon, with strong, well-stitched attachment points. It is always useful to have separate pockets, especially on tank bags, for items which will be needed on the journey. When purchasing a tank bag, look for one which has a separate, well-padded, base. This will protect the tank's paintwork and permit easy access to the filler cap at petrol stations.

Hard luggage is confined to two types: panniers, and top boxes or tail trunks. Most hard luggage manufacturers produce matching sets of these items, the basis of which is generally that manufacturer's own heavy-duty luggage rack. Variations on this theme occur in the form of separate frames for the better quality panniers, fixed or quickly-detachable luggage, and in size and carrying capacity. Hard luggage offers a reasonable degree of security against theft and good protection against weather and accident damage. Carrying capacity is greater than that of soft luggage, around 15 – 20 lb in the case of panniers, although top boxes should never be loaded as much as their apparent capacity might imply. A top box should only be used for lightweight items, because one that is heavily laden can have a serious effect on the stability of the machine. When purchasing hard luggage look for the same good points as mentioned under fairings and windscreens, ie good quality mounting brackets and fittings, and well-finished fibreglass or ABS plastic cases. Again as with fairings, always purchase luggage made specifically for your motorcycle, using as few separate brackets as possible, to ensure that everything remains securely bolted in place. When fitting hard luggage, be careful to check that the rear suspension and brake operation will not be impaired in any way and remember that many pannier kits require re-siting of the indicators. Remember also that a non-standard exhaust system may make fitting extremely difficult.

Handlebars

The occupation of fitting alternative types of handlebar is extremely popular with modern motorcyclists, whose motives may vary from the purely practical, wishing to improve the comfort of their machines, to the purely aesthetic, where form is more important than function. Whatever the reason, there are several considerations to be borne in mind when changing the handlebars of your machine. If fitting lower bars, check carefully that the switches and cables do not foul the petrol tank on full lock and that the surplus length of cable, brake pipe, and electrical wiring are smoothly and tidily disposed of. Avoid tight kinks in cable or brake pipes which will produce stiff controls or the premature and disastrous failure of an overstressed component. If necessary, remove the petrol tank and re-route the cable from the engine/gearbox unit upwards, ensuring smooth gentle curves are produced. In extreme cases, it will be necessary to purchase a shorter brake pipe to overcome this problem. In the case of higher handlebars than standard it will almost certainly be necessary to purchase extended cables and brake pipes. Fortunately, many standard motorcycles have a custom version which will be equipped with higher handlebars and, therefore, factory-built extended components will be available from your local dealer. It is not usually necessary to extend electrical wiring, as switch clusters may be used on several different motorcycles, some being custom versions. This point should be borne in mind however when fitting extremely high or wide handlebars.

When fitting different types of handlebar, ensure that the mounting clamps are correctly tightened to the manufacturer's specifications and that cables and wiring, as previously mentioned, have smooth easy runs and do not snag on any part of the motorcycle throughout the full steering lock. Ensure that the fluid level in the front brake master cylinder remains level to avoid any chance of air entering the hydraulic system. Also check that the cables are adjusted correctly and that all handlebar controls operate correctly and can be easily reached when riding.

Crashbars

Crashbars, also known as engine protector bars, engine guards, or case savers, are extremely useful items of equipment which can contribute protection to the machine's structure if a crash occurs. They do not, as has been inferred in the US, prevent the rider from crashing, or necessarily prevent rider injury should a crash occur.

It is recommended that only the smaller, neater, engine protector type of crashbar is considered. This type will offer protection while restricting, as little as is possible, access to the engine and the machine's ground clearance. The crashbars should be designed for use specifically on your machine, and should be constructed of heavy-gauge tubing with strong, integral mounting brackets. Where possible, they should bolt to a strong lug on the frame, usually at the engine mounting bolts.

The alternative type of crashbar is the larger cage type. This type is not recommended in spite of their appearance which promises some protection to the rider as well as to the machine. The larger amount of leverage imposed by the size of this type of crashbar increases the risk of severe frame damage in the event of an accident. This type also decreases the machine's ground clearance and restricts access to the engine. The amount of protection afforded the rider is open to some doubt as the design is based on the premise that the rider will stay in the normally seated position during an accident, and the crash bar structure will not itself fail. Neither result can in any way be guaranteed.

As a general rule, always purchase the best, ie usually the most expensive, set of crashbars you an afford. The investment will be repaid by minimising the amount of damage incurred, should the machine be involved in an accident. Finally, avoid the universal type of crashbar. This should be regarded only as a last resort to be used if no alternative exists. With its usual multitude of separate brackets and spacers, the universal crashbar is far too weak in design and construction to be of any practical value.

Exhaust systems

The fitting of aftermarket exhaust systems is another extremely popular pastime amongst motorcyclists. The usual motive is to gain more performance from the engine but other considerations are to gain more ground clearance, to lose weight from the motorcycle, to obtain a more distinctive exhaust note or to find a cheaper alternative to the manufacturer's original equipment exhaust system. Original equipment exhaust systems often cost more and may well have a relatively short life. It should be noted that it is rare for an aftermarket exhaust system alone to give a noticeable increase in the engine's power output. Modern motorcycles are designed to give the highest power output possible allowing for factors such as quietness, fuel economy, spread of power, and long-term reliability. If there were a magic formula which allowed the exhaust system to produce more power without affecting these other considerations you can be sure that the manufacturers, with their large research and development facilities, would have found it and made use of it. Performance increases of a worthwhile and noticeable nature only come from well-tried and properly matched modifications to the entire engine, from the air filter, through the carburettors, port timing or camshaft and valve design, combustion chamber shape, compression ratio, and the exhaust system. Such modifications are well outside the scope of this manual but interested owners might refer to specialist books produced by the publisher of this manual which go into the whole subject in great detail.

Whatever your motive for wishing to fit an alternative exhaust system, be sure to seek expert advice before doing so. Changes to the carburettor jetting will almost certainly be required for which you must consult the exhaust system manufacturer. If he cannot supply adequately specific information it is reasonable to assume that insufficient development work has been carried out, and that particular make should be avoided. Other factors to be borne in mind are whether the exhaust system allows the use of both centre and side stands, whether it allows sufficient access to permit oil and filter changing and whether modifications are necessary to the standard exhaust system. Many two-stroke expansion chamber systems require the use of the standard exhaust pipe; this is all very well if the standard exhaust pipe and silencer are separate units but can cause problems if the two, as with so many modern two-strokes, are a one-piece unit. While the exhaust pipe can be removed easily by means of a hacksaw it is not so easy to refit the original silencer should you at any time wish to return the machine to standard trim. The same applies to several four-stroke systems.

On the subject of the finish of aftermarket exhausts, avoid black-painted systems unless you enjoy painting. As any trail-bike owner will tell you, rust has a great affinity for black exhausts and re-painting or rust removal becomes a task which must be carried out with monotonous regularity. A bright chrome finish is, as a general rule, a far better proposition as it is much easier to keep clean and to prevent rusting. Although the general finish of aftermarket exhaust systems is not always up to the standard of the original equipment the lower cost of such systems does at least reflect this fact.

When fitting an alternative system always purchase a full set of new exhaust gaskets, to prevent leaks. Fit the exhaust first to the cylinder head or barrel, as appropriate, tightening the retaining nuts or bolts by hand only and then line up the exhaust rear mountings. If the new system is a one-piece unit and the rear mountings do not line up exactly, spacers must be fabricated to take up the difference. Do not force the system into place as the stress thus imposed will rapidly cause cracks and splits to appear. Once all the mountings are loosely fixed, tighten the retaining nuts or bolts securely, being careful not to overtighten them. Where the motorcycle manufacturer's torque

settings are available, these should be used. Do not forget to carry out any carburation changes recommended by the exhaust system's manufacturer.

Electrical equipment

The vast range of electrical equipment available to motorcyclists is so large and so diverse that only the most general outline can be given here. Electrical accessories vary from electric ignition kits fitted to replace contact breaker points, to additional lighting at the front and rear, more powerful horns, various instruments and gauges, clocks, anti-theft systems, heated clothing, CB radios, radio-cassette players, and intercom systems, to name but a few of the more popular items of equipment.

As will be evident, it would require a separate manual to cover this subject alone and this section is therefore restricted to outlining a few basic rules which must be borne in mind when fitting electrical equipment. The first consideration is whether your machine's electrical system has enough reserve capacity to cope with the added demand of the accessories you wish to fit. The motorcycle's manufacturer or importer should be able to furnish this sort of information and may also be able to offer advice on uprating the electrical system. Failing this, a good dealer or the accessory manufacturer may be able to help. In some cases, more powerful generator components may be available, perhaps from another motorcycle in the manufacturer's range. The second consideration is the legal requirements in force in your area. The local police may be prepared to help with this point. In the UK for example, there are strict regulations governing the position and use of auxiliary riding lamps and fog lamps.

When fitting electrical equipment always disconnect the battery first to prevent the risk of a short-circuit, and be careful to ensure that all connections are properly made and that they are waterproof. Remember that many electrical accessories are designed primarily for use in cars and that they cannot easily withstand the exposure to vibration and to the weather. Delicate components must be rubber-mounted to insulate them from vibration, and sealed carefully to prevent the entry of rainwater and dirt. Be careful to follow exactly the accessory manufacturer's instructions in conjunction with the wiring diagram at the back of this manual.

Accessories – general

Accessories fitted to your motorcycle will rapidly deteriorate if not cared for. Regular washing and polishing will maintain the finish and will provide an opportunity to check that all mounting bolts and nuts are securely fastened. Any signs of chafing or wear should be watched for, and the cause cured as soon as possible before serious damage occurs.

As a general rule, do not expect the re-sale value of your motorcycle to increase by an amount proportional to the amount of money and effort put into fitting accessories. It is usually the case that an absolutely standard motorcycle will sell more easily at a better price than one that has been modified. If you are in the habit of exchanging your machine for another at frequent intervals, this factor should be borne in mind to avoid loss of money.

Fault diagnosis

Contents

1   Introduction

This Section provides an easy reference-guide to the more common ailments that are likely to afflict your machine. Obviously, the opportunities are almost limitless for faults to occur as a result of obscure failures, and to try and cover all eventualities would require a book. Indeed, a number have been written on the subject.

Successful fault diagnosis is not a mysterious 'black art' but the application of a bit of knowledge combined with a systematic and logical approach to the problem. Approach any fault diagnosis by first accurately identifying the symptom and then checking through the list of possible causes, starting with the simplest or most obvious and progressing in stages to the most complex. Take nothing for granted, but above all apply liberal quantities of common sense.

The main symptom of a fault is given in the text as a major heading below which are listed, as Section headings, the various systems or areas which may contain the fault. Details of each possible cause for a fault and the remedial action to be taken are given, in brief, in the paragraphs below each Section heading. Further information should be sought in the relevant Chapter.

In some cases reference will be made in the singular to a component, for example a carburettor, where in fact more than one item is fitted to the machine. The particular reference should be applied to all those components.

Starter motor problems

2   Starter motor not rotating

Engine stop switch off.

Fuse blown. Check the main fuse located behind the battery side cover.

Battery voltage low. Switching on the headlamp and operating the horn will give a good indication of the charge level. If necessary recharge the battery from an external source.

Neutral gear not selected. Where a neutral indicator switch is fitted.

Faulty neutral indicator switch or starter interlock switch (where fitted). Check the switch wiring and switches for correct operation.

Ignition switch defective. Check switch for continuity and connections for security.

Engine stop switch defective. Check switch for continuity in 'Run' position. Fault will be caused by broken, wet or corroded switch contacts. Clean or renew as necessary.

Starter button switch faulty. Check continuity of switch. Faults as for engine stop switch.

Starter relay (solenoid) faulty. If the switch is functioning correctly a pronounced click should be heard when the starter button is depressed. This presupposes that current is flowing to the solenoid when the button is depressed.

Wiring open or shorted. Check first that the battery terminal connections are tight and corrosion free. Follow this by checking that all wiring connections are dry, tight and corrosion free. Check also for frayed or broken wiring. Occasionally a wire may become trapped between two moving components, particularly in the vicinity of the steering head, leading to breakage of the internal core but leaving the softer but more resilient outer cover intact. This can cause mysterious intermittent or total power loss.

Starter motor defective. A badly worn starter motor may cause high current drain from a battery without the motor rotating. If current is found to be reaching the motor, after checking the starter button and starter relay, suspect a damaged motor. The motor should be removed for inspection.

3   Starter motor rotates but engine does not turn over

Starter motor clutch defective. Suspect jammed or worn engagement rollers, plungers and springs.

Damaged starter motor drive train. Inspect and renew component where necessary. Failure in this area is unlikely.

4   Starter motor and clutch function but engine will not turn over

Engine seized. Seizure of the engine is always a result of damage to internal components due to lubrication failure, or component breakage resulting from abuse, neglect or old age. A seizing or partially seized component may go un-noticed until the engine has cooled down and an attempt is made to restart the engine. Suspect first seizure of the valves, valve gear and the pistons. Instantaneous seizure whilst the engine is running indicates component breakage. In either case major dismantling and inspection will be required.

Engine does not start when turned over

5   No fuel flow to carburettor

No fuel or insufficient fuel in tank.

Fuel tap lever position incorrectly selected.

Float chambers require priming after running dry.

Tank filler cap air vent obstructed. Usually caused by dirt or water. Clean the vent orifice.

Fuel tap or filter blocked. Blockage may be due to accumulation of rust or paint flakes from the tank's inner surface or of foreign matter from contaminated fuel. Remove the tap and clean it and the filter. Look also for water droplets in the fuel.

Fuel line blocked. Blockage of the fuel line is more likely to result from a kink in the line rather than the accumulation of debris.

6   Fuel not reaching cylinder

Float chamber not filling. Caused by float needle or floats sticking in up position. This may occur after the machine has been left standing for an extended length of time allowing the fuel to evaporate. When this occurs a gummy residue is often left which hardens to a varnish-like substance. This condition may be worsened by corrosion and crystaline deposits produced prior to the total evaporation of contaminated fuel. Sticking of the float needle may also be caused by wear. In any case removal of the float chamber will be necessary for inspection and cleaning.

Blockage in starting circuit, slow running circuit or jets. Blockage of these items may be attributable to debris from the fuel tank by-passing the filter system or to gumming up as described in paragraph 1. Water droplets in the fuel will also block jets and passages. The carburettor should be dismantled for cleaning.

Fuel level too low. The fuel level in the float chamber is controlled by float height. The float height may increase with wear or damage but will never reduce, thus a low float height is an inherent rather than developing condition. Check the float height and make any necessary adjustment.

7   Engine flooding

Float valve needle worn or stuck open. A piece of rust or other debris can prevent correct seating of the needle against the valve seat thereby permitting an uncontrolled flow of fuel. Similarly, a worn needle or needle seat will prevent valve closure. Dismantle the carburettor float bowl for cleaning and, if necessary, renewal of the worn components.

Fuel level too high. The fuel level is controlled by the float height which may increase due to wear of the float needle, pivot pin or operating tang. Check the float height, and make any necessary adjustment. A leaking float will cause an increase in fuel level, and thus should be renewed.

Cold starting mechanism. Check the choke (starter mechanism) for correct operation. If the mechanism jams in the 'On' position subsequent starting of a hot engine will be difficult.

Blocked air filter. A badly restricted air filter will cause flooding. Check the filter and clean or renew as required. A collapsed inlet hose will have a similar effect.

8 No spark at plug

Ignition switch not on.
Engine stop switch off.
Fuse blown. Check fuse for ignition circuit. See wiring diagram.
Battery voltage low. The current draw required by a starter motor is sufficiently high that an under-charged battery may not have enough spare capacity to provide power for the ignition circuit during starting.
Starter motor inefficient. A starter motor with worn brushes and a worn or dirty commutator will draw excessive amounts of current causing power starvation in the ignition system. See the preceding paragraph. Starter motor overhaul will be required.
Spark plug failure. Clean the spark plug thoroughly and reset the electrode gap. Refer to the spark plug section and the   condition guide in Chapter 3. If the spark plug shorts internally or has sustained visible damage to the electrodes, core or ceramic insulator it should be renewed. On rare occasions a plug that appears to spark vigorously will fail to do so when refitted to the engine and subjected to the compression pressure in the cylinder.
Spark plug cap or high tension (HT) lead faulty. Check condition and security. Replace if deterioration is evident.
Spark plug cap loose. Check that the spark plug cap fits securely over the plug and, where fitted, the screwed terminal on the plug end is secure.
Shorting due to moisture. Certain parts of the ignition system are susceptible to shorting when the machine is ridden or parked in wet weather. Check particularly the area from the spark plug cap back to the ignition coil. A water dispersant spray may be used to dry out waterlogged components. Recurrence of the problem can be prevented by using an ignition sealant spray after drying out and cleaning.
Ignition, starter interlock or stop switch shorted. May be caused by water, corrosion or wear. Water dispersant and contact cleaning sprays may be used. If this fails to overcome the problem dismantling and visual inspection of the switches will be required.
Shorting or open circuit in wiring. Failure in any wire connecting any of the ignition components will cause ignition malfunction. Check also that all connections are clean, dry and tight.
Ignition coil failure. Check the coil, referring to Chapter 3.
Pickup coil failure. Check each coil, referring to Chapter 3.
IC ignitor failure. See Chapter 3.

9 Weak spark at plug

Feeble sparking at the plug may be caused by any of the faults mentioned in the preceding Section other than those items in paragraphs 1 and 2.

10 Compression low

Spark plug loose. This will be self-evident on inspection, and may be accompanied by a hissing noise when the engine is turned over. Remove the plug and check that the threads in the cylinder head are not damaged. Check also that the plug sealing washer is in good condition.
Cylinder head gasket leaking. This condition is often accompanied by a high pitched squeak from around the cylinder head and oil loss, and may be caused by insufficiently tightened cylinder head fasteners, a warped cylinder head or mechanical failure of the gasket material. Re-torqueing the fasteners to the correct specification may seal the leak in some instances but if damage has occurred this course of action will provide, at best, only a temporary cure.
Valve not seating correctly. The failure of a valve to seat may be caused by insufficient valve clearance, pitting of the valve seat or face, carbon deposits on the valve seat or seizure of the valve stem or valve gear components. Valve spring breakage will also prevent correct valve closure. The valve clearances should be checked first and then, if these are found to be in order, further dismantling will be required to inspect the relevant components for failure.
Cylinder, piston and ring wear. Compression pressure will be lost if any of these components are badly worn. Wear in one component is invariably accompanied by wear in another. A top end overhaul will be required.

Piston rings sticking or broken. Sticking of the piston rings may be caused by seizure due to lack of lubrication or heating as a result of poor carburation or incorrect fuel type. Gumming of the rings may result from lack of use, or carbon deposits in the ring grooves. Broken rings result from over-revving, overheating or general wear. In either case a top-end overhaul will be required.

Engine stalls after starting

11 General causes

Improper cold start mechanism operation. Check that the operating controls function smoothly and, where applicable, are correctly adjusted. A cold engine may not require application of an enriched mixture to start initially but may baulk without choke once firing. Likewise a hot engine may start with an enriched mixture but will stop almost immediately if the choke is inadvertently in operation.
Ignition malfunction. See Section 9, 'Weak spark at plug'.
Carburettor incorrectly adjusted. Maladjustment of the mixture strength or idle speed may cause the engine to stop immediately after starting. See Chapter 2.
Fuel contamination. Check for filter blockage by debris or water which reduces, but does not completely stop, fuel flow or blockage of the slow speed circuit in the carburettor by the same agents. If water is present it can often be seen as droplets in the bottom of the float bowl. Clean the filter and, where water is in evidence, drain and flush the fuel tank and float bowl.
Intake air leak. Check for security of the carburettor mounting and hose connections, and for cracks or splits in the hoses. Check also that the carburettor top is secure and that the vacuum gauge adaptor plug (where fitted) is tight.
Air filter blocked or omitted. A blocked filter will cause an over-rich mixture; the omission of a filter will cause an excessively weak mixture. Both conditions will have a detrimental effect on carburation. Clean or renew the filter as necessary.
Fuel filler cap air vent blocked. Usually caused by dirt or water. Clean the vent orifice.

Poor running at idle and low speed

12 Weak spark at plug or erratic firing

Battery voltage low. In certain conditions low battery charge, especially when coupled with a badly sulphated battery, may result in misfiring. If the battery is in good general condition it should be recharged; an old battery suffering from sulphated plates should be renewed.
Spark plug fouled, faulty or incorrectly adjusted. See Section 8 or refer to Chapter 3.
Spark plug cap or high tension lead shorting. Check the condition of both these items ensuring that they are in good condition and dry and that the cap is fitted correctly.
Spark plug type incorrect. Fit plug of correct type and heat range as given in Specifications. In certain conditions a plug of hotter or colder type may be required for normal running.
Pickup coil faulty. Partial failure of a pickup coil internal insulation will diminish the performace of the coil. No repair is possible; fit a new item.
Faulty ignition coil. Partial failure of the coil internal insulation will diminish the performance of the coil. No repair is possible, a new component must be fitted.
IC ignitor malfunction. See Chapter 3 for details.

13 Fuel/air mixture incorrect

Intake air leak. See Section 11.
Mixture strength incorrect. Adjust slow running mixture strength using pilot adjustment screw.
Carburettor synchronisation.
Pilot jet or slow running circuit blocked. The carburettor should be removed and dismantled for thorough cleaning. Blow through all jets and air passages with compressed air to clear obstructions.

Air cleaner clogged or omitted. Clean or fit air cleaner element as necessary. Check also that the element and air filter cover are correctly seated.

Cold start mechanism in operation. Check that the choke has not been left on inadvertently and the operation is correct. Where applicable check the operating cable free play.

Fuel level too high or too low. Check the float height and adjust as necessary. See Section 7.

Fuel tank air vent obstructed. Obstruction usually caused by dirt or water. Clean vent orifice.

Valve clearance incorrect. Check, and if necessary, adjust, the clearances.

14 Compression low

See Section 10.

Acceleration poor

15 General causes

All items as for previous Section.

Timing not advancing. This is caused by a sticking or damaged automatic timing unit (ATU). Cleaning and lubrication of the ATU will usually overcome sticking, failing this, and in any event if damage is evident, renewal of the ATU will be required.

Sticking throttle vacuum piston.

Brakes binding. Usually caused by maladjustment or partial seizure of the operating mechanism due to poor maintenance. Check brake adjustment (where applicable). A bent wheel spindle or warped brake disc can produce similar symptoms.

Poor running or lack of power at high speeds

16 Weak spark at plug or erratic firing

All items as for Section 12.

HT lead insulation failure. Insulation failure of the HT lead and spark plug cap due to old age or damage can cause shorting when the engine is driven hard. This condition may be less noticeable, or not noticeable at all at lower engine speeds.

17 Fuel/air mixture incorrect

All items as for Section 13, with the exception of items 2 and 4.

Main jet blocked. Debris from contaminated fuel, or from the fuel tank, and water in the fuel can block the main jet. Clean the fuel filter, the float bowl area, and if water is present, flush and refill the fuel tank.

Main jet is the wrong size. The standard carburettor jetting is for sea level atmospheric pressure. For high altitudes, usually above 5000 ft, a smaller main jet will be required.

Jet needle and needle jet worn. These can be renewed individually but should be renewed as a pair. Renewal of both items requires partial dismantling of the carburettor.

Air bleed holes blocked. Dismantle carburettor and use compressed air to blow out all air passages.

Reduced fuel flow. A reduction in the maximum fuel flow from the fuel tank to the carburettor will cause fuel starvation, proportionate to the engine speed. Check for blockages through debris or a kinked fuel line.

Vacuum diaphragm split. Renew.

18 Compression low

See Section 10.

Knocking or pinking

19 General causes

Carbon build-up in combustion chamber. After high mileages have been covered large accumulation of carbon may occur. This may glow red hot and cause premature ignition of the fuel/air mixture, in advance of normal firing by the spark plug. Cylinder head removal will be required to allow inspection and cleaning.

Fuel incorrect. A low grade fuel, or one of poor quality may result in compression induced detonation of the fuel resulting in knocking and pinking noises. Old fuel can cause similar problems. A too highly leaded fuel will reduce detonation but will accelerate deposit formation in the combustion chamber and may lead to early pre-ignition as described in item 1.

Spark plug heat range incorrect. Uncontrolled pre-ignition can result from the use of a spark plug the heat range of which is too hot.

Weak mixture. Overheating of the engine due to a weak mixture can result in pre-ignition occurring where it would not occur when engine temperature was within normal limits. Maladjustment, blocked jets or passages and air leaks can cause this condition.

Overheating

20 Firing incorrect

Spark plug fouled, defective or maladjusted. See Section 6.

Spark plug type incorrect. Refer to the Specifications and ensure that the correct plug type is fitted.

Incorrect ignition timing. Timing that is far too much advanced or far too much retarded will cause overheating. Check the ignition timing is correct and that the advance mechanism is functioning.

21 Fuel/air mixture incorrect

Slow speed mixture strength incorrect. Adjust pilot air screw.

Main jet wrong size. The carburettor is jetted for sea level atmospheric conditions. For high altitudes, usually above 5000 ft, a smaller main jet will be required.

Air filter badly fitted or omitted. Check that the filter element is in place and that it and the air filter box cover are sealing correctly. Any leaks will cause a weak mixture.

Induction air leaks. Check the security of the carburettor mountings and hose connections, and for cracks and splits in the hoses. Check also that the carburettor top is secure and that the vacuum gauge adaptor plug (where fitted) is tight.

Fuel level too low. See Section 6.

Fuel tank filler cap air vent obstructed. Clear blockage.

22 Lubrication inadequate

Engine oil too low. Not only does the oil serve as a lubricant by preventing friction between moving components, but it also acts as a coolant. Check the oil level and replenish.

Engine oil overworked. The lubricating properties of oil are lost slowly during use as a result of changes resulting from heat and also contamination. Always change the oil at the recommended interval.

Engine oil of incorrect viscosity or poor quality. Always use the recommended viscosity and type of oil.

Oil filter and filter by-pass valve blocked. Renew filter and clean the by-pass valve.

23 Miscellaneous causes

Engine fins clogged. A build-up of mud in the cylinder head and cylinder barrel cooling fins will decrease the cooling capabilities of the fins. Clean the fins as required.

Clutch operating problems

24 Clutch slip

No clutch lever play. Adjust clutch lever end play according to the procedure in Routine Maintenance.

Friction plates worn or warped. Overhaul clutch assembly, replacing plates out of specification.

Steel plates worn or warped. Overhaul clutch assembly, replacing plates out of specification.

Clutch springs broken or wear. Old or heat-damaged (from slipping clutch) springs should be replaced with new ones.

Clutch release not adjusted properly.

Clutch inner cable snagging. Caused by a frayed cable or kinked outer cable. Replace the cable with a new one. Repair of a frayed cable is not advised.

Clutch release mechanism defective. Worn or damaged parts in the clutch release mechanism could include the shaft, cam, actuating arm or pivot. Replace parts as necessary.

Clutch hub and outer drum worn. Severe indentation by the clutch plate tangs of the channels in the hub and drum will cause snagging of the plates preventing correct engagement. If this damage occurs, renewal of the worn components is required.

Lubricant incorrect. Use of a transmission lubricant other than that specified may allow the plates to slip.

25 Clutch drag

Clutch lever play excessive. Adjust lever at bars or at cable end if necessary.

Clutch plates warped or damaged. This will cause a drag on the clutch, causing the machine to creep. Overhaul clutch assembly.

Clutch spring tension uneven. Usually caused by a sagged or broken spring. Check and replace springs.

Engine oil deteriorated. Badly contaminated engine oil and a heavy deposit of oil sludge and carbon on the plates will cause plate sticking. The oil recommended for this machine is of the detergent type, therefore it is unlikely that this problem will arise unless regular oil changes are neglected.

Engine oil viscosity too high. Drag in the plates will result from the use of an oil with too high a viscosity. In very cold weather clutch drag may occur until the engine has reached operating temperature.

Clutch hub and outer drum worn. Indentation by the clutch plate tangs of the channels in the hub and drum will prevent easy plate disengagement. If the damage is light the affected areas may be dressed with a fine file. More pronounced damage will necessitate renewal of the components.

Clutch release mechanism defective. Worn or damaged release mechanism parts can stick and fail to provide leverage. Overhaul clutch cover components.

Loose clutch hub nut. Causes drum and hub misalignment, putting a drag on the engine. Engagement adjustment continually varies. Overhaul clutch assembly.

Gear selection problems

26 Gear lever does not return

Weak or broken centraliser spring. Renew the spring.

Gearchange shaft bent or seized. Distortion of the gearchange shaft often occurs if the machine is dropped heavily on the gear lever. Provided that damage is not severe straightening of the shaft is permissible.

27 Gear selection difficult or impossible

Clutch not disengaging fully. See Section 25.

Gearchange shaft bent. This often occurs if the machine is dropped heavily on the gear lever. Straightening of the shaft is permissible if the damage is not too great.

Gearchange arms, pawls or pins worn or damaged. Wear or breakage of any of these items may cause difficulty in selecting one or more gears. Overhaul the selector mechanism.

Gearchange arm spring broken. Renew spring.

Gearchange drum cam detent plunger damaged. Failure, rather than wear, of these items may jam the drum thereby preventing gearchanging. The damaged items must be renewed.

Selector forks bent or seized. This can be caused by dropping the machine heavily on the gearchange lever or as a result of lack of lubrication. Though rare, bending of a shaft can result from a missed gearchange or false selection at high speed.

Selector fork end and pin wear. Pronounced wear of these items and the grooves in the gearchange drum can lead to imprecise selection and, eventually, no selection. Renewal of the worn components will be required.

Structural failure. Failure of any one component of the selector rod and change mechanism will result in improper or fouled gear selection.

28 Jumping out of gear

Detent plunger assembly worn or damaged. Wear of the plunger and the cam with which it locates and breakage of the detent spring can cause imprecise gear selection resulting in jumping out of gear. Renew the damaged components.

Gear pinion dogs worn or damaged. Rounding off the dog edges and the mating recesses in adjacent pinion can lead to jumping out of gear when under load. The gears should be inspected and renewed. Attempting to reprofile the dogs is not recommended.

Selector forks, gearchange drum and pinion grooves worn. Extreme wear of these interconnected items can occur after high mileages especially when lubrication has been neglected. The worn components must be renewed.

Gear pinions, bushes and shafts worn. Renew the worn components.

Bent gearchange shaft. Often caused by dropping the machine on the gear lever.

Gear pinion tooth broken. Chipped teeth are unlikely to cause jumping out of gear once the gear has been selected fully; a tooth which is completely broken off, however, may cause problems in this respect and in any event will cause transmission noise.

29 Overselection

Pawl spring weak or broken. Renew the spring.

Detent plunger worn or broken. Renew the damaged items.

Gearchange arm stop pads worn. Repairs can be made by welding and reprofiling with a file.

Selector limiter claw components (where fitted) worn or damaged. Renew the damaged items.

Abnormal engine noise.

30 Knocking or pinking

See Section 19.

31 Piston slap or rattling from cylinder

Cylinder bore/piston clearance excessive. Resulting from wear, partial seizure or improper boring during overhaul. This condition can often be heard as a high, rapid tapping noise when the engine is under little or no load, particularly when power is just beginning to be applied. Reboring to the next correct oversize should be carried out and a new oversize piston fitted.

Connecting rod bent. This can be caused by over-revving, trying to start a very badly flooded engine (resulting in a hydraulic lock in the cylinder) or by earlier mechanical failure such as a dropped valve. Attempts at straightening a bent connecting rod from a high performance engine are not recommended. Careful inspection of the crankshaft should be made before renewing the damaged connecting rod.

Gudgeon pin, piston boss bore or small-end bearing wear or seizure. Excess clearance or partial seizure between normal moving parts of these items can cause continuous or intermittent tapping noises. Rapid wear or seizure is caused by lubrication starvation resulting from an insufficient engine oil level or oilway blockage.

Piston rings worn, broken or sticking. Renew the rings after careful inspection of the piston and bore.

32 Valve noise or tapping from the cylinder head

Valve clearance incorrect. Adjust the clearances with the engine cold.

Valve spring broken or weak. Renew the spring set.

Camshaft or cylinder head worn or damaged. The camshaft lobes are the most highly stressed of all components in the engine and are subject to high wear if lubrication becomes inadequate. The bearing surfaces on the camshaft and cylinder head are also sensitive to a lack of lubrication. Lubrication failure due to blocked oilways can occur, but over-enthusiastic revving before engine warm-up is complete is the usual cause.

Worn camshaft drive components. A rustling noise or light tapping which is not improved by correct re-adjustment of the cam chain tension can be emitted by a worn cam chain or worn sprockets and chain. If uncorrected, subsequent cam chain breakage may cause extensive damage. The worn components must be renewed before wear becomes too far advanced.

33 Other noises

Big-end bearing wear. A pronounced knock from within the crankcase which worsens rapidly is indicative of big-end bearing failure as a result of extreme normal wear or lubrication failure. Remedial action in the form of a bottom end overhaul should be taken; continuing to run the engine will lead to further damage including the possibility of connecting rod breakage.

Main bearing failure. Extreme normal wear or failure of the main bearings is characteristically accompanied by a rumble from the crankcase and vibration felt through the frame and footrests. Renew the worn bearings and carry out a very careful examination of the crankshaft.

Crankshaft excessively out of true. A bent crank may result from over-revving or damage from an upper cylinder component or gearbox failure. Damage can also result from dropping the machine on either crankshaft end. Straightening of the crankshaft is not possible in normal circumstances; a replacement item should be fitted.

Engine mounting loose. Tighten all the engine mounting nuts and bolts.

Cylinder head gasket leaking. The noise most often associated with a leaking head gasket is a high pitched squeaking, although any other noise consistent with gas being forced out under pressure from a small orifice can also be emitted. Gasket leakage is often accompanied by oil seepage from around the mating joint or from the cylinder head holding down bolts and nuts. Leakage into the cam chain tunnel or oil return passages will increase crankcase pressure and may cause oil leakage at joints and oil seals. Also, oil contamination will be accelerated. Leakage results from insufficient or uneven tightening of the cylinder head fasteners, or from random mechanical failure. Retightening to the correct torque figure will, at best, only provide a temporary cure. The gasket should be renewed at the earliest opportunity.

Exhaust system leakage. Popping or crackling in the exhaust system, particularly when it occurs with the engine on the overrun, indicates a poor joint either at the cylinder port or at the exhaust pipe/silencer connection. Failure of the gasket or looseness of the clamp should be looked for.

Abnormal transmission noise

34 Clutch noise

Clutch outer drum/friction plate tang clearance excessive.
Clutch outer drum/spacer clearance excessive.
Clutch outer drum/bearing clearance excessive.

Primary drive gear teeth worn or damaged.
Clutch shock absorber assembly worn or damaged.

35 Transmission noise

Bearing or bushes worn or damaged. Renew the affected components.

Gear pinions worn or chipped. Renew the gear pinions.

Metal chips jammed in gear teeth. This can occur when pieces of metal from any failed component are picked up by a meshing pinion. The condition will lead to rapid bearing wear or early gear failure.

Engine/transmission oil level too low. Top up immediately to prevent damage to gearbox and engine.

Gearchange mechanism worn or damaged. Wear or failure of certain items in the selection and change components can induce mis-selection of gears (see Section 27) where incipient engagement of more than one gear set is promoted. Remedial action, by the overhaul of the gearbox, should be taken without delay.

Chain drive models

Loose gearbox chain sprocket. Remove the sprocket and check for impact damage to the splines of the sprocket and shaft. Excessive slack between the splines will promote loosening of the securing nut; renewal of the worn components is required. When retightening the nut ensure that it is tightened fully and that, where fitted, the lock washer is bent up against one flat of the nut.

Chain snagging on cases or cycle parts. A badly worn chain or one that is excessively loose may snag or smack against adjacent components.

Shaft drive models

Worn or damaged bevel gear sets. A whine emitted from either bevel gear set is indicative of improper meshing. This may increase progressively as wear develops or suddenly due to mechanical failure. Drain the lubricant and inspect for metal chips prior to dismantling.

Output shaft joint failure. This can cause vibration and noise. Renew the affected component.

Exhaust smokes excessively

36 White/blue smoke (caused by oil burning)

Piston rings worn or broken. Breakage or wear of any ring, but particularly the oil control ring, will allow engine oil past the piston into the combustion chamber. Overhaul the cylinder barrel and piston.

Cylinder cracked, worn or scored. These conditions may be caused by overheating, lack of lubrication, component failure or advanced normal wear. The cylinder barrel should be renewed or rebored and the next oversize piston fitted.

Valve oil seal damaged or worn. This can occur as a result of valve guide failure or old age. The emission of smoke is likely to occur when the throttle is closed rapidly after acceleration, for instance, when changing gear. Renew the valve oil seals and, if necessary, the valve guides.

Valve guides worn. See the preceding paragraph.

Engine oil level too high. This increases the crankcase pressure and allows oil to be forced past the piston rings. Often accompanied by seepage of oil at joints and oil seals.

Cylinder head gasket blown between cam chain tunnel or oil return passage. Renew the cylinder head gasket.

Abnormal crankcase pressure. This may be caused by blocked breather passages or hoses causing back-pressure at high engine revolutions.

37 Black smoke (caused by over-rich mixture)

Air filter element clogged. Clean or renew the element.

Main jet loose or too large. Remove the float chamber to check for tightness of the jet. If the machine is used at high altitudes rejetting will be required to compensate for the lower atmospheric pressure.

Cold start mechanism jammed on. Check that the mechanism works smoothly and correctly and that, where fitted, the operating cable is lubricated and not snagged.

Fuel level too high. The fuel level is controlled by the float height which can increase as a result of wear or damage. Remove the float bowl and check the float height. Check also that floats have not punctured; a punctured float will loose buoyancy and allow an increased fuel level.

Float valve needle stuck open. Caused by dirt or a worn valve. Clean the float chamber or renew the needle and, if necessary, the valve seat.

Oil pressure level indicator lamp goes on

38 Engine lubrication system failure

Engine oil defective. Oil pump shaft or locating pin sheared off from ingesting debris or seizing from lack of lubrication (low oil level).

Engine oil screen clogged. Change oil and filter and service pickup screen.

Engine oil level too low. Inspect for leak or other problem causing low oil level and add recommended lubricant.

Engine oil viscosity too low. Very old, thin oil, or an improper weight of oil used in engine. Change to correct lubricant.

Camshaft or journals worn. High wear causing drop in oil pressure. Replace cam and/or head. Abnormal wear could be caused by oil starvation at high rpm from low oil level, improper oil weight or type, or loose oil fitting on upper cylinder oil line.

Crankshaft and/or bearings worn. Same problems as paragraph 5. Overhaul lower end.

Relief valve stuck open. This causes the oil to be dumped back into the sump. Repair or replace.

39 Electrical system failure

Oil pressure level switch defective. Check switch according to the procedures in Chapter 6. Replace if defective.

Oil pressure level indicator lamp wiring system defective. Check for pinched, shorted, disconnected or damaged wiring.

Poor handling or roadholding

40 Directional instability

Steering head bearing adjustment too tight. This will cause rolling or weaving at low speeds. Re-adjust the bearings.

Steering head bearing worn or damaged. Correct adjustment of the bearing will prove impossible to achieve if wear or damage has occurred. Inconsistent handling will occur including rolling or weaving at low speed and poor directional control at indeterminate higher speeds. The steering head bearing should be dismantled for inspection and renewed if required. Lubrication should also be carried out.

Bearing races pitted or dented. Impact damage caused, perhaps, by an accident or riding over a pot-hole can cause indentation of the bearing, usually in one position. This should be noted as notchiness when the handlebars are turned. Renew and lubricate the bearings.

Steering stem bent. This will occur only if the machine is subjected to a high impact such as hitting a curb or a pot-hole. The lower yoke/stem should be renewed; do not attempt to straighten the stem.

Front or rear tyre pressures too low.

Front or rear tyre worn. General instability, high speed wobbles and skipping over white lines indicates that tyre renewal may be required. Tyre induced problems, in some machine/tyre combinations, can occur even when the tyre in question is by no means fully worn.

Swinging arm bearing badly adjusted (shaft drive models). Adjust as required.

Swinging arm bearings worn. Difficulties in holding line, particularly when cornering or when changing power settings indicates wear in the swinging arm bearings. The swinging arm should be removed from the machine and the bearings renewed.

Swinging arm flexing. The symptoms given in the preceding paragraph will also occur if the swinging arm fork flexes badly. This can be caused by structural weakness as a result of corrosion, fatigue or impact damage, or because the rear wheel spindle is slack.

Wheel bearings worn. Renew the worn bearings.

Tyres unsuitable for machine. Not all available tyres will suit the characteristics of the frame and suspension, indeed, some tyres or tyre combinations may cause a transformation in the handling characteristics. If handling problems occur immediately after changing to a new tyre type or make, revert to the original tyres to see whether an improvement can be noted. In some instances a change to what are, in fact, suitable tyres may give rise to handling deficiences. In this case a thorough check should be made of all frame and suspension items which affect stability.

41 Steering bias to left or right

Rear wheel out of alignment. On chain drive machines, caused by uneven adjustment of chain tensioner adjusters allowing the wheel to be askew in the fork ends. A bent rear wheel spindle will also misalign the wheel in the swinging arm.

Wheels out of alignment. This can be caused by impact damage to the frame, swinging arm, wheel spindles or front forks. Although occasionally a result of material failure or corrosion it is usually as a result of a crash.

Front forks twisted in the steering yokes. A light impact, for instance with a pot-hole or low curb, can twist the fork legs in the steering yokes without causing structural damage to the fork legs or the yokes themselves. Re-alignment can be made by loosening the yoke pinch bolts, wheel spindle and mudguard bolts. Re-align the wheel with the handlebars and tighten the bolts working upwards from the wheel spindle. This action should be carried out only when there is no chance that structural damage has occurred.

42 Handlebar vibrates or oscillates

Tyres worn or out of balance. Either condition, particularly in the front tyre, will promote shaking of the fork assembly and thus the handlebars. A sudden onset of shaking can result if a balance weight is displaced during use.

Tyres badly positioned on the wheel rims. A moulded line on each wall of a tyre is provided to allow visual verification that the tyre is correctly positioned on the rim. A check can be made by rotating the tyre; any misalignment will be immediately obvious.

Wheels rims warped or damaged. Inspect the wheels for runout as described in Chapter 5.

Swinging arm bearings badly adjusted (shaft drive models). Readjust.

Swinging arm bearings worn. Renew the bearings.

Wheel bearings worn. Renew the bearings.

Steering head bearings incorrectly adjusted. Vibration is more likely to result from bearings which are too loose rather than too tight. Re-adjust the bearings.

Loose fork component fasteners. Loose nuts and bolts holding the fork legs, wheel spindle, mudguards or steering stem can promote shaking at the handlebars. Fasteners on running gear such as the forks and suspension should be check tightened occasionally to prevent dangerous looseness of components occurring.

Engine mounting bolts loose. Tighten all fasteners.

43 Poor front fork performance

Damping fluid level incorrect. If the fluid level is too low poor suspension control will occur resulting in a general impairment of roadholding and early loss of tyre adhesion when cornering and braking. Too much oil is unlikely to change the fork characteristics unless severe overfilling occurs when the fork action will become stiffer and oil seal failure may occur.

Damping oil viscosity incorrect. The damping action of the fork is directly related to the viscosity of the damping oil. The lighter the oil used, the less will be the damping action imparted. For general use, use the recommended viscosity of oil, changing to a slightly higher or heavier oil only when a change in damping characteristic is required. Overworked oil, or oil contaminated with water which has found its way past the seals, should be renewed to restore the correct damping performance and to prevent bottoming of the forks.

Air pressure incorrect. An imbalance in the air pressure between the fork legs can give rise to poor fork performance. Similarly, if the air pressure is outside the recommended range, problems can occur.

Damping components worn or corroded. Advanced normal wear of the fork internals is unlikely to occur until a very high mileage has been covered. Continual use of the machine with damaged oil seals which allows the ingress of water, or neglect, will lead to rapid corrosion and wear. Dismantle the forks for inspection and overhaul.

Weak fork springs. Progressive fatigue of the fork springs, resulting in a reduced spring free length, will occur after extensive use. This condition will promote excessive fork dive under braking, and in its advanced form will reduce the at-rest extended length of the forks and thus the fork geometry. Renewal of the springs as a pair is the only satisfactory course of action.

Bent stanchions or corroded stanchions. Both conditions will prevent correct telescoping of the fork legs, and in an advanced state can cause sticking of the fork in one position. In a mild form corrosion will cause stiction of the fork thereby increasing the time the suspension takes to react to an uneven road surface. Bent fork stanchions should be attended to immediately because they indicate that impact damage has occurred, and there is a danger that the forks will fail with disastrous consequences.

44 Front fork judder when braking (see also Section 56)

Wear between the fork stanchions and the fork legs. Renewal of the affected components is required.

Slack steering head bearings. Re-adjust the bearings.

Warped brake disc. If irregular braking action occurs fork judder can be induced in what are normally serviceable forks. Renew the damaged brake components.

45 Poor rear suspension performance

Rear suspension unit damper worn out or leaking. The damping performance of most rear suspension units falls off with age. This is a gradual process, and thus may not be immediately obvious. Indications of poor damping include hopping of the rear end when cornering or braking, and a general loss of positive stability. See Chapter 4.

Weak rear springs. If the suspension unit springs fatigue they will promote excessive pitching of the machine and reduce the ground clearance when cornering. Although replacement springs are available separately from the rear suspension damper unit it is probable that if spring fatigue has occurred the damper units will also require renewal.

Swinging arm flexing or bearings worn. See Sections 40 and 41.

Bent suspension unit damper rod. This is likely to occur only if the machine is dropped or if seizure of the piston occurs. If either happens the suspension units should be renewed as a pair.

Abnormal frame and suspension noise

46 Front end noise

Oil level low or too thin. This can cause a 'spurting' sound and is usually accompanied by irregular fork action.

Spring weak or broken. Makes a clicking or scraping sound. Fork oil will have a lot of metal particles in it.

Steering head bearings loose or damaged. Clicks when braking. Check, adjust or replace.

Fork clamps loose. Make sure all fork clamp pinch bolts are tight.

Fork stanchion bent. Good possibility if machine has been dropped. Repair or replace tube.

47 Rear suspension noise

Fluid level too low. Leakage of a suspension unit, usually evident by oil on the outer surfaces, can cause a spurting noise. The suspension units should be renewed as a pair.

Defective rear suspension unit with internal damage. Renew the suspension units as a pair.

Brake problems

48 Brakes are spongy or ineffective

Air in brake circuit. This is only likely to happen in service due to neglect in checking the fluid level or because a leak has developed. The problem should be identified and the brake system bled of air.

Pad worn. Check the pad wear against the wear lines provided and renew the pads if necessary.

Contaminated pads. Cleaning pads which have been contaminated with oil, grease or brake fluid is unlikely to prove successful; the pads should be renewed.

Pads glazed. This is usually caused by overheating. The surface of the pads may be roughened using glass-paper or a fine file.

Brake fluid deterioration. A brake which on initial operation is firm but rapidly becomes spongy in use may be failing due to water contamination of the fluid. The fluid should be drained and then the system refilled and bled.

Master cylinder seal failure. Wear or damage of master cylinder internal parts will prevent pressurisation of the brake fluid. Overhaul the master cylinder unit.

Caliper seal failure. This will almost certainly be obvious by loss of fluid, a lowering of fluid in the master cylinder reservoir and contamination of the brake pads and caliper. Overhaul the caliper assembly.

Brake lever or pedal improperly adjusted. Adjust the clearance between the lever end and master cylinder plunger to take up lost motion, as recommended in Routine maintenance.

49 Brakes drag

Disc warped. The disc must be renewed.

Caliper piston, caliper or pads corroded. The brake caliper assembly is vulnerable to corrosion due to water and dirt, and unless cleaned at regular intervals and lubricated in the recommended manner, will become sticky in operation.

Piston seal deteriorated. The seal is designed to return the piston in the caliper to the retracted position when the brake is released. Wear or old age can affect this function. The caliper should be overhauled if this occurs.

Brake pad damaged. Pad material separating from the backing plate due to wear or faulty manufacture. Renew the pads. Faulty installation of a pad also will cause dragging.

Wheel spindle bent. The spindle may be straightened if no structural damage has occurred.

Brake lever or pedal not returning. Check that the lever or pedal works smoothly throughout its operating range and does not snag on any adjacent cycle parts. Lubricate the pivot if necessary.

Twisted caliper support bracket. This is likely to occur only after impact in an accident. No attempt should be made to re-align the caliper; the bracket should be renewed.

50 Brake lever or pedal pulsates in operation

Disc warped or irregularly worn. The disc must be renewed.

Wheel spindle bent. The spindle may be straightened provided no structural damage has occurred.

51 Disc brake noise

Brake squeal. This can be caused by the omission or incorrect installation of the anti-squeal shim fitted to the rear of one pad. The arrow on the shim should face the direction of wheel normal rotation. Squealing can also be caused by dust on the pads, usually in combination with glazed pads, or other contamination from oil, grease, brake fluid or corrosion. Persistent squealing which cannot be traced to any of the normal causes can often be cured by applying a thin layer of high temperature silicone grease to the rear of the pads. Make absolutely certain that no grease is allowed to contaminate the braking surface of the pads.

Glazed pads. This is usually caused by high temperatures or contamination. The pad surfaces may be roughened using glass-paper or a fine file. If this approach does not effect a cure the pads should be renewed.

Disc warped. This can cause a chattering, clicking or intermittent squeal and is usually accompanied by a pulsating brake lever or pedal or uneven braking. The disc must be renewed.

Brake pads fitted incorrectly or undersize. Longitudinal play in the pads due to omission of the locating springs (where fitted) or because pads of the wrong size have been fitted will cause a single tapping noise every time the brake is operated. Inspect the pads for correct installation and security.

52 Brake induced fork judder

Worn front fork stanchions and legs, or worn or badly adjusted steering head bearings. These conditions, combined with uneven or pulsating braking as described in Sections 50 and 54 will induce more or less judder when the brakes are applied, dependent on the degree of wear and poor brake operation. Attention should be given to both areas of malfunction. See the relevant Sections.

Electrical problems

53 Battery dead or weak

Battery faulty. Battery life should not be expected to exceed 3 to 4 years, particularly where a starter motor is used regularly. Gradual sulphation of the plates and sediment deposits will reduce the battery performance. Plate and insulator damage can often occur as a result of vibration. Complete power failure, or intermittent failure, may be due to a broken battery terminal. Lack of electrolyte will prevent the battery maintaining charge.

Battery leads making poor contact. Remove the battery leads and clean them and the terminals, removing all traces of corrosion and tarnish. Reconnect the leads and apply a coating of petroleum jelly to the terminals.

Load excessive. If additional items such as spot lamps, are fitted, which increase the total electrical load above the maximum alternator output, the battery will fail to maintain full charge. Reduce the electrical load to suit the electrical capacity.

Regulator/rectifier failure.

Alternator generating coils open-circuit or shorted.

Charging circuit shorting or open circuit. This may be caused by frayed or broken wiring, dirty connectors or a faulty ignition switch. The system should be tested in a logical manner. See Section 60.

54 Battery overcharged

Rectifier/regulator faulty. Overcharging is indicated if the battery becomes hot or it is noticed that the electrolyte level falls repeatedly between checks. In extreme cases the battery will boil causing corrosive gases and electrolyte to be emitted through the vent pipes.

Battery wrongly matched to the electrical circuit. Ensure that the specified battery is fitted to the machine.

55 Total electrical failure

Fuse blown. Check the main fuse. If a fault has occurred, it must be rectified before a new fuse is fitted.

Battery faulty. See Section 53.

Earth failure. Check that the frame main earth strap from the battery is securely affixed to the frame and is making a good contact.

Ignition switch or power circuit failure. Check for current flow through the battery positive lead (red) to the ignition switch. Check the ignition switch for continuity.

56 Circuit failure

Cable failure. Refer to the machine's wiring diagram and check the circuit for continuity. Open circuits are a result of loose or corroded connections, either at terminals or in-line connectors, or because of broken wires. Occasionally, the core of a wire will break without there being any apparent damage to the outer plastic cover.

Switch failure. All switches may be checked for continuity in each switch position, after referring to the switch position boxes incorporated in the wiring diagram for the machine. Switch failure may be a result of mechanical breakage, corrosion or water.

Fuse blown. Refer to the wiring diagram to check whether or not a circuit fuse is fitted. Replace the fuse, if blown, only after the fault has been identified and rectified.

57 Bulbs blowing repeatedly

Vibration failure. This is often an inherent fault related to the natural vibration characteristics of the engine and frame and is, thus, difficult to resolve. Modifications of the lamp mounting, to change the damping characteristics may help.

Intermittent earth. Repeated failure of one bulb, particularly where the bulb is fed directly from the generator, indicates that a poor earth exists somewhere in the circuit. Check that a good contact is available at each earthing point in the circuit.

Reduced voltage. Where a quartz-halogen bulb is fitted the voltage to the bulb should be maintained or early failure of the bulb will occur. Do not overload the system with additional electrical equipment in excess of the system's power capacity and ensure that all circuit connections are maintained clean and tight.

KAWASAKI 750 FOURS

Check list

Daily checks

1 Check the engine/gearbox oil level
2 Check the tyre pressures
3 Check the level of fluid in the master cylinder

Weekly checks

1 Lubricate the final drive chain
2 Check the battery electrolye level
3 Lubricate the exposed portions of control cables
4 Check around the machine for loose fittings and check that the electrical components are functioning correctly

3 monthly or every 3000 miles (5000 km)

1 Check the brake pads for wear and inspect the hydraulic system for leakage
2 Adjust the clutch
3 Adjust the carburettor
4 Check the operation of the steering and suspension
5 Clean and adjust the spark plugs
6 Check and adjust the valve clearances
7 Check the emission control components – US models
8 Clean and inspect the air filter element
9 Clean the fuel system
10 Change the engine/transmission oil
11 Lubricate the control cables and stand pivots

6 monthly or every 6000 miles (10 000 km)

1 Change the oil filter
2 Change the front fork oil
3 Lubricate the automatic timing unit (ATU)
4 Lubricate the swinging arm pivot

Two yearly or every 12 000 miles (20 000 km)

1 Grease the wheel bearings
2 Overhaul the steering head bearings

Adjustment data

Tyre pressures	750 E, L, R and P	KZ750 H
Front	28 psi	25 psi
Rear – up to 215 lb load	32 psi	22 psi
Rear – 215–364 lb load	36 psi	25 psi
Z750 H	Up to 110 mph (180 kph)	Above 110 mph (180 kph)
Front	25 psi	28 psi
Rear – up to 210 lb load	25 psi	28 psi
Rear – 210–300 lb load	28 psi	32 psi
Rear – 300–397 lb load	32 psi	36 psi
Z750 R	Up to 130 mph (210 kph)	Above 130 mph (210 kph)
Front	28 psi	32 psi
Rear – up to 215 lb load	32 psi	41 psi
Rear – 215–397 lb load	36 psi	41 psi
KZ750 N		
Front	25 psi	
Rear – up to 215 lb load	25 psi	
Rear – 215–397 lb load	32 psi	

	All models
Valve clearances	0.08 – 0.18 mm (0003 – 0.007 in)

Spark plug type	NGK	ND
US models	B8ES	W24ES-U
UK and Canadian models	BR8ES	W24ESR-U

Spark plug gap	0.7 – 0.8 mm (0.028 – 0.031 in)
Idle speed	1000 – 1100 rpm

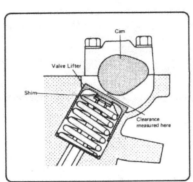

Valve clearance measurement

Recommended lubricants

① Component	Quantity	Type/viscosity
Engine/transmission		SAE 10W/40, 10W/50
Oil change	3.0 lit (5.3 Imp pt/ 3.2 US qt)	20W/40 or 20W/50 SE class or equivalent
Oil and filter change	3.5 lit (6.2 Imp pt, 3.7 US qt)	
✱ ② Final drive box	190cc (6.9/6.4 Imp/ US fl oz)	AP1 GL-5 hypoid gear oil, SAE 90 (above 5°C) SAE 80 (below 5°C)

③ Front forks	Oil change	Level	
E1,E2,L1, E3, L2 (US)	230cc	355 ± 4 mm	
E1,E2,L1, E3,L2 (UK)	215 cc	382 ± 4 mm	SAE 10W
H1,H2,H3	260cc	436 ± 4 mm	
R1	240cc	168 ± 4 mm	
N1	250cc	457 ± 2 mm	SAE 5W/20
P1	260cc	408.5 ± 2 mm	

④ Steering head bearings	As required	High melting point grease
⑤ Hydraulic disc brake	As required	Hydraulic brake fluid DOT 3 (US) or SAE J1703 (UK)
⑥ Swinging arm	As required	High melting point grease
⑦ Final drive chain	As required	Aerosol type chain lubricant
⑧ Wheel bearings	As required	High melting point grease
⑨ Control cables	As required	Light machine oil
⑩ Pivot points	As required	General purpose grease

✱ shaft drive models only

ROUTINE MAINTENANCE GUIDE

Routine maintenance

Refer to Chapter 7 for information relating to the 1983 on models

Introduction

Periodic routine maintenance is a continuous process which should commence immediately the machine is used. It must be carried out at specified mileage recording, or on a calendar basis if the machine is not used frequently, whichever is the sooner. Maintenance should be regarded as an insurance policy, to help keep the machine in the peak of condition and to ensure long, trouble-free service. It has the additional benefit of giving early warning of any faults that may develop and will act as a regular safety check, to the obvious advantage of both rider and machine alike.

The various maintenance tasks are described under their respective mileage and calendar headings. Accompanying diagrams are provided, where necessary. It should be remembered that the interval between the various maintenance tasks serves only as a guide. As the machine gets older or is used under particularly adverse conditions, it would be advisable to reduce the period between each check.

For ease of reference each service operation is described in detail under the relevant heading. However, if further general information is required, it can be found within the manual under the pertinent section heading in the relevant Chapter.

In order that the routine maintenance tasks are carried out with as much ease as possible, it is essential that a good selection of general workshop tools is available.

Included in the kit must be a range of metric ring or combination spanners, a selection of crosshead screwdrivers and at least one pair of circlip pliers.

Additionally, owing to the extreme tightness of most casing screws on Japanese machines, an impact screwdriver, together with a choice of large or small crosshead screw bits, is absolutely indispensable. This is particularly so if the engine has not been dismantled since leaving the factory.

It will be noted that Allen screws are used extensively on the engine outer covers, and it follows that a set of metric Allen keys (wrenches) will be required. These are not expensive and can be obtained from most auto accessory or tool suppliers.

Daily

A daily check of the motorcycle is essential both from mechanical and safety aspects. It is a good idea to develop this checking procedure in a specific sequence so that it will ultimately become as instinctive as actually riding the machine. Done properly, this simple checking sequence will give advanced warning of impending mechanical failures and any condition which may jeopardise the safety of the rider.

1 Oil level

The level of the engine oil is quickly checked by way of the oil sight glass set in the right-hand outer casing. With the machine standing on level ground, the oil should be visible half way up the plastic window. Marks are provided on the rim of the window, indicating the maximum and minimum oil levels. If necessary, top up the oil by way of the filler cap at the rear of the casing. Should too much oil have been added, it should be removed, using a syringe or an empty plastic squeeze pack such as that used for gear oils.

Oil level should be kept between upper and lower level lines in sight glass

Top up as required using a good quality engine oil

2 Tyre pressures

Check the tyre pressures with a pressure gauge that is known to be accurate. Always check the pressure when the tyres are cold. If the machine has travelled a number of miles, the tyres will have become hot and consequently the pressure will have increased. A false reading will therefore result.

It is well worth purchasing a small pocket pressure gauge which can be relied on to give consistent readings, and which will remove any reliance on garage forecourt gauges which tend to be less dependable.

3 Hydraulic fluid level

Check the level of the fluid in the master cylinder reservoir. This may be observed through the window in the side of the reservoir. During normal service, it is unlikely that the hydraulic fluid level will fall dramatically, unless a leak has developed in the system. If this occurs, the fault should be remedied **at once**. The level will fall slowly as the brake linings wear and the fluid defiency should be corrected, when required. Always use an hydraulic fluid of DOT 3 or SAE J1703 specification, and do not mix different types of fluid, even if the specifications appear the same. This will preclude the possibility of two incompatible fluids being mixed and the resultant chemical reaction damaging the seals.

If the level in the reservoir has been allowed to fall below the specified limit, and air has entered the system, the brake in question must be bled, as described in Chapter 5.

4 General checks

In addition to the above points, a running check of the machine in general should be made. It will be found that conditions such as control cables becoming slack will soon make themselves apparent during riding, necessitating adjustment as soon as possible. The electrical system should also be fully functional, noting that in the UK and in many other countries, it is illegal to use the machine with a defective horn or lights, even if they are not in use.

Weekly

1 Final drive chain cleaning and lubrication

The final drive chain is of the endless type, having no joining link in an effort to eliminate any tendency towards breakage. The rollers are equipped with an 'O-ring at each end which seals the lubricant inside and prevents the ingress of water or abrasive grit. It should not, however, be supposed that the need for lubrication is lessened. On the contrary, frequent but sparse lubrication is essential to minimise wear between the chain and sprockets. This can be accomplished by using one of the aerosol chain lubricants but make sure that they are suitable for use on O-ring chains. Many of the chain lubricants available in aerosol form employ a propellant which will damage the O-rings. It is advisable to check the can label carefully before buying. At a pinch, a heavy gear oil may be used, but be prepared to wipe excess oil off the rear of the machine and the rider or passenger. The chain and sprockets should be wiped clean before the lubricant is applied, to ensure adequate penetration.

In particularly adverse weather conditions, or when touring, lubrication should be undertaken more frequently.

A final word of caution; the importance of chain lubrication cannot be overstressed in view of the cost of replacement, and the fact that a considerable amount of dismantling work, including swinging arm removal, will need to be undertaken should replacement be necessary.

Adjust the chain after lubrication, so that there is approximately 25 mm (1 in) slack in the middle of the lower run. Always check with the chain at the tightest point as a chain rarely wears evenly during service.

Adjustment is accomplished after placing the machine on the centre stand and slackening the wheel nut, so that the wheel can be drawn backwards by means of the drawbolt adjusters in the fork ends.

Adjust the drawbolts an equal amount to preserve wheel alignment. The fork ends are clearly marked with a series of parallel lines above the adjusters, to provide a simple visual check.

Front brake fluid level should lie between level marks in sight glass

Rear reservoir has level marks in translucent body

When adjusting chain, check that alignment marks on each side are in the same position

2 Topping up the battery

All models except the 'P' and 'N' are fitted with a 12 volt 12Ah battery, the latter models having a 12 volt 14Ah unit. The battery is housed beneath the dual seat in a compartment to the rear of the air cleaner casing. It is just possible to see the electrolyte upper and lower level lines with the battery in position, but it should be removed for topping up. With the dualseat removed or lifted as appropriate, the battery compartment can be seen just to the rear of the tank. On KZ750 R1 and N1 models (GPz750 and GT750) remove the seat hook bracket bolts. On all models, release the single screw which secures the battery holder strap. Disconnect the battery sensor lead, where fitted. Disconnect the battery negative (−) lead followed by the positive (+) lead. The battery can now be lifted clear of its compartment. Note the routing of the battery vent tube. This must be refitted in its original position during installation.

The transparent plastic case of the battery permits the upper and lower levels of the electrolyte to be observed when the battery is lifted from its housing below the dualseat. Maintenance is normally limited to keeping the electrolyte level between the prescribed upper and lower limits and by making sure that the vent pipe is not blocked. The lead plates and their separators can be seen through the translucent case, a further guide to the general condition of the battery.

Unless acid is split, as may occur if the machine falls over, the electrolyte should always be topped up with distilled water, to restore the correct level. If acid is spilt on any part of the machine, it should be neutralised with an alkaline such as washing soda and washed away with plenty of water, otherwise serious corrosion will occur. Top up with sulphuric acid of the correct specific gravity (1.260-1.280) only when spillage has occurred. Check that the vent pipe is well clear of the frame tubes or any other cycle parts, for obvious reasons.

3 Control cable lubrication

Apply a few drops of motor oil to the exposed inner portion of each control cable. This will prevent drying-off of the cables between the more thorough lubrication that should be carried out during the 3000 mile/3 monthly service.

4 Safety check

Give the machine a close visual inspection, checking for loose nuts and fittings, frayed control cables etc. Check the tyres for damage, especially splitting of the sidewalls. Remove any stones or other objects caught between the treads.

5 Legal check

Ensure that the lights, horn and flashing indicators function correctly, also the speedometer.

3 Monthly, or every 3000 miles (5000 km)

1 Checking the braking system and renewing the pads
Front brake

Check the front brake master cylinder, hose and caliper unit for signs of fluid leakage. Pay particular attention to the condition of the synthetic rubber hose, which should be renewed without question if there are signs of cracking, splitting or other exterior damage.

Check the level of hydraulic fluid by removing the cap on the brake fluid reservoir and lifting out the diaphragm and diaphragm plate. This is one of the maintenance tasks which should never be neglected. Make certain that the handlebars are in the central position when removing the reservoir cap, because if the fluid level is high, the fluid will spill over the reservoir brim. If the level is particularly low, the fluid delivery passage will be allowed direct contact with the air and may necessitate the bleeding of the system at a later date. A level window is provided on the side of the cylinder; if the level is below the mark, brake fluid of the correct grade must be added. **NEVER USE ENGINE OIL** or anything other than the recommended fluid. Other fluids have unsatisfactory characteristics and will quickly destroy the seals.

The brake pads should be inspected for wear. Each pad is marked with a scribed line, indicating the maximum wear limit. If worn beyond this point, **both** brake pads should be renewed as a set. The calipers are provided with a small inspection window through which the pad condition can be checked. The window is very small, allowing only part of the pads to be seen, and is often obscured by brake dust, so it is

Separate R/H switch to lubricate throttle cable and twistgrip

Small caliper inspection window gives restricted view of pads

preferable to lift away the caliper body by removing the two bolts which retain the caliper to its mounting bracket. The caliper body can now be lifted clear, leaving the bracket and pads in situ. If the maximum wear mark cannot be distinguished, renew both pads if less than 1.0 mm (0.0394 in) of friction material remain.

If new pads are to be fitted, wash off any accumulated road dirt and brake dust using a non-greasy solvent. **Do not** blow out brake dust with an air line or inhale the dust because it is toxic. Lift out the old pads and place the new items in position ensuring that they locate properly. Check that the pad shims are correctly positioned on the sides of the caliper bracket and that the anti-rattle spring is in place inside the caliper bridge.

If new pads are to be installed it will be necessary to make room for them in the caliper, which will have compensated for the old worn pads. To this end, attach a length of plastic tubing to the bleed nipple, placing the end in a jar or drain tray. Open the nipple very slightly and push the caliper piston inwards as far as it will go, then secure the nipple to 0.80 kgf m (69 lbf in). Remove the drain tube and wipe up any spilled fluid. **Note:** If done carefully, the above procedure will not introduce air into the hydraulic system, If, however, there is **any** suspicion that air may have entered the system, or if the fluid has not been changed for some time, carry out the bleeding operation described in Chapter 5 after assembly has been completed.

Fit the caliper unit over the pads and bracket and fit the two retaining bolts. These should be tightened to 1.8 kgf m (13.0 lbf ft).

Remove caliper bolts (arrowed) and lift caliper clear of bracket and pads

Pads can now be examined and caliper cleaned

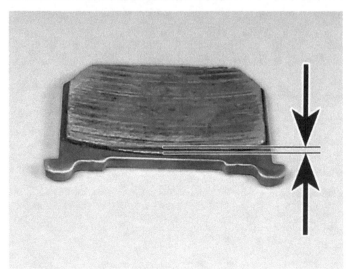

Stepped area of friction material indicates usable range

Repeat the procedure on the remaining caliper, noting that it is preferable to renew the pads in both calipers at the same time to ensure even braking force on the two discs. When work is complete, check that the brakes work properly before riding the machine. Note

that the first few strokes of the brake lever will serve only to adjust the piston to the new pads. This will almost certainly cause a drop in the reservoir fluid level, so check this and top up as required using new DOT 3 or SAEJ1703 hydraulic fluid only.

Rear brake
Refer to the procedure described above, noting that the rear master cylinder is connected to a remote reservoir via a short synthetic rubber hose. The translucent reservoir body has upper and lower level marks and is closed by a screw cap. The caliper support bracket is of a different design, but the associated shims and the caliper unit itself is as described for the front brake. Pay particular attention to accumulated dirt and oil around the rear brake components which are far more susceptible to contamination than front brake items.

2 Adjusting the clutch
There should be 2-3 mm (0.08-0.12 in) of free play in the clutch cable, measured between the clutch lever and the handlebar stock. If the free play is not correct, the clutch should be adjusted as described below.

Chain drive models
Start by locating the in-line clutch cable adjuster which is clipped to the left-hand front down tube. Slacken the locknut and turn the adjuster fully clockwise to give maximum free play in the cable.

Slacken the handlebar adjuster knurled locknut and set the adjuster to leave a gap of 5-6 mm between the underside of the adjuster and the knurled locknut.

Remove the small inspection cover to reveal the clutch pushrod adjuster and locknut. Slacken the locknut and unscrew the adjuster by about two turns. Holding the locknut with a spanner, turn the adjuster screw inwards until resistance is felt, indicating that all free play in the clutch pushrod has been taken up. Back the adjuster off by $\frac{1}{2}$ a turn and hold it in this position until the locknut has been secured.

Using the in-line adjuster, remove all free play from the cable and secure the locknut. Make sure that the cable ends are located properly in the various adjusters. To prevent the ingress of water and road dirt into the clutch cable itself it is good practice to seal the open ends of the adjuster with grease.

Finally, using the adjuster at the handlebar lever, set the cable free play to give a gap of 2-3 mm measured between the lever stock and blade.

Shaft drive models
Make adjustment by first using the knurled adjuster set in the handlebar lever stock. If this does not allow sufficient adjustment to give the correct free play, screw it fully inwards and make adjustment using the in-line adjuster midway down the cable; minor adjustment can then be made at the knurled adjuster. When the free play has been correctly set, tighten both adjuster locknuts securely.

3 Carburettor adjustment
Before attempting to perform any form of carburettor adjustment it should be realised that uneven running may be caused by a number

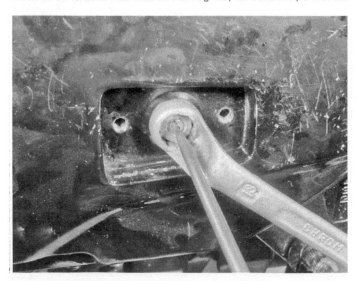

Remove inspection cover to reveal clutch pushrod adjuster and locknut

of other factors, many of which are unrelated to carburation. In particular, attention should be given to spark plug and air filter condition, camshaft chain tensioner and valve clearance adjustment, and on US models, the condition of the Clean Air System.

Start by ensuring that the throttle cable operates smoothly and is adjusted properly. There should be 2-3 mm of free play in the cable, and this can be adjusted as required by turning the in-line adjuster below the twistgrip housing. Once adjusted, tighten the locknut securely and check that the prescribed amount of free play is evident with the handlebar turned to full lock in both directions. If the amount of free play varies with the position of the handlebar, check the routing of the throttle cable and improve it as required.

Idle speed should be checked with the engine at its normal operating temperature, and thus is best dealt with after a run. Allow the engine to idle and check that it runs at 1000-1100 rpm. If adjustment is necessary, use the black plastic adjusting knob which projects between the float bowls of the two centre carburettors.

If, after idle speed adjustment, the engine is not running smoothly then it will be necessary to check carburettor synchronsation. This requires the use of a set of vacuum gauges; refer to the Section on synchronisation in Chapter 2. Failing this, have the synchronisation checked by a Kawasaki dealer.

4 Steering and suspension check

Sit astride the machine and 'bounce' it up and down to check suspension operation. Movement should be smooth and progressive with no signs of jerkiness or unusual noises. Place the machine on its centre stand and check the front forks for signs of oil leakage or scoring on the stanchions. Remove any dirt trapped between the stanchions and dust seals because this will score the chromium plating and cause damage to the fork oil seals. Note that the use of fork gaiters obviates this risk and should result in less wear on the stanchions and seals. It follows that close inspection of these areas should not be necessary on Spectre models.

Grasp the bottom of the fork stanchions, pushing and pulling them

to check for play in the steering head bearings. If movement is found, refer to Chapter 4 for further information on adjustment. Turn the handlebar from lock to lock noting any feeling of resistance or roughness which might indicate worn or badly lubricated steering head bearings. Again, Chapter 4 provides information on overhaul should this prove necessary.

5 Cleaning and adjusting spark plugs

Remove the spark plugs and clean them using a wire brush. Clean the electrode points using emery paper or a fine file and then reset the gaps. To reset the gap, bend the outer electrode to bring it closer to or further from the central electrode, until a feeler gauge of the correct size can just be slid between the gap. Never bend the central electrode or the insulator will crack, causing engine damage if the particles fall in whilst the engine is running. The correct plug gap is 0.7-0.8 mm (0.028-0.032 in). Before replacing the plugs, smear the threads with a small quantity of graphite grease to aid subsequent removal.

Note that where fouling of the electrodes occurs frequently and the machine is used solely for short runs or at low speed in an ambient temperature of 10°C (50°F) or less, it may prove advisable to use slightly hotter plugs than normal. The recommended plug grades and cold weather alternatives are given in Chapter 3.

6 Checking and adjusting the valve clearances

It is advisable to check the valve clearances at the prescribed intervals or at any time that valve train noise, a light regular tapping, becomes evident. Note that incorrect valve clearances will result in poor performance, uneven running and may cause damage to the valves and valve seats if left uncorrected.

Lift or remove the dualseat, depending on the model, to gain access to the rear of the fuel tank. Check that the fuel tap is set to the 'On' or 'Res' position, then release the fuel and vacuum pipes. These are secured by wire clips which can be slid clear of the fuel tap stubs after squeezing the clip tangs together. The pipes can now be worked off the stubs with the aid of a screwdriver. Remove the tank fixing bolt which will be found at or under the rear of the tank. Lift the tank

PART NUMBER 192025	1090	1091	1092	1093	1094	1095	1096	1097	1098	1099	1100	1101	1102	1103	1104	1105	1106	1107	1108	1109	1110	1111	1112	1113	1114
THICKNESS (mm)	2.00	2.05	2.10	2.15	2.20	2.25	2.30	2.35	2.40	2.45	2.50	2.55	2.60	2.65	2.70	2.75	2.80	2.85	2.90	2.95	3.00	3.05	3.10	3.15	3.20

VALVE CLEARANCE (mm) / PRESENT SHIM SIZE — INSTALL THE SHIM OF THIS THICKNESS (mm)

VALVE CLEARANCE (mm)	1090	1091	1092	1093	1094	1095	1096	1097	1098	1099	1100	1101	1102	1103	1104	1105	1106	1107	1108	1109	1110	1111	1112	1113	1114
0.00 ~ 0.03			2.00	2.00	2.05	2.10	2.15	2.20	2.25	2.30	2.35	2.40	2.45	2.50	2.55	2.60	2.65	2.70	2.75	2.80	2.85	2.90	2.95	3.00	3.05
0.04 ~ 0.07		2.00	2.00	2.05	2.10	2.15	2.20	2.25	2.30	2.35	2.40	2.45	2.50	2.55	2.60	2.65	2.70	2.75	2.80	2.85	2.90	2.95	3.00	3.05	3.10
0.08 ~ 0.18	SPECIFIED CLEARANCE / NO CHANGE REQUIRED																								
0.19 ~ 0.22	2.05	2.10	2.15	2.20	2.25	2.30	2.35	2.40	2.45	2.50	2.55	2.60	2.65	2.70	2.75	2.80	2.85	2.90	2.95	3.00	3.05	3.10	3.15	3.20	
0.23 ~ 0.27	2.10	2.15	2.20	2.25	2.30	2.35	2.40	2.45	2.50	2.55	2.60	2.65	2.70	2.75	2.80	2.85	2.90	2.95	3.00	3.05	3.10	3.15	3.20		
0.28 ~ 0.32	2.15	2.20	2.25	2.30	2.35	2.40	2.45	2.50	2.55	2.60	2.65	2.70	2.75	2.80	2.85	2.90	2.95	3.00	3.05	3.10	3.15	3.20			
0.33 ~ 0.37	2.20	2.25	2.30	2.35	2.40	2.45	2.50	2.55	2.60	2.65	2.70	2.75	2.80	2.85	2.90	2.95	3.00	3.05	3.10	3.15	3.20				
0.38 ~ 0.42	2.25	2.30	2.35	2.40	2.45	2.50	2.55	2.60	2.65	2.70	2.75	2.80	2.85	2.90	2.95	3.00	3.05	3.10	3.15	3.20					
0.43 ~ 0.47	2.30	2.35	2.40	2.45	2.50	2.55	2.60	2.65	2.70	2.75	2.80	2.85	2.90	2.95	3.00	3.05	3.10	3.15	3.20						
0.48 ~ 0.52	2.35	2.40	2.45	2.50	2.55	2.60	2.65	2.70	2.75	2.80	2.85	2.90	2.95	3.00	3.05	3.10	3.15	3.20							
0.53 ~ 0.57	2.40	2.45	2.50	2.55	2.60	2.65	2.70	2.75	2.80	2.85	2.90	2.95	3.00	3.05	3.10	3.15	3.20								
0.58 ~ 0.62	2.45	2.50	2.55	2.60	2.65	2.70	2.75	2.80	2.85	2.90	2.95	3.00	3.05	3.10	3.15	3.20									
0.63 ~ 0.67	2.50	2.55	2.60	2.65	2.70	2.75	2.80	2.85	2.90	2.95	3.00	3.05	3.10	3.15	3.20										
0.68 ~ 0.72	2.55	2.60	2.65	2.70	2.75	2.80	2.85	2.90	2.95	3.00	3.05	3.10	3.15	3.20											
0.73 ~ 0.77	2.60	2.65	2.70	2.75	2.80	2.85	2.90	2.95	3.00	3.05	3.10	3.15	3.20												
0.78 ~ 0.82	2.65	2.70	2.75	2.80	2.85	2.90	2.95	3.00	3.05	3.10	3.15	3.20													
0.83 ~ 0.87	2.70	2.75	2.80	2.85	2.90	2.95	3.00	3.05	3.10	3.15	3.20														
0.88 ~ 0.92	2.75	2.80	2.85	2.90	2.95	3.00	3.05	3.10	3.15	3.20															
0.93 ~ 0.97	2.80	2.85	2.90	2.95	3.00	3.05	3.10	3.15	3.20																
0.98 ~ 1.02	2.85	2.90	2.95	3.00	3.05	3.10	3.15	3.20																	
1.03 ~ 1.07	2.90	2.95	3.00	3.05	3.10	3.15	3.20																		
1.08 ~ 1.12	2.95	3.00	3.05	3.10	3.15	3.20																			
1.13 ~ 1.17	3.00	3.05	3.10	3.15	3.20																				
1.18 ~ 1.22	3.05	3.10	3.15	3.20																					
1.23 ~ 1.27	3.10	3.15	3.20																						
1.28 ~ 1.32	3.15	3.20																							
1.33 ~ 1.38	3.20																								

1 Measure the clearance (when cold)
2 Check present shim size.
3 Match clearance in vertical column with present shim size in horizontal column.
4 The shim specified where the lines intersect is the one that will give you the proper clearance.

NOTE: If there is no clearance, select a shim which is several sizes smaller and then measure the clearance.

CAUTION 1. Do not put shim stock under the shim. This may cause the shim to pop out at high rpm causing extensive engine damage.
2. Do not grind the shim. This may cause it to fracture, causing extensive engine damage.
3. Check the valve clearance with the proper method in the text. Checking the clearance at any other cam position may result in improper valve clearance.

Cam

Valve Lifter

Shim

Clearance measured here

Valve clearances shim selection chart

slightly and, where applicable, detach the connector from the fuel level sender unit. The tank may now be pulled back and lifted clear of the frame.

Remove the spark plug leads and release the ignition coils from the frame. On US models only, disconnect and remove as an assembly the air suction valve hoses, vacuum switch valve and silencer. On all models, slacken the cylinder head cover securing screws, then remove the cover. Release the inspection cover at the right-hand end of the crankshaft to expose the ignition pickup assembly.

The crankcase index mark, and the various timing marks on the automatic timing unit, are visible through the oval window in the pickup coil plate. Using the large hexagon provided for this purpose, turn the crankshaft until the '1 4' T mark aligns with the fixed index mark.

Using feeler gauges, measure the clearance between each cam follower and the camshaft, noting that the lobe of the cam should be pointing away from the follower where measurements are being taken. Note down the clearances obtained, turning the crankshaft as necessary to complete the check. The correct clearance for both inlet and exhaust valves is 0.08-0.18 mm (0.003-0.007 in).

If the valve clearances require adjustment it will be necessary to remove the camshafts to gain access to the cam followers and the valve shims. Refer to Chapter 1 Section 7 for details on camshaft removal. Lift the cam follower from its bore in the cylinder head to expose the shim beneath. A magnet may prove useful in removing the cam follower.

Check the size of the existing shim, this being etched on the shim surface. If the marking is indistinct, use a micrometer to check the size. Refer to the accompanying chart which shows the sequence for calculating the size of shim required to restore the correct clearance. Repeat this procedure for the remaining valves which require adjustment. Once all the clearances are set, refit the camshafts as described in Chapter 1, Section 44. Re-check all valve clearances, then complete reassembly by reversing the dismantling sequence, but clean the air filter element prior to fitting the fuel tank and seat (See Section 8).

7 Checking the emission control components – US models

The US versions of the KZ750 are equipped with a system which ensures more complete combustion of the exhaust gases. The system comprises an air suction valve, a vacuum switch valve and a silencer unit, connected by rubber hoses to the air cleaner casing. Reference should be made to Chapter 1, Section 27, where a full description of the system is given, together with the procedure for checking its operation.

The system should be checked at the specified intervals to ensure that the engine is running correctly and that the system is performing its function. In addition, check the system whenever:

a) Idle speed is reduced or unstable
b) Engine power is lacking
c) Unusual engine noises are apparent
d) Frequent backfiring occurs

8 Air filter element cleaning and renewal

To gain access to the air filter casing lid, it is necessary to remove the fuel tank and seat. Where this has not already been done refer to Section 6 for details. Unscrew and remove the large circular lid together with the element.

Check the element closely for signs of splits or holes. If damage of this nature is discovered it will be necessary to renew the element promptly. Remember that unfiltered air entering the engine will carry abrasive dust, leading to accelerated wear of the engine components.

The element can be cleaned by washing it in a non-oily high flash point solvent. The manufacturer warns against the use of paraffin (kerosene) because of the oily residue that this leaves, and against petrol (gasoline) because of the attendant fire risk. With all solvents, use in a well ventilated area, preferably outside, to avoid inhaling the vapour, and take great care to guard against explosion or fire.

Dry the element after cleaning, either by blowing it dry with compressed air or by leaving it overnight for the solvent to evaporate. Check that the foam sealing ring is securely attached to each end of the element, gluing it back in place if necessary. Refit the element, ensuring that the lid clicks into position.

Note that the element will deteriorate with age and repeated

cleanings, and will eventually cease to function effectively. To prevent any risk of damage the element should be renewed at every fifth cleaning or after 6000 miles (10000 km), whichever comes first.

Use **large** hexagon to align 'T' mark with index mark (pickup assembly removed for clarity only)

Unscrew air cleaner lid

Element can now be removed for cleaning

9 Cleaning the fuel system

Although the manufacturer recommends that the fuel system is flushed at the above intervals, it is difficult to advise specific, regular cleaning intervals. In most cases flushing the system will prove unnecessary, unless fuel contaminated with dirt or water has been used. It follows that if a spate of running problems or blocked carburettor jets suggest that such contaminants may be present in the tank, the system should be flushed through as a precaution.

Place the machine on its centre stand in a well ventilated position, preferably out of doors, and well away from any possible sources of fire. Note also that the engine should be **cold**. Pull the overflow tubes clear of the machine and lead them into a suitable container. If a large glass jar is available this will allow any dirt or water to be seen. Set the fuel tap to the 'Pri' position.

Each float bowl incorporates a drain screw at its base. When this is unscrewed by a few turns it will allow the fuel in the float bowl to drain out through the overflow pipe. Slacken each screw in turn and check to see whether dirt or water appears in the drain container. If contamination is evident it will be necessary to drain and clean out the fuel tank, fuel tap and the carburettors. Refer to Chapter 2 for details. When work is complete, remember to reconnect the overflow pipes at the air cleaner casing.

10 Changing the engine/transmission oil

It is important that the engine/transmission oil is changed at the recommended intervals to ensure adequate lubrication of the engine components. If regular oil changes are overlooked the prolonged use of degraded and contaminated oil will lead to premature engine wear. Note that where mileages are unusually low, the engine oil should be changed annually irrespective of mileage readings.

The oil should be changed with the engine at its normal operating temperature, preferably after a run. This ensures that the oil is relatively thin and will drain more quickly and completely. Obtain a container of at least 3.5 litres (6.2 Imp pint/3.7 US qt) capacity, and arrange it beneath the crankcase drain plug. Slacken and remove the plug, noting that the oil filter cover should be left undisturbed, and allow the oil to drain.

When the crankcase is completely emptied, clean the drain plug orifice and refit the plug, tightening it to 27 lbf ft (3.8 kgf m). Remove

the filler plug, and add the specified motor oil to bring the level half way up the window in the outer cover. This will normally take about 3 litres, the oil filter system requiring about ½ litre if this has been renewed.

Start the engine and allow it to idle for a few minutes to distribute the new oil through the system. Switch off and allow the machine to stand for a while, then check that the oil level comes between the upper and lower level marks on the sight glass.

11 General lubrication

Carry out general lubrication at regular intervals. This routine will prolong the life of many of the controls and cables and will usually draw attention to worn or frayed cables before they fail in use.

Use engine oil or machine oil applied with a suitable oil can for lever and pedal pivots. Check the operation of each one prior to lubrication, and where necessary dismantle and clean the pivot first. In the case of the centre and side stands, check for cracks or distortion and ensure that the clamp bolts are secure and that the return spring is in good condition.

The clutch and throttle cables will benefit from full lubrication. This can be accomplished with the aid of a proprietary hydraulic cable oiler, or less quickly but economically using the method shown in the accompanying illustration. The free end of the cable is pushed through the bottom of a plastic bag and the bag taped to the cable outer. Suspend the cable vertically and pour a small quantity of oil into the funnel formed by the bag. Leave the oil to find its way through the cable, preferably overnight. While the cables are detached, grease the clutch lever cable nipple recess and the sliding surfaces of the throttle twistgrip drum, including the end of the handlebar over which the twistgrip fits.

The speedometer and tachometer drive cables can be lubricated by releasing the lower end and partially withdrawing the cable inner. Lubricate sparingly with grease, but avoid greasing the upper end of the cable to prevent grease from finding its way up into the instrument heads.

---

6 Monthly, or every 6000 miles (10 000 km)

1 Changing the oil filter

Carry out the normal engine oil change as described under the 3 monthly/3000 miles heading. After removing the sump drain plug, slacken the centre bolt which secures the oil filter cover to the underside of the unit. Remember that the oil filter chamber will be full of oil, so be ready to catch it.

Clean around the oil filter chamber and the inside of the cover and check the condition of the large O-ring which seals the latter, renewing it if in any doubt as to its serviceability. Fit the new element and the filter cover, tightening the retaining bolt to 14.5 lbf ft (2.0 kgf m). **Do not** overtighten the bolt. Fit and tighten the sump drain plug to 27 lbf ft (3.8 kgf m). Top up with engine oil, noting that approximately 3.5 litre will be required after a filter change.

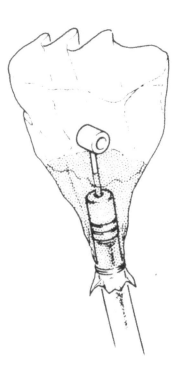

Oiling a control cable

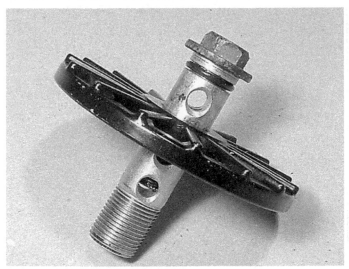

Check O-ring near bolt head, then fit cover

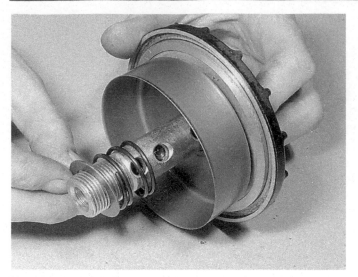

Metal cup fits over bolt as shown; do not omit spring and plain washer

Check cover O-ring, then install as shown

Although not specified as a maintenance item, the oil pump filter gauze should be removed periodically for cleaning. To gain access, drain the oil and remove the sump as described in Chapter 1. Pick the filter gauze out of the pump inlet. Clean the gauze in solvent, removing any deposits with a soft-bristled brush, then refit the gauze to the pump inlet. Wipe out the inside of the sump and clean the gasket faces. Refit all disturbed components using a new gasket and O-rings, as described in Chapter 1, and refill the crankcases with oil.

2 Changing the front fork oil

The machine should be placed on its centre stand on level ground. Place some thick cloth across the fuel tank to protect the paintwork, and remove the handlebar clamps. The handlebar assembly can be lodged on the protected tank while attention is turned to the forks. Where separate cast alloy bars are fitted, slacken the fork top bolts and slide the bars clear of the top yoke, tying them to the frame or lower yoke.

Release air pressure from the fork legs by depressing the air valve core(s). Remove each fork drain screw in turn, allowing the oil to drain into a suitable tray or bowl. Do not let the oil run onto the brake discs. To assist in draining the oil, pump the fork legs up and down to expel the contents. When all of the old oil is drained, clean and dry the plugs and their threaded holes. The plugs should be refitted using a smear of instant gasket to make sure that they are oil tight.

Where necessary, unscrew the inner fork plugs, then lift out the fork springs and place them to one side. Using a measuring jug or similar, add the prescribed quantity and grade of oil to each leg. To check that the oil quantities are correct, first work the fork legs up and down to expel any trapped air. Raise the front of the machine until the wheel is just clear of the ground. This can be done with a jack or by placing blocks under the crankcase, but a strip of wood should be placed against the sump to spread the load.

Using a measuring rod – a length of wire or wooden dowel will suffice – check the level of the oil below the top of the stanchion as shown in the accompanying figure. If necessary, add or subtract oil until the level is correct, and make sure that the level in each fork is identical. When refitting the fork components and handlebar assembly, remember to check and adjust fork air pressure. For further general information on the front fork arrangement fitted to the various models, refer to Chapter 4.

3 Lubricating the automatic timing unit (ATU)

To gain access to the ATU, remove the inspection cover on the right-hand end of the crankcase. Slacken and remove the three screws which retain the ignition pickup baseplate and lift it clear of the casing recess. Note that the screws which secure the pickup coils to the baseplate must not be disturbed.

Slacken the central securing bolt and lift the ATU away from the crankshaft end. The ATU should be clean and should move smoothly when the rotor is turned against spring tension. Make sure that the weights move freely and without excessive free play. Lubricate the pivots with a drop of engine oil. Check that the springs are unbroken

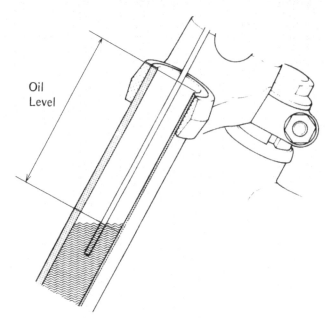

Oil Level

Measuring the fork leg oil level

and that they have not weakened. If in any doubt as to their condition, compare the ATU with a new component to ensure that it is in serviceable condition. Before refitting the unit, fill the internal groove of the rotor with grease.

Refit the ATU by reversing the removal sequence, noting that a pin in the crankshaft end should engage in a slot in the ATU to ensure correct alignment. Refit the pickup baseplate and install the inspection cover using a new gasket where necessary.

4 Lubricating the swinging arm pivot

The swinging arm rear fork is carried on needle roller bearings which should last for many miles if kept properly lubricated. A grease nipple is located at the centre of the pivot tube through which the bearings and pivot shaft can be greased. Using a normal lever type grease gun, clean the nipple and then pump grease into the pivot tube until it is forced out at both ends of the swinging arm. Wipe off any excess grease when the greasing operation is completed.

Should the nipple become blocked or if grease does not exude from the seals the pivot shaft should be removed and the bearings greased manually. Refer to Chapter 4 for further details on the procedure for swinging arm removal and replacement. It is worthwhile taking the extra time required to adopt this approach every two years, so that the old grease can be removed and the bearings and shaft condition checked.

5 Checking the final gear case oil level and greasing the driveshaft joint

Place the machine on its centre stand on level ground and remove the larger level/filler plug from the rear of the casing; oil should be seen level with the bottom threads of the plug orifice. If necessary, top up with the recommended type of oil and use only the same brand as that already in the case. Note that oil is not consumed, it can only leak out; if the level drops noticeably at any time, the gear case should be checked thoroughly for signs of oil leakage and the fault rectified by a competent Kawasaki dealer. Tighten the plug securely.

To allow for the changes in its effective length as the swinging arm moves up and down, the driveshaft is fitted with a sliding joint at its rear end. Both this joint and the splines at the shaft forward end must be greased regularly as follows:

Remove the rear wheel as described in Chapter 5, then remove the left-hand suspension unit bottom mounting nut and pull the unit off its mounting stud. Remove its four retaining nuts and withdraw the final drive casing from the swinging arm, taking care not to lose the coil spring. Slacken its clamp screw and pull the gaiter back off the front gear case unit. Rotate the shaft until a small hole is located in the shaft end, then insert a metal rod into the hole to depress the locking pin inside. The shaft can then be pulled rearwards off its splines.

Wipe all old grease off the shaft male and female splines and apply a thin coat of new high melting-point grease to all of them. The universal joint is sealed for life; check that there are no signs of free play, indicating wear, or a lack of lubrication. If any of this is found, the propeller shaft unit must be renewed. If all is well, check that the locking pin and its hole are aligned and press the shaft on to its splines; the pin should be heard to click into place. Refit the rubber gaiter and tighten its clamp. At the rear, wipe all old grease from the gear case splines and from the inside of the sliding joint, then pack it with high melting-point grease. A specific amount, 17 ml, is recommended, to be packed around the outside of the joint, as shown in the accompanying illustration.

Refit the final drive casing, placing the coil spring over the pinion nut as shown in the accompanying illustration. If smears of grease have appeared at the swinging arm/drive casing joint face, apply a thin smear of jointing compound to the mating surfaces. Apply thread locking compound to the stud threads and tighten the nuts to a torque setting of 2.3 kgf m (16.5 lbf ft). Refit the suspension unit, tightening its nut to 2.5 kgf m (18 lbf ft), then refit the rear wheel as described in Chapter 5.

6 Checking the tightness of all fasteners

The manufacturer recommends that the cylinder head nuts and bolts be check tightened at this interval. Refer to Chapter 1, Section 43 and check that all are secured to the specified torque setting; take note of the tightening sequence in Fig. 1.19.

In addition, move around the machine checking the tightness of all securing nuts and bolts, particularly those on the cycle parts. Refer to the torque settings given at the start of each Chapter.

Two yearly, or every 12 000 miles (20 000 km)

1 Greasing the wheel bearings

Remove each wheel in turn, driving out the bearings and seals for examination. Clean the hub and bearings in a degreasing solution or solvent, then check the latter for wear or damage. If the bearings are sound, pack the races with high melting point grease and refit them into the hub. It is usual for the seals to be damaged during removal, and these should be renewed unless they are in good condition. For further information on wheel bearing removal and replacement refer to Chapter 5, Sections 6 and 7.

Before the front wheel is refitted it is advisable to repack the speedometer drive gearbox. Grease can usually be pushed in between the central sleeve and the seal, but if it is wished to dismantle the gearbox to check the condition of the gears, take care when removing the seal. With care it can be worked out of the gearbox without damage.

2 Overhauling the steering head bearings

The steering head assembly should be dismantled at the prescribed intervals to allow the bearings to be cleaned, examined and greased prior to adjustment. Neglecting this task is a common cause of bearing failure and deteriorating handling, and should not be omitted. For full details covering steering head dismantling examination and reassembly, refer to Chapter 4.

Every 18 000 miles (30 000 km)

Repeat the tasks listed under the previous mileage/time headings, then carry out the following:

Change the final gear case oil

When ready, take the machine for a journey of sufficient length to warm up fully the oil in the final drive unit. The oil is thick and will not drain quickly or remove any impurities until it is fully warmed up.

Place the machine on its centre stand and place a container under the gear case with a sheet of paper or cardboard to keep oil off the wheel and tyre. Remove the oil level/filler and drain plugs and allow the oil to drain fully. Renew the plug sealing washer (or O-ring, as appropriate) if it is damaged or worn. When draining is complete, refit the drain plug, tightening it to 2.0 kgf m (14.5 lbf ft). Add sufficient oil of the recommended grade and viscosity to bring the level up to the bottom of the filler/level plug orifice; the amount required should be 190 cc (6.9 Imp fl oz, 6.4 US fl oz). Refit the filler/level plug, wash off any surplus oil and take the machine for a short journey to warm up the oil and distribute it , then stop the engine and allow a few minutes for the level to settle before rechecking it, top up as necessary. Tighten the filler/level plug securely and wash off all traces of oil from the outside of the swinging arm and casing.

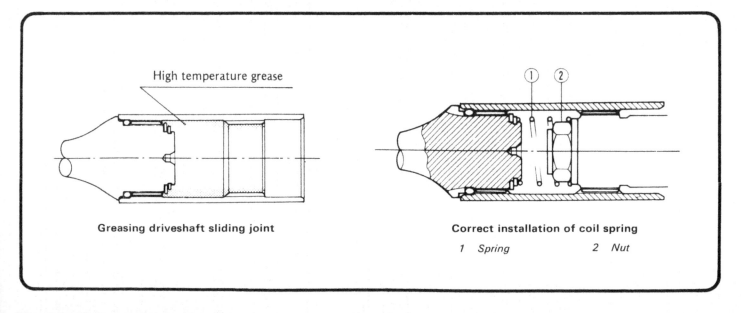

High temperature grease

Greasing driveshaft sliding joint

Correct installation of coil spring

1 Spring 2 Nut

Conversion factors

Length (distance)

Inches (in)	X 25.4	= Millimetres (mm)	X 0.0394	= Inches (in)
Feet (ft)	X 0.305	= Metres (m)	X 3.281	= Feet (ft)
Miles	X 1.609	= Kilometres (km)	X 0.621	= Miles

Volume (capacity)

Cubic inches (cu in; in³)	X 16.387	= Cubic centimetres (cc; cm³)	X 0.061	= Cubic inches (cu in; in³)
Imperial pints (Imp pt)	X 0.568	= Litres (l)	X 1.76	= Imperial pints (Imp pt)
Imperial quarts (Imp qt)	X 1.137	= Litres (l)	X 0.88	= Imperial quarts (Imp qt)
Imperial quarts (Imp qt)	X 1.201	= US quarts (US qt)	X 0.833	= Imperial quarts (Imp qt)
US quarts (US qt)	X 0.946	= Litres (l)	X 1.057	= US quarts (US qt)
Imperial gallons (Imp gal)	X 4.546	= Litres (l)	X 0.22	= Imperial gallons (Imp gal)
Imperial gallons (Imp gal)	X 1.201	= US gallons (US gal)	X 0.833	= Imperial gallons (Imp gal)
US gallons (US gal)	X 3.785	= Litres (l)	X 0.264	= US gallons (US gal)

Mass (weight)

Ounces (oz)	X 28.35	= Grams (g)	X 0.035	= Ounces (oz)
Pounds (lb)	X 0.454	= Kilograms (kg)	X 2.205	= Pounds (lb)

Force

Ounces-force (ozf; oz)	X 0.278	= Newtons (N)	X 3.6	= Ounces-force (ozf; oz)
Pounds-force (lbf; lb)	X 4.448	= Newtons (N)	X 0.225	= Pounds-force (lbf; lb)
Newtons (N)	X 0.1	= Kilograms-force (kgf; kg)	X 9.81	= Newtons (N)

Pressure

Pounds-force per square inch (psi; lbf/in²; lb/in²)	X 0.070	= Kilograms-force per square centimetre (kgf/cm²; kg/cm²)	X 14.223	= Pounds-force per square inch (psi; lbf/in²; lb/in²)
Pounds-force per square inch (psi; lbf/in²; lb/in²)	X 0.068	= Atmospheres (atm)	X 14.696	= Pounds-force per square inch (psi; lbf/in²; lb/in²)
Pounds-force per square inch (psi; lbf/in²; lb/in²)	X 0.069	= Bars	X 14.5	= Pounds-force per square inch (psi; lbf/in²; lb/in²)
Pounds-force per square inch (psi; lbf/in²; lb/in²)	X 6.895	= Kilopascals (kPa)	X 0.145	= Pounds-force per square inch (psi; lbf/in²; lb/in²)
Kilopascals (kPa)	X 0.01	= Kilograms-force per square centimetre (kgf/cm²; kg/cm²)	X 98.1	= Kilopascals (kPa)
Millibar (mbar)	X 100	= Pascals (Pa)	X 0.01	= Millibar (mbar)
Millibar (mbar)	X 0.0145	= Pounds-force per square inch (psi; lbf/in²; lb/in²)	X 68.947	= Millibar (mbar)
Millibar (mbar)	X 0.75	= Millimetres of mercury (mmHg)	X 1.333	= Millibar (mbar)
Millibar (mbar)	X 0.401	= Inches of water (inH₂O)	X 2.491	= Millibar (mbar)
Millimetres of mercury (mmHg)	X 0.535	= Inches of water (inH₂O)	X 1.868	= Millimetres of mercury (mmHg)
Inches of water (inH₂O)	X 0.036	= Pounds-force per square inch (psi; lbf/in²; lb/in²)	X 27.68	= Inches of water (inH₂O)

Torque (moment of force)

Pounds-force inches (lbf in; lb in)	X 1.152	= Kilograms-force centimetre (kgf cm; kg cm)	X 0.868	= Pounds-force inches (lbf in; lb in)
Pounds-force inches (lbf in; lb in)	X 0.113	= Newton metres (Nm)	X 8.85	= Pounds-force inches (lbf in; lb in)
Pounds-force inches (lbf in; lb in)	X 0.083	= Pounds-force feet (lbf ft; lb ft)	X 12	= Pounds-force inches (lbf in; lb in)
Pounds-force feet (lbf ft; lb ft)	X 0.138	= Kilograms-force metres (kgf m; kg m)	X 7.233	= Pounds-force feet (lbf ft; lb ft)
Pounds-force feet (lbf ft; lb ft)	X 1.356	= Newton metres (Nm)	X 0.738	= Pounds-force feet (lbf ft; lb ft)
Newton metres (Nm)	X 0.102	= Kilograms-force metres (kgf m; kg m)	X 9.804	= Newton metres (Nm)

Power

Horsepower (hp)	X 745.7	= Watts (W)	X 0.0013	= Horsepower (hp)

Velocity (speed)

Miles per hour (miles/hr; mph)	X 1.609	= Kilometres per hour (km/hr; kph)	X 0.621	= Miles per hour (miles/hr; mph)

Fuel consumption*

Miles per gallon, Imperial (mpg)	X 0.354	= Kilometres per litre (km/l)	X 2.825	= Miles per gallon, Imperial (mpg)
Miles per gallon, US (mpg)	X 0.425	= Kilometres per litre (km/l)	X 2.352	= Miles per gallon, US (mpg)

Temperature

Degrees Fahrenheit = (°C x 1.8) + 32   Degrees Celsius (Degrees Centigrade; °C) = (°F - 32) x 0.56

*It is common practice to convert from miles per gallon (mpg) to litres/100 kilometres (l/100km), where mpg (Imperial) x l/100 km = 282 and mpg (US) x l/100 km = 235

Chapter 1 Engine, clutch and gearbox

Refer to Chapter 7 for information relating to the 1983 on models

Contents

Specifications

Engine

Type	DOHC, 4-cylinder, air-cooled, four-stroke
Bore	66 mm (2.598 in)
Stroke	54 mm (2.126 in)
Displacement	738 cc (45 cu in)
Compression ratio:	
R1	9.5:1
All others	9.0:1

Maximum horsepower:
  R1 .... 80 hp @ 9500 rpm
  N1 .... 75 hp @ 9500 rpm
  P1 .... 78 hp @ 9500 rpm
  All others .... 74 hp @ 9000 rpm
Maximum torque:
  R1 .... 6.7 kgf m @ 7500 rpm
  All others .... 6.4 kgf m @ 7500 rpm

Pistons

Type .... Aluminium alloy, solid skirt
OD service limit .... 65.81 mm (2.590 in)
Piston ring groove width service limit:
  Top — N1 and P1 .... 1.12 mm (0.044 in)
  Top — all others .... 1.60 mm (0.063 in)
  2nd — N1 and P1 .... 1.31 mm (0.051 in)
  2nd — all others .... 1.60 mm (0.063 in)
  Oil .... 2.60 mm (0.102 in)

Piston rings

End gap service limit (installed):
  Top and 2nd .... 0.70 mm (0.028 in)
Ring/groove clearance service limit:
  Top — N1 and P1 .... 0.17 mm (0.0066 in)
  Top — all others .... 0.15 mm (0.0060 in)
  2nd — N1 and P1 .... 0.16 mm (0.0062 in)
  2nd — all others .... 0.15 mm (0.0060 in)
Ring thickness service limit:
  Top — N1 and P1 .... 0.90 mm (0.035 in)
  Top — all others .... 1.40 mm (0.055 in)
  2nd — N1 and P1 .... 1.10 mm (0.043 in)
  2nd — all others .... 1.40 mm (0.055 in)
Ring width service limit (not N1)
  Top and 2nd .... 1.60 mm (0.63 in)
  Oil .... 2.60 mm (0.102 in)

Cylinder bores

Standard bore diameter .... 66.005 – 66.017 mm (2.5986 – 2.5990 in)
Service limit .... 66.10 mm (2.6024 in)
Ovality (nominal) .... less than 0.01 mm (0.0004 in)
Service limit .... more than 0.05 mm (0.0020 in)
Piston/bore clearance .... 0.040 – 0.067 mm (0.0016 – 0.0026 in)

Small-end assembly (service limits)

Gudgeon pin OD .... 14.96 mm (0.5890 in)
Piston bore ID .... 15.07 mm (0.5933 in)
Small end ID .... 15.05 mm (0.5925 in)

Crankshaft and big-end assembly

Big-end to journal clearance service limit .... 0.10 mm (0.0039 in)
Big-end journal diameter:
  Unmarked .... 34.984 – 34.994 mm (1.3773 – 1.3777 in)
  Marked '0' .... 34.995 – 35.000 mm (1.3777 – 1.3779 in)
  Service limit .... 34.97 mm (1.3767 in)
Connecting rod big-end diameter:
  Marked '0' .... 38.009 – 38.016 mm (1.4964 – 1.4966 in)
  Unmarked .... 38.000 – 38.008 mm (1.4960 – 1.4963 in)
Bearing insert sizes:
  Green .... 1.485 – 1.490 mm (0.0584 – 0.5866 in)
  Black .... 1.480 – 1.485 mm (0.0582 – 0.0584 in)
  Brown .... 1.475 – 1.480 mm (0.0580 – 0.0582 in)
Big-end side clearance service limit .... 0.45 mm (0.0177 in)
Crankshaft runout service limit .... 0.05 mm (0.0019 in)
Main bearing to journal clearance service limit .... 0.11 mm (0.0043 in)
Main bearing journal diameter:
  Unmarked .... 35.984 – 35.991 mm (1.4166 – 1.4169 in)
  Marked '1' .... 35.992 – 36.000 mm (1.4170 – 1.4173 in)
  Service limit .... 35.96 mm (1.4157 in)
Main bearing boss ID:
  Marked '0' .... 39.000 – 39.008 mm (1.5354 – 1.5357 in)
  Unmarked .... 39.009 – 39.016 mm (1.5357 – 1.5360 in)
Main bearing insert size:
  Brown .... 1.490 – 1.494 mm (0.0586 – 0.0588 in)
  Black .... 1.494 – 1.498 mm (0.0588 – 0.0589 in)
  Blue .... 1.498 – 1.502 mm (0.0589 – 0.0591 in)
Crankshaft end float service limit .... 0.40 mm (0.0157 in)

Valves

Stem diameter service limit:	
N1 and P1	6.94 mm (0.2732 in)
All others	6.89 mm (0.2692 in)
Stem warpage service limit	0.05 mm (0.0019 in)
Guide inside diameter service limit	7.08 mm (0.2787 in)
Valve/guide clearance service limit:	
Inlet – N1 and P1	0.33 mm (0.0118 in)
Inlet – all others	0.24 mm (0.0094 in)
Exhaust – N1 and P1	0.33 mm (0.0118 in)
Exhaust – all others	0.19 mm (0.0074 in)
Valve seat width	0.5 – 1.0 mm (0.0196 – 0.0393 in)
Valve clearances (cold)	0.08 – 0.18 mm (0.0031 – 0.0070 in)

Valve timing

Inlet opens at	30° BTDC
Inlet closes at	60° ABDC
Duration	270°
Exhaust opens at	60° BBDC
Exhaust closes at	30° ATDC
Duration	270°

Camshafts

Cam height service limit:	
N1 and P1	36.15 mm (1.4232 in)
All others	35.65 mm (1.4035 in)
Journal diameter service limit:	
N1 and P1	21.91 mm (0.8625 in)
All others	21.93 mm (0.8633 in)
Bearing surface ID service limit	22.12 mm (0.8708 in)
Bearing/journal clearance service limit	0.19 mm (0.0074 in)
Camshaft runout service limit	0.10 mm (0.0039 in)
Cam chain length, (20 pin length) service limit	128.9 mm (5.0747 in)
Chain guide groove depth service limit:	
Upper	3.5 mm (0.1377 in)
Front	2.2 mm (0.0866 in)
Rear	3.5 mm (0.1377 in)

Clutch

Number of plain plates	6
Number of friction plates	7
Friction plate thickness service limit	3.5 mm (0.1377 in)
Plain plate warpage service limit	0.4 mm (0.0157 in)
Friction plate tang/clutch housing clearance service limit	1.0 mm (0.0393 in)
Clutch housing gear/secondary shaft gear backlash service limit	0.12 mm (0.0078 in)
Clutch housing bore service limit	37.03 mm (1.4578 in)
Clutch bearing sleeve OD service limit	31.96 mm (1.2440 in)
Clutch spring free length service limit:	
N1 and P1	33.9 mm (1.3346 in)
All others	N/Av

Transmission

Gearbox type	5-speed, constant mesh
Gear ratios:	
1st	2.33:1 (35/15)
2nd	1.63:1 (31/19)
3rd	1.27:1 (28/22)
4th	1.04:1 (26/25)
Top	0.88:1 (21/24)
Primary reduction ratio:	
N1 and P1	N/App
All others	2.55:1 (27/23 x 63/29)
Final reduction ratio:	
N1 and P1	2.52:1 (15/22 x 37/10)
H1, H2 and H3	2.46:1 (32/13)
All others	2.54:1 (33/13)
Overall drive ratio (top):	
N1 and P1	5.62:1
H1, H2 and H3	5.49:1
All others	5.66:1

Torque settings – E, H and L models

Component	kgf m	lbf ft	lbf in
Air suction valve cover bolts	0.80	–	69.0
Alternator rotor bolt – 10 mm (see Section 39)	7.00	51.0	–
Alternator rotor bolt – 12 mm (see Section 39)	13.0	94.0	–
Alternator stator bolts	0.80	–	69.0

Breather cover bolt	0.60	–	52.0
Camshaft cap bolts	1.20	–	104.0
Camshaft sprocket bolts	1.50	11.0	–
Clutch centre nut	13.50	98.0	–
Clutch spring bolts	0.90	–	78.0
Connecting rod big-end nuts	3.70	27.0	–
Crankcase bolts:			
6 mm	1.00	–	87.0
8 mm	2.50	18.0	–
Cylinder head:			
bolts	3.00	22.0	–
nuts	4.00	29.0	–
Cylinder head cover bolts	0.80	–	69.0
Engine mounting bolts	4.00	29.0	–
Engine mounting bracket bolts	2.40	17.5	–
Gearbox sprocket nut	8.00	58.0	–
Neutral switch	1.50	11.0	–
Oil filter bolt	2.00	14.5	–
Oil pressure switch	1.00	–	87.0
Oil pressure relief valve	1.50	11.0	–
Sump retaining bolts	1.00	–	87.0
Sump drain plug	3.80	27.0	–
Secondary shaft nut	6.00	43.0	–
Spark plugs	2.80	20.0	–
Starter clutch Allen bolts	3.50	25.0	–
Automatic timing unit	2.50	18.0	–
Cam chain tensioner cap (cross-wedge type)	2.50	18.0	–

Additional torque settings* – R1 model

Carburettor adaptor Allen bolts	1.40	10.0	–
Oil passage plug	1.30	–	113.0
Oil cooler gland nuts	2.20	16.0	–

Additional torque settings* – P1 and N1 models

Front gear case components:			
Bearing housing nuts	1.0	–	87.0
Gear case mounting bolts	1.0	–	87.0
Damper cam nut	12.0	87.0	–
Drive pinion nut	12.0	87.0	–
Driven pinion nut	12.0	87.0	–
Engine mounting bracket bolts	2.6	19.0	–

* For main torque settings refer to E, H and L models above.

1 General description

The engine/gearbox unit fitted to the Kawasaki Z750 fours is of the 4 cylinder in-line type, fitted transversely across the frame. The valves are operated by double overhead camshafts driven off the crankshaft by a centre chain. The engine/gear unit is of aluminium alloy construction, with the crankcase divided horizontally.

The crankcase incorporates a wet sump, pressure fed lubrication system, which incorporates a gear driven dual rotor oil pump, an oil filter, a safety by-pass valve, and an oil level or pressure switch.

The engine is built in unit with the gearbox. This means that when the engine is completely dismantled, the clutch and gearbox are dismantled too. This task is made easy by arranging the crankcase to separate horizontally.

Power from the crankshaft is transmitted via a Hy-Vo chain to the primary shaft which runs to the rear of the crankshaft. The primary shaft carries the starter clutch assembly and provides gear drive to the clutch. From the clutch, power is transmitted to the input shaft of the five-speed constant mesh gearbox.

On most of the models covered in this manual the output shaft carries an external sprocket which drives a final drive chain. In the case of the N1 (Spectre) and P1 (GT 750) models, the output shaft engages a bolted-on bevel gearbox unit, transferring drive to the rear wheel by shaft.

2 Operations with the engine/gearbox unit in the frame

The components and assemblies listed below can be removed without having to remove the engine unit first. If, however, a number of areas must be attended to then removal of the engine is recommended. Note that although the cylinder block and pistons can be attended to with the engine in situ, this is inadvisable because accumulation of road dirt around the cylinder barrel holding studs is likely to drop into the crankcase as the block is lifted. Refer to Section 7 for more details.

The following components/assemblies can be removed and refitted with the engine unit in place:

a) Sprocket cover/output bevel gearbox cover
b) Sprocket/output bevel gearbox
c) Clutch release mechanism
d) Neutral switch
e) Gear selector mechanism (external)
f) Alternator
g) Secondary shaft assembly and starter motor
h) Ignition pickup assembly and ATU
i) Clutch assembly
j) Oil filter
k) Bypass valve, sump and oil pump
l) Cylinder head cover, camshafts and tensioner
m) Cylinder head
n) Cylinder block and pistons*

*Not recommended – see above

3 Operations with the engine/gearbox unit removed from the frame

Removal of any component or assembly contained within the crankcase halves, except the secondary shaft components and the oil pump, will require the removal of the engine unit and separation of the crankcase halves. These can be summarised as follows:

- a) Crankshaft assembly
- b) Main and big-end bearings
- c) Connecting rods
- d) Gearbox shafts and pinions
- e) Gear selector drum and forks
- f) All crankcase bearings
- g) Primary drive chain
- h) Camshaft chain

4 Likely problem areas: general

1 In the course of the workshop project which provided the photographs which illustrate this manual, a number of minor problems were encountered which would normally require the use of a factory service tool to overcome. It is, however, possible to avoid this provided that the correct approach is taken.

Clutch centre nut
2 This nut will be **very** tight having been tightened to 98 lbf ft (13.5 kgf m) and in addition has a flattened section which will exert considerable drag on the shaft threads. Kawasaki produce a universal holding tool Part Number 57001-305 with which the clutch centre can be held during removal of the nut. A good home made alternative is described in Section 10 of this Chapter. The only other possible course of action is to strip the clutch while the engine unit is in the frame. This will allow the clutch centre to be held by selecting top gear and applying the rear brake while it is slackened.

Secondary shaft nut
3 This may also prove to be rather tight, and will require the crankshaft to be held while it is removed. To this end, remove the nut **before** removing the alternator rotor, using a strap wrench around the outside of the rotor to hold the crankshaft.

5 Removing the engine/gearbox unit

1 If possible, raise the machine to a comfortable working height on a hydraulic ramp or a platform made from stout planks and strong blocks. Place the machine on its centre stand and check that it is in no danger of toppling over. Place a drain tray or bowl of about one gallon (5 litre) capacity beneath the sump and remove the drain plug. Leave the machine until most of the oil has drained, then remove the centre bolt which retains the circular oil filter cover to the underside of the unit. Remove and discard the filter element but check that the cup, spring and plain washer are retained.
2 On all models except the R1 and P1 versions lift the seat, remove the split pin from the end of the seat stay, release the two retaining screws which secure each hinge and the seat. On R1 and P1 models remove the seat by unlocking the combined helmet and seat lock and pushing the catch lever forward.
3 The fuel tank on all models is retained by a single bolt fixing at the rear and is located on rubber buffers at the front. Slacken and remove the fuel tank rear fixing bolt, and check that the fuel tap is turned to the 'ON' or 'RES' position. Free the fuel pipe and the smaller diameter vacuum pipe at the tab stubs by squeezing together the ears of the retaining clip and sliding it down the pipe. The pipes can now be worked off the stubs with the aid of a small screwdriver. Where fitted, trace the fuel gauge or low level warning lamp wiring from the sender unit on the underside of the tank and disconnect it at the connector block. Lift the rear of the tank by an inch or two and remove it by pulling it rearwards. Remember that the tank represents a significant fire hazard; store it accordingly.
4 Pull off the plastic side panels. Disconnect the battery leads to prevent accidental short circuits and, where fitted disconnect the battery sensor lead. Working from the left-hand side of the machine, release the single screw which retains the plastic cover over the electrical panel. Open the hinged cover and unscrew the two rubber bushed mounting bolts which secure the electrical panel to the frame. Move the panel away from the frame to allow access to the air cleaner bracket bolt. Remove the bolt, followed by the corresponding bolt on the opposite side of the bracket assembly.

5 Where fitted, remove the seat front mounting bracket after releasing its two retaining bolts (applies to machines with lift-off seat units only). Unscrew the air cleaner casing cap and remove it and the air filter element. Remove the screw which secures the battery strap and remove the battery to a safe place for storage.
6 Slacken the carburettor clamp screws. Roll the spring retainers rearwards along the carburettor to air cleaner hoses. Pull the carburettor drain hoses clear of the frame. Disconnect the vacuum hoses from their stubs.
7 The throttle cable arrangement varies slightly according to the model. Where a cable adjuster is fitted, slacken the adjuster locknut so that the adjuster can be disengaged from the support bracket. Where no lower adjuster is fitted, lift the cable outer clear of its locating recess and slide the inner cable clear of the bracket via the slot provided. In both cases the cable nipple can be released from the pulley by aligning the cable inner with the slot and sliding it clear. Since access is limited this operation can be left until the carburettors are partly removed if desired.
8 The carburettors are a tight fit between the two sets of mounting rubbers and a certain amount of careful manipulation will be required to extricate them. Pull the carburettor bank rearwards, twisting it clear of the intake rubbers. Once these have been freed, manoeuvre the assembly clear of the air cleaner rubbers and withdraw from the right-hand side. As the carburettor bank clears the mountings, unhook the throttle cable inner.
9 Displace the clip which secures the breather hose to the breather cover by squeezing the clip ears together and sliding the clip up the hose. The hose can now be pulled off its stud. Pull the air cleaner casing forward and remove it from the frame.
10 Release the two domed nuts which secure the left-hand footrest and lift it away. Note that where a side stand interlock switch is fitted, the operating spring should be unhooked as the footrest assembly is drawn clear. Remove the gearchange lever, either by removing its pinch bolt and sliding the lever off the splined shaft, or in the case of machines fitted with a rearset linkage, by removing the pinch bolt which removes the operating link and removing the circlip which secures the pedal itself, allowing the complete assembly to be drawn off. Note that a plastic cap is clipped over the circlip fitting.
11 Release the two bolts and plain washers which retain the starter motor cover to the crankcase and lift the cover away. The sprocket cover incorporates the clutch release mechanism and is removed together with the clutch cable. Start by slackening the clutch cable adjuster at the handlebar lever, aligning the slots in the lever stock, adjuster and locknut so that the cable can be disengaged. Follow the cable down the frame tube, releasing the cable clips to free it. (Note that in some circumstances it may not be necessary to remove the cover completely, in which case the cable may be left in place).

Chain drive models only
12 Slacken the four bolts which secure the sprocket cover to the crankcase. The cover can now be pulled clear. Pull the cover locating dowels out of the crankcase. Slide the clutch pushrod out of its bore and place it with the sprocket cover.

Shaft drive models
13 Where shaft final drive is fitted then engine removal procedure differs somewhat at this point, and should be continued as described below. Start by removing the rear wheel as described in Chapter 5, Section 4. Release the domed nuts which secure the left-hand rear suspension unit and tie it clear of the swinging arm/final drive casing. Slacken and remove the four nuts which secure the final drive casing to the swinging arm, then lift the casing away, taking care not to lose the spring which will be released as the casing comes free.
14 Working from the left-hand side of the engine unit, remove the rear section of the engine casing to reveal the front gear case unit. Note that the clutch release mechanism is incorporated here and thus the clutch cable should be disconnected as described in paragraph 14 above. Slacken the screw clip which retains the rubber gaiter (boot) to the front gear case. Pull the gaiter back to expose the plain cylindrical end of the drive shaft. Rotate the shaft, by turning the crankshaft with a gear engaged, until a small hole is located in the shaft end. Insert a thin metal rod or a small screwdriver into the hole, depress the locking pin inside and pull the drive shaft rearwards until it disengages with the front gear case output splines.
15 Slacken evenly and remove the eight bolts which secure the front gear case to the crankcase, and lift the unit away. To remove the shock

absorber cam half from the output shaft, straighten the staked portion of the retaining nut, then remove the nut whilst holding the cam with the holding tool, Part Number 57001-1025, or a chain wrench.

16 Pull back the rubber boot which covers the starter motor terminal and disconnect the lead. Disconnect the neutral switch lead at the switch, and the alternator output leads at their connectors. Make sure all leads are lodged away from the engine.

17 The procedure for removing the exhaust system differs in the case of the R1. For details refer to paragraph 19, otherwise follow the procedure detailed here. Slacken the clamps which hold each of the four exhaust pipes to the balance pipe/silencer assembly. These will almost always be rusty and may come apart easier if they are soaked in penetrating oil or WD40 and left for an hour or so to free off. Moving to the front of the engine, unscrew the flanged nuts which secure the finned exhaust pipe retainers to the two centre exhaust ports. Slide the retainers clear and remove the split collet halves to avoid loss. Pull the centre exhaust pipes forward and clear of the machine. It is normal for the pipes to be firmly rusted in place, in which case a hide mallet or a hammer and a wooden block can be used.

18 Release the two outer exhaust pipe retainers as described above, sliding the retainers away from the ports and removing the split collets. Remove the pillion footrest mounting bolts to free the silencer support brackets. The system can now be pulled forward to free the exhaust pipes, then lowered and manoeuvred clear of the machine.

19 The exhaust arrangement on the R1 differs from that of the remaining models, and is dismantled as follows. Dismantle the two inner exhaust pipe retainers as described above. Slacken the clamps at the lower ends of the pipes, using penetrating oil or WD40 as required, then pull off the inner pipes. Remove the single central bolt which secures the system to the crankcase. Remove the remaining exhaust pipe retainers and then remove the pillion footrest mounting bolts to free the rear of the system, which can now be manoeuvred clear of the machine.

20 The R1 models sold in the UK are equipped with an oil cooler, and it is essential that the hose unions are released at the crankcase prior to engine removal. Place a drain tray or bowl beneath the front of the crankcase to catch the oil which will be released when the unions are freed. Each union is retained by two bolts. Once the residual oil has drained the hoses can be tied clear of the engine. Better still, remove the four rubber-bushed mounting bolts and lift the assembly clear of the frame, for safe keeping.

21 On the right-hand side of the machine, release the engine earth lead from the crankcase. It is secured by a small hexagon-headed screw just to the rear of the oil filler cap. Unhook the brake light switch operating spring and disconnect the switch leads. Remove the brake pedal pinch bolt and remove the pedal, then dismantle the right-hand footrest assembly.

22 Trace the ignition pickup leads back to the four-pin connector. Disconnect the ignition wiring and coil it up clear of the frame. Where

fitted, disconnect the oil pressure switch lead. Where an oil level switch is fitted, disconnect its lead. Where a mechanical tachometer is fitted, release the drive cable from the cylinder head.

23 Knock back the locking tab which secures the gearbox sprocket retaining nut. With the aid of an assistant, lock the rear wheel, either by holding it or by applying the rear brake with the operating arm at the hub. Slacken and remove the sprocket retaining nut and tab washer and slide the sprocket off the shaft. If there is not sufficient play in the chain, slacken the rear wheel spindle and adjusters and push the wheel forward to obtain the necessary clearance.

24 Check around the unit for any hoses, electrical leads or cables that might impede removal. Before attempting to remove the engine note that a **minimum** of two people will be required to ensure that the operation is accomplished without incident. For most purposes all models are similar, but it should be noted that the front mountings on R1 models, and all mountings on the shaft drive models, are rubber bushed.

25 Place a jack beneath the crankcase to take the weight of the engine, using a wooden block to spread the loading on the casting. Alternatively, have ready a pair of large tyre levers or similar levers so that the engine can be lifted sufficiently to allow the bolts to be removed. Remove the nuts from the left-hand end of the rear mounting bolts. Remove the four front engine mounting bolts. Remove the front brackets from the frame.

26 Withdraw the engine rear mounting bolts, noting that the upper bolt runs through a spacer which should be displaced and removed. Take care not to damage any of the bolt threads as they are withdrawn. Dismantle the upper rear mounting bracket and remove it together with the brake light switch.

27 Lift the engine upwards just enough to ensure that the crankcase clears the front and rear lower mounting lugs. Move the unit to the right and rest it on the frame tube. Establish a firm purchase on the engine and lift it out to the right.

---

6   Dismantling the engine/gearbox: general

1   Before commencing work on the engine unit, the external surfaces should be cleaned thoroughly. There are a number of proprietary cleaning solvents on the market including Jizer and Gunk. It is best to soak the parts in one of these solvents, using a cheap paint brush. Allow the solvent to penetrate the dirt, and afterwards wash down with water, making sure not to let water penetrate the electrical system or get into the engine, as many parts are now more exposed.

2   Never use force to remove any stubborn part unless specific mention is made of this requirement in the text. There is invariably good reason why a part is difficult to remove, often because the dismantling procedure has been tackled out of sequence.

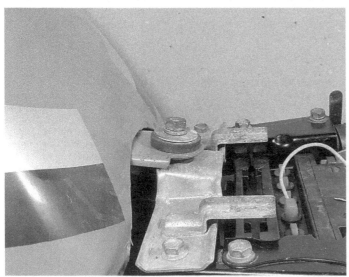

5.3a Slacken and remove the single tank fixing bolt

5.3b Turn tap to 'On' and detach fuel and vacuum hoses

5.3c Trace and disconnect fuel gauge sender wiring

5.4a Remove electrical panel and air cleaner bracket bolt ...

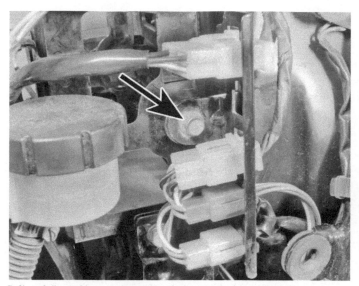

5.4b ... followed by corresponding bolt on right-hand side

5.5 Remove seat bracket, battery strap, battery and sensor leads and lift battery out of tray

5.8 Disconnect clamps, retainers and hoses to allow carburettor assembly to be manoeuvred clear

5.10 Remove domed nuts to free left-hand footrest

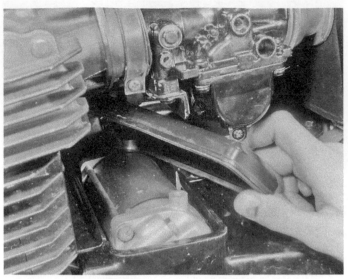
5.11a Remove bolts and lift starter motor cover clear

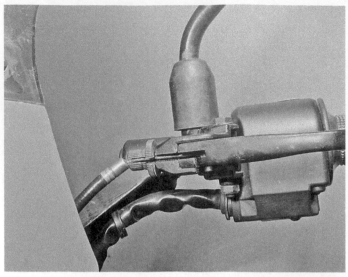
5.11b Slacken clutch cable adjuster and disconnect cable as shown

5.12 Release cover bolts and remove complete with cable

5.19 Note central exhaust mounting to crankcase (GPz750 model)

5.20a Release oil cooler connections (where fitted)

5.20b Remove oil cooler bolts and lift assembly clear of frame

5.21 Do not forget to release engine earth strap

7 Dismantling the engine/gearbox unit: removing the camshafts, cylinder head and cylinder block

1 The above components can be removed with the engine installed in the frame, noting that it will first be necessary to remove the fuel tank, carburettors, exhaust system, ignition coils, and on US models, the vacuum switch valve and silencer. For details on this refer to Section 5 of this Chapter.

2 Remove the circular inspection cover at the right-hand end of the crankcase to expose the ignition pickup assembly. Using the large hexagon provided (do **not** use the smaller retaining bolt hexagon) rotate the crankshaft until the pistons of cylinders 1 and 4 are at TDC (top dead centre). A small oval inspection hole in the pickup baseplate allows the timing marks and the fixed index mark to be seen. Turn the crankshaft clockwise just past the '1 4' and 'F' marks until the 'T' mark is aligned with the index mark.

3 On models with a mechanically driven tachometer, remove the single Allen headed screw which retains the stop plate to the cylinder head and withdraw the tachometer gear. Slacken evenly and remove the 24 cylinder head cover bolts and remove the cover. If the cover is stuck to the gasket, tap around the jointing face with a soft faced mallet or a hammer and wooden block to jar it free. On no account try to lever the cover off.

4 On 1980 models lock the camshaft chain tensioner by removing the short lock bolt and washer located on the right-hand side of the tensioner body, and fitting in its place a longer (16 mm or more) 6 mm bolt or screw which should be tightened down until it clamps the tensioner plunger. Detach the camshaft chain tensioner assembly by releasing its two retaining bolts, noting that the upper one has an aluminium washer.

5 For 1981 and later models an automatic tensioner was fitted. This type can be distinguished from the earlier version by the cylindrical extension of the casting in the form of a cap pointing to the right-hand side of the machine, and is known by Kawasaki as the 'Cross-wedge' type. Remove it by first releasing spring pressure by unscrewing the cap. The two retaining bolts can now be removed and the tensioner body lifted away.

6 Slacken each of the camshaft cap bolts evenly by about one turn at a time. The camshafts are under pressure from the valve springs and will be pushed clear of the bearing surfaces in the cylinder head. Once valve spring pressure has been relieved, remove the bolts and place them with the bearing caps in a safe place. Place a bar or a length of

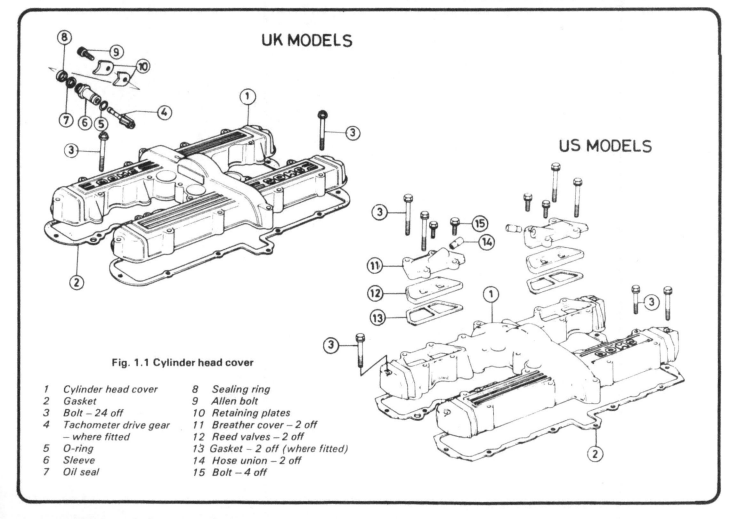

Fig. 1.1 Cylinder head cover

1	Cylinder head cover
2	Gasket
3	Bolt – 24 off
4	Tachometer drive gear – where fitted
5	O-ring
6	Sleeve
7	Oil seal
8	Sealing ring
9	Allen bolt
10	Retaining plates
11	Breather cover – 2 off
12	Reed valves – 2 off
13	Gasket – 2 off (where fitted)
14	Hose union – 2 off
15	Bolt – 4 off

wire between the camshaft sprockets and under the chain to prevent the latter from dropping into the crankcase when the camshafts are removed. Lift each camshaft clear of the cylinder head and disengage it from the chain. The exhaust camshaft on machines with mechanical tachometers can be identified by its tachometer drive gear. Although on electronic tachometer models a raised blank will be found in place of the machined gear, identification is far from obvious. To avoid any possible mistake during assembly tie a marked label to the exhaust camshaft to identify it.

7   The cylinder head is retained by a total of twelve nuts plus two smaller bolts located inside the camshaft chain tunnel. Remove the two bolts, then release the cylinder head nuts in the reverse order of the tightening sequence shown in Fig. 1.19. Slacken each nut by about ½ turn at a time to release evenly pressure on the head thus avoiding any risk of warpage. Once the nuts are slack, run them off the studs. The cylinder head should now lift off the holding studs, but may prove to be stuck in place by the gasket. Avoid the temptation to lever between the cylinder head and block fins; they are brittle and are easily broken off if overstressed. Tap around the joint face using a wooden block and a hammer, or a soft-faced mallet, to jar the head free. Once the seal has been broken, lift the head clear, feeding the camshaft chain through the tunnel.

8   **Important note:** Before removing the cylinder block, it should be noted that there is likely to be an accumulation of road dirt around the base of the holding down studs. Unless great care is taken, this will drop down into the crankcase during removal, necessitating crankcase separation. Try to arrange the unit so that the block faces downwards, permitting the debris to drop clear of the crankcase mouths. Clean the studs carefully before turning the unit up the right way again. If the cylinder block is to be removed with the engine unit in the frame the above approach will obviously be impracticable. The only alternative here is to clean the area around the front studs as carefully as possible, using a vacuum cleaner to remove as much of the loose dirt as can be reached. Remove the small rubber bungs in the drillings on each side of the stud holes, then spray the area with aerosol chain lubricant and leave it to solidify. The grease-like consistency of the chain lubricant should engulf and hold the dirt particles, but great care should be taken when lifting the block. If any of the debris enters the crankcase it is vital that it is removed, even if it proves necessary to strip the engine bottom end to be certain.

9   Lift the cylinder block by an inch or two and carefully remove any residual road dirt from the vicinity of the studs. Before the pistons are allowed to emerge from the bores, guard gainst pieces of broken ring or any further debris entering the crankcase by packing clean rag into the crankcase mouths. The block can now be lifted clear of the holding studs and pistons.

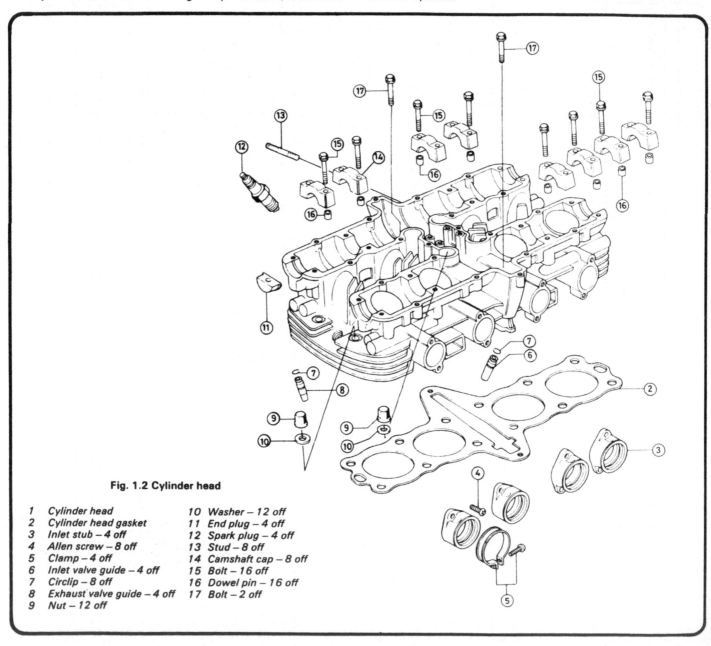

Fig. 1.2 Cylinder head

1	Cylinder head	10	Washer – 12 off
2	Cylinder head gasket	11	End plug – 4 off
3	Inlet stub – 4 off	12	Spark plug – 4 off
4	Allen screw – 8 off	13	Stud – 8 off
5	Clamp – 4 off	14	Camshaft cap – 8 off
6	Inlet valve guide – 4 off	15	Bolt – 16 off
7	Circlip – 8 off	16	Dowel pin – 16 off
8	Exhaust valve guide – 4 off	17	Bolt – 2 off
9	Nut – 12 off		

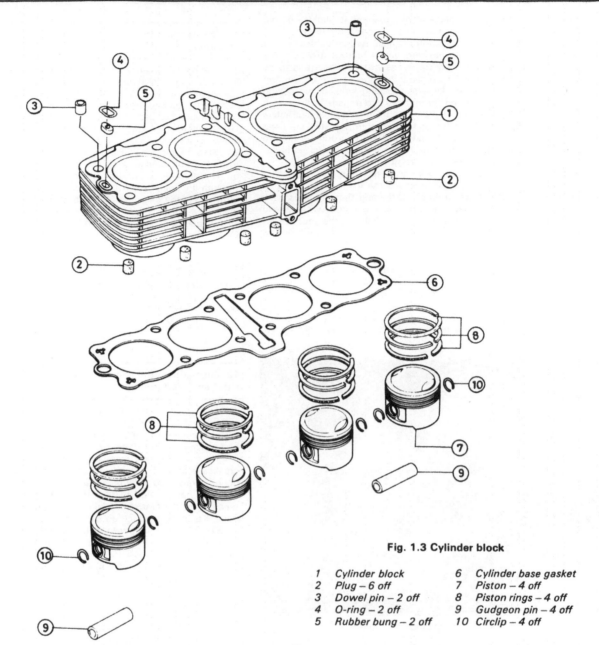

Fig. 1.3 Cylinder block

1	Cylinder block	6	Cylinder base gasket
2	Plug – 6 off	7	Piston – 4 off
3	Dowel pin – 2 off	8	Piston rings – 4 off
4	O-ring – 2 off	9	Gudgeon pin – 4 off
5	Rubber bung – 2 off	10	Circlip – 4 off

8 Dismantling the engine/gearbox unit: removing the pistons and piston rings

1 Remove the circlips from the pistons by inserting a screwdriver (or a piece of welding rod chamfered one end) through the groove in each piston boss. Discard them. Never re-use old circlips during the rebuild.

2 Using a drift of suitable diameter, tap each gudgeon pin out of position, supporting each piston and connecting rod in turn. Using a spirit-based marker, mark each piston inside the skirt so that it is replaced in the appropriate bore. If the gudgeon pins are a tight fit in the piston bosses, it is advisable to warm the pistons. One way is to soak a rag in very hot water, wring the water out and wrap the rag round the piston very quickly. The resultant expansion should ease the grip of the piston bosses on the steel pins.

9 Dismantling the engine/gearbox unit: removing the ignition pickup assembly

1 The ignition pickup components are mounted at the right-hand end of the crankshaft and are housed beneath a circular inspection cover. The assembly can be removed with the engine installed, noting that if it is necessary to remove it completely the pickup leads should be traced back to the connector block and disconnected.

2 Remove the three cross head screws which retain the ignition pickup baseplate to the crankcase. **Do not** disturb the inner screws which retain the pickup coils to the baseplate. Lift the assembly clear of the crankcase, pulling the cable grommet clear of its slot. Slacken and remove the single bolt which secures the automatic timing unit to the crankshaft end and lift it away.

3 Most of the 750 models are fitted with an oil pressure switch located at the bottom of the pickup recess. The switch lead runs through the same length of sleeving as the pickup leads, and thus should be disconnected by releasing its single terminal screw.

10 Dismantling the engine/gearbox unit: removing the clutch

1 The clutch can be removed for inspection or overhaul with the engine unit in or out of the frame. In the former case it will first be necessary to drain the engine oil, and to remove the right-hand footrest assembly to gain working space around the clutch cover.

2 **Note**: If it is necessary to remove the clutch centre nut, some sort of holding tool will be required. The nut is designed to resist loosening

in service by the simple expedient of crushing it slightly during manufacture. Added to the drag that this produces is its 98 lbf ft (13.5 kgf m) torque figure, the combination of these two factors calling for a lot of force to be applied during removal. It follows that a secure method of holding the clutch centre is essential if damage is to be avoided. Kawasaki produce an excellent holding tool, Part Number 57001-3—5, which is rather like a self-locking wrench, but having extended blade-like jaws turned through 90° to engage in the clutch centre splines. In the absence of the correct tool, a simple alternative is shown in photograph 38.1h. It was fabricated from ⅛ steel strip and uses a nut and bolt as a pivot. The jaws should be ground or filed to suit the clutch centre splines, and the handles should be left at about 2 – 3 feet in length to provide a secure grip.

3   Remove the clutch cover screws and detach the cover, catching any residual oil which is released. Remove the five hexagon-headed bolts which retain the clutch springs, slackening them progressively until spring pressure has been released. Remove the washers and springs, then lift the pressure plate away. Pull out the mushroom-headed pushrod from the centre of the input shaft and displace and remove the single steel ball which is fitted behind it. Remove the clutch plates as an assembly.

4   Using the Kawasaki holding tool or substitute, hold the clutch centre stationary while the clutch centre nut is slackened. Once the nut has been slackened, remove it and the large plain washer or Belville washer beneath it.

5   Pull off the clutch centre. Remove the clutch outer drum and plain thrust washer, followed by its needle roller bearing and inner sleeve. Finally, remove the thick thrust washer.

10.3 Remove bolts, springs, clutch cover and plates

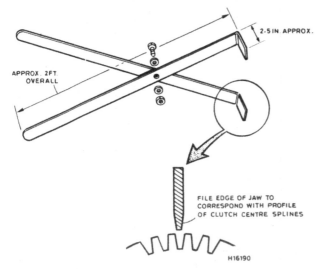

2·5 IN. APPROX.

APPROX. 2FT. OVERALL

FILE EDGE OF JAW TO CORRESPOND WITH PROFILE OF CLUTCH CENTRE SPLINES

H16190

Fig. 1.4 Fabricated clutch holding tool

11 Dismantling the engine/gearbox unit: removing the sump and oil pump

1   The oil pump can be removed with the engine unit installed in the frame or on the workbench. If the operation is to be carried out with the engine installed, it will be necessary to drain the engine oil and remove the oil filter first. The clutch cover and clutch assembly should be dismantled as described in Section 10 of this Chapter. Where an oil level switch is incorporated, trace and disconnect its lead to permit sump removal.

2   Slacken and remove the sump retaining screws and lift the sump away from the crankcase. Make provision to catch the residual engine oil. The pump is secured by two screws and a single bolt which pass through the crankcase wall and screw into the pump body. The screws form two of the three mounting points for the secondary shaft retainer, whilst the bolt is located just to the rear of the retainer. Remove the screws and bolt, noting that the former are staked against the retainer for security. Use an impact driver to free these screws to avoid damage to the screw heads. The third screw in the retainer does not affect the oil pump and can be left undisturbed if no further dismantling is necessary. Otherwise remove it and place it with the pump mounting screws and bolt for safe keeping.

3   The pump can now be removed by disengaging it from the two dowel pins which locate it. These may come away with the pump, or can be pushed out of the crankcase.

12 Dismantling the engine/gearbox unit: removing the secondary shaft and starter clutch

1   The secondary shaft, incorporating the starter clutch unit can be removed with the engine installed in the frame after the exhaust system, engine sprocket cover (or front gear case), clutch, oil pump and alternator cover have been removed.

2   Remove the two screws which secure the bearing cap to the crankcase at the left-hand end of the secondary shaft, noting that the upper screw also retains a wiring clip. The secondary shaft nut is removed next, noting that its torque setting of 43 lbf ft (6.0 kgf m) means that the crankshaft will have to be held securely while this is done. Kawasaki produce an alternator rotor holding tool, Part Number 57001-308, for this purpose, but most owners will have to find an alternative. If the cylinder head, cylinder block and pistons have been removed, it is permissible to run a smooth round metal bar through one of the connecting rod small ends, provided that the ends of the bar are supported on wooden blocks to avoid damage to the crankcase mouth. Alternatively, use a heavy duty strap wrench around the alternator rotor, noting that the lightweight types sold for oil filter removal will probably be too weak for this job.

3   Check that the bearing retainer at the right-hand end of the shaft has been removed, then tap the shaft through from the left-hand side to displace the right-hand bearing. (Note: On the model used for the workshop project, the shaft had been coated with Loctite or some similar compound, presumably to eliminate clutch chatter. This made removal of the shaft more difficult, but may not have been done on other models). Support the starter clutch/sprocket assembly with one hand and withdraw the shaft and bearing. The clutch unit can now be disengaged from the primary chain and lifted out of the crankcase.

13 Dismantling the engine/gearbox unit: removing the alternator rotor

1   Remove the screws securing the alternator cover to the left-hand side of the engine, and lift away the cover complete with the alternator stator. The alternator rotor is secured to the crankshaft end by a single retaining bolt. To remove the rotor from the tapered shaft end it is necessary to employ an extractor. Use only the Kawasaki rotor removal tool, Part Number 57001-254 or 57001-1099. An alternative is to obtain a large metric bolt having the necessary fine-pitched thread – we found that a Ford Transit wheel stud worked well.

2   Hold the rotor using either the Kawasaki rotor holding tool, Part Number 57001-308, or a heavy duty strap wrench. Slacken and

remove the retaining bolt and screw the extractor into the large thread provided. Hold the rotor, and tighten the extractor firmly. If the rotor does not draw off the shaft, strike smartly the end of the extractor to jar it free. If necessary, tighten the extractor a little more and repeat the above, but beware of overtightening.

14 Dismantling the engine/gearbox unit: removing the external gearchange components

1 Access to the external components of the gearchange mechanism is gained after removal of the engine left-hand cover and sprocket (chain drive models) or front gear case (shaft drive models), and may be carried out with the engine unit in or out of the frame.
2 In the case of chain drive machines, remove the three bolts which secure the final drive chain guard to the crankcase, followed by the gearchange mechanism cover bolts. Detach the cover and remove the two dowels. On shaft drive models, the front gear case takes the place of the cover and will already have been removed.
3 Holding the selector and overshift limiter claws clear of the end of the selector drum, grasp the end of the selector shaft and pull the assembly clear of the crankcase.

12.3a Remove retainer screws – note that they are staked

12.3b Tap shaft through from left-hand side to displace bearing ...

12.3c ... then withdraw shaft and disengage clutch unit

13.1 Slacken and remove rotor holding bolt

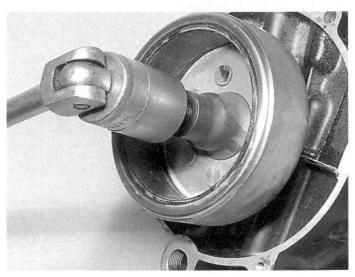

13.2 Use bolt or extractor to draw rotor off

15 Dismantling the engine/gearbox unit: separating the crankcase halves

1   The crankcase halves are secured by thirteen 6 mm bolts fitted from the upper crankcase, plus a further seven 6 mm bolts fitted from the underside. In addition, the area around the main bearings is closed by ten 8 mm bolts, fitted from the underside. These should be removed progressively and evenly to avoid any risk of warpage.

2   Leverage points are provided at the jointing face, and a screwdriver can be used here to break the joint. Alternatively, a hammer may be used with a block of wood interposed between it and the casing, to jar the casing halves apart. Usually the joint will break fairly easily, but it may be found that separation will be impaired during the first half inch or so of removal. Check carefully to determine which component or components is sticking, and take steps to release it before proceeding further. Separate the crankcase halves with the unit inverted on the bench, drawing the lower half off the upper half. The crankshaft assembly and gearshafts and clusters will remain in position in the upper half.

16 Dismantling the engine and gearbox unit: removing the upper crankcase half components

1   The crankshaft assembly can be lifted out of the upper casing half, and the camshaft drive chain and Morse secondary shaft drive chain disengaged and removed.

2   The gearbox mainshaft and layshaft can also be lifted out of the casing, noting that these have half ring retainers fitted to the bearing grooves. The camshaft drive chain guide sprocket may be left in position unless there is specific need to remove it.

17 Dismantling the engine and gearbox unit: removing the lower crankcase half components

1   The selector mechanism has three selector forks, two of which are supported by a selector fork shaft. Support the two forks, withdraw the shaft from the left-hand side and lift the forks from position. The third gear selector fork fits round the selector drum itself, and has a locating pin which runs in the selector drum track. Remove and discard the split pin which retains the locating pin, then remove the latter using a pair of pointed-nose pliers.

2   The selector drum is located by a special bolt which screws into the casing and engages in a locating groove. This bolt should be removed after bending back the locking tab. The detent plunger is located at the opposite end of the selector drum, and should also be removed. Detach the large external circlip which retains the detent plate to the end of the selector drum. The plate can now be removed, and the selector drum displaced and withdrawn, leaving the selector fork to be lifted clear of the casing half.

18 Examination and renovation: general

1   Before examining the parts of the dismantled engine unit clean them using a petrol/paraffin mix to remove all traces of old oil and sludge which may have accumulated within the engine.

2   Examine the crankcase castings for damage. If a crack is discovered it will require a specialist repair.

3   Examine carefully each part to determine the extent of wear, checking with the tolerance figures listed in the Specifications section of this Chapter. If there is any question of doubt play safe and renew.

4   Should any internal threads be damaged, they can be repaired by fitting an insert of the Helicoil type.

19 Examination and renovation: crankcase and fittings

1   The crankcase halves should be thoroughly degreased, using one of the proprietary water-soluble degreasing solutions such as Gunk and then inspected for damage. Any fault will probably require either professional repair or renewal of the crankcases as a pair. It is important to check crankcase condition at the earliest opportunity, because this will permit any necessary machining or welding to be

done whilst attention is turned to the remaining engine parts.

2   Badly worn or damaged threads can be reclaimed by fitting a thread insert of the Helicoil type. Many motorcycle repair specialists offer this type of reclamation service.

3   When handling the crankcase halves, be careful not to lose the various loose pins and location rings. Most importantly, remove and retain the small rubber ball which closes the oilway leading into the left-hand input shaft journal.

4   Locate and retrieve the two oil nozzles from the lower crankcase half. One of these will be found in the right-hand output shaft boss, and the other near the centre main bearing cap. Check that these are secure, in which case they can be cleaned and left in position.

5   It should be noted that apparent cylinder head or cylinder base oil leakage can often be attributed to oil leakage past the lower threads of the cylinder head studs. The oil works around the threads, then up through the passages in the cylinder block to emerge at the cylinder base or the cylinder head joints.

6   If such leakage has been noted, or to prevent any risk of leakage after assembly, proceed as follows. Remove each of the studs in turn, pushing it through a piece of card marked to denote its position. If possible, use a proprietary stud remover to release the stud from the crankcase. Clean the stud threads thoroughly with electrical contact cleaner or paint thinners, being sure to remove all traces of oil or dirt.

7   Clean the stud ends in the same way, after wire-brushing them to remove loose dirt, then dry the studs and their threads with compressed air. When all threads are completely clean and dry, refit the studs using Locktite Lock 'N Seal or a similar compound to ensure a sound and leak-proof fit.

15.2a Use prising points to assist crankcase separation ...

15.2b ... after all retaining bolts have been removed

20 Examination and renovation: connecting rods and big-end bearings

1 Examine the connecting rods for signs of cracking or distortion, renewing any rod which is not in perfect condition. Check the connecting rod side clearance using feeler gauges. If the clearance exceeds the service limit of 0.45 mm (0.0177 in) it will be necessary to renew the crankshaft assembly.

2 Examine closely the main and big-end bearing inserts (shells). The bearing surface should be smooth and of even texture, with no sign of scoring or streaking on its surface. If any insert is in less than perfect condition the complete set should be renewed. In practice, it is advisable to renew the bearing inserts during a major overhaul as a precautionary measure. The inserts are relatively cheap and it should be considered a false economy to reuse part worn components.

3 The crankshaft journals should be given a close visual examination, paying particular attention where damaged bearing inserts were discovered. If the journals are scored or pitted in any way, a new crankshaft will be required. Note that undersized inserts are not available, thus precluding re-grinding the crankshaft.

4 Measure the big-end journals using a micrometer, making a written note of each. Note that each journal should be checked in several different places and the smallest diameter noted. If significant ovality is found, renew the crankshaft. Compare the readings obtained with those shown below. If any journal falls below the service limit,

renew the crankshaft, otherwise mark the flywheel edges as shown in the accompanying figure to indicate its diameter range. Disregard the existing flywheel marks.

Connecting rod journal diameter

Mark	Size
No mark	34.984 – 34.994 mm (1.3773 – 1.3777 in)
O mark	34.995 – 35.000 mm (1.3778 – 1.3780 in)
Service limit	34.97 mm (1.3768 in)

5 If necessary, the big-end eye diameter can be checked, with each connecting rod and cap assembled without inserts, using a bore micrometer. It is unlikely that the diameter will have changed from that marked on the assemblies' machined faces because this area is not subjected to mechanical wear.

Connecting rod big-end eye diameter

Mark	Size
0	38.009 – 38.016 mm (1.4964 – 1.4967 in)
No mark	38.000 – 38.008 mm (1.4961 – 1.4964 in)

6 Using the sizings obtained, select the appropriate insert size from the figures shown below.

Bearing insert selection

Connecting rod mark	Crankshaft mark	Insert colour
0	Unmarked	Green
0	0	Black
Unmarked	Unmarked	Black
Unmarked	0	Brown

20.2a Note locating tabs on shells (inserts)

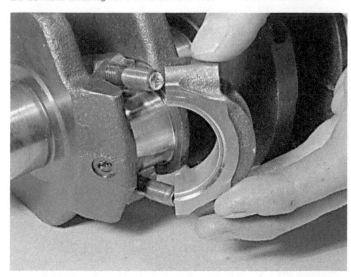

20.2b Lubricate and assemble connecting rod and cap, tightening nuts to specified torque

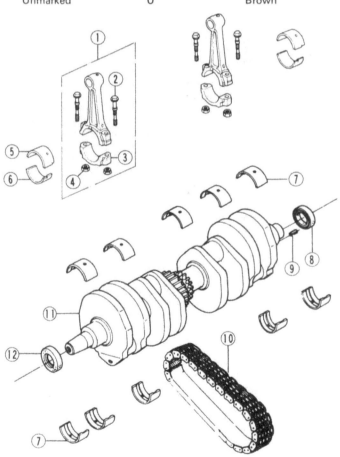

Fig. 1.5 Crankshaft

1	Connecting rod – 4 off	7	Main bearing – 10 off
2	Bolt – 8 off	8	Oil seal
3	Big-end cap – 4 off	9	Key
4	Nut – 8 off	10	Primary chain
5	Big-end bearing with oilway – 4 off	11	Crankshaft
6	Big-end bearing without oilway – 4 off	12	Oil seal

Bearing insert thickness

Colour	Thickness
Green or blue	1.485 – 1.490 mm (0.0585 – 0.0587 in)
Black	1.480 – 1.485 mm (0.0582 – 0.0585 in)
Brown	1.475 – 1.480 mm (0.0581 – 0.0582 in)

When ordering new bearing inserts quote the relevant colour code and provide the machine's engine and frame numbers.

7 If the connecting rods are to be renewed for any reason, it is important that rods of the correct weight are fitted. If this is not done the balance of the crankshaft will be affected and excessive vibration will result. Each connecting rod is marked with a code letter electro-etched across the machined face of the rod and cap. It is this code which indicates the weight range of the rod assembly. The difference in weight must not exceed 12 grammes, and the replacement rod should be selected on that basis from the following table:

Code letter (std)	Weight range including bolts and nuts	Acceptable replacement code letter
H	347 – 352 gram	H or I
I	353 – 358 gram	H, I or J
J	359 – 364 gram	I or J

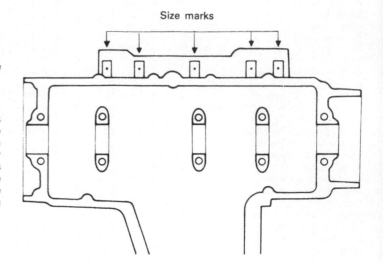

Size marks

Fig. 1.6 Crankcase main bearing size range marks

21 Examination and renovation: main bearings

1 Measure each main bearing journal using a micrometer, as described above for the big-end journals. If any one journal is worn beyond the service limit, renew the crankshaft, otherwise mark the appropriate flywheel face to show the diameter range.

Crankshaft main bearing journal diameter

Mark	Size
No mark	35.984 – 35.991 mm (1.4167 – 1.4170 in)
1	35.992 – 36.000 mm (1.4170 – 1.4173 in)

2 Assemble the crankcase halves, without main bearing inserts, and check each main bearing boss inside diameter using a bore micrometer. It is unlikely that the readings obtained will differ from the original sized noted during manufacture. Check the readings against the figures shown below, and where necessary re-mark the crankcase to denote the appropriate size range in the positions shown in the accompanying figure.

Main bearing boss diameter

Mark	Size
0	39.000 – 39.008 mm (1.5354 – 1.5357 in)
No mark	39.009 – 39.016 mm (1.5358 – 1.5361 in)

3 Using the measurements obtained above, select new bearing inserts according to the table shown below.

Bearing insert selection

Crankcase mark	Crankshaft mark	Insert colour
0	1	Brown
No mark	No mark	Blue
0	No mark	Black
No mark	1	Black

Bearing insert thickness

Colour	Thickness	Part number
Brown	1.490 – 1.494 mm (0.0587 – 0.0588 in)	92028-1102
Black	1.494 – 1.498 mm (0.0588 – 0.0590 in)	92028-1101
Blue	1.498 – 1.502 mm (0.0590 – 0.0591 in)	92028-1100

22 Examination and renovation: secondary shaft components

1 Power from the crankshaft is transmitted by way of a Morse chain to a sprocket on the secondary shaft, which in turn drives the clutch. The secondary shaft also incorporates a rubber-segment shock absorber which counteracts any engine vibration. It is not normally necessary to dismantle the secondary shaft components unless one of the following symptoms has been apparent.

a) Starter motor not engaging or disengaging correctly, indicating wear in the clutch rollers, weak or broken springs or damaged clutch housing.
b) Snatch in primary transmission, indicating wear or damage to the shock absorber rubbers or hub.

2 If starter clutch problems are in evidence, dismantle the unit by sliding the starter gear off the end of the shaft, together with its needle roller bearing. The clutch body will now be exposed, and will be seen to contain three sets of springs, caps and rollers. These should be removed and examined for wear or damage. Look for signs of flats appearing on the roller faces, and for signs of wear in the clutch body and on the gear boss on which the rollers act. The only safe course of action, if wear is evident, is to renew the parts concerned. If the rollers are to be renewed, fit new springs as a matter of course.

3 Like the starter clutch, the shock absorber components rarely give any trouble. The main body, which incorporates the Morse drive sprocket, is retained by a circlip. When this has been released, the body can be slid off and the rubber segments removed for examination. Any damage will be self-evident and normally will be confined to the rubber segments. These will tend to become compressed and rounded off after a very high mileage, and should be renewed if this is the case.

4 Examine the teeth on the outside on the housing, looking for chips and signs of wear. If the teeth are only slightly marked, they may be reclaimed, using a fine oilstone. More severe damage will necessitate renewal.

5 The journal ball bearings which support the shaft should be checked for signs of roughness and free play after they have been washed in clean petrol and dried off. Any sign of grittiness or slop is indicative of the need for renewal. The bearing which is still in place in the casing can be driven out using a large diameter socket as a drift. The remaining bearing can be pulled off the shaft by way of a bearing extractor or small sprocket puller.

6 The Morse primary chain has no provision for adjustment, but normally wears very slowly. Wear can be checked by temporarily reinstalling the crankshaft and secondary shaft in the casing half, with the chain fitted in its normal position. Free play should be measured at the middle of the run, and should not exceed 27 mm (1.063 in). If worn beyond this amount, a new chain must be fitted.

22.3a Free circlip to dismantle shock absorber unit

22.3b Renew rubbers if worn or damaged

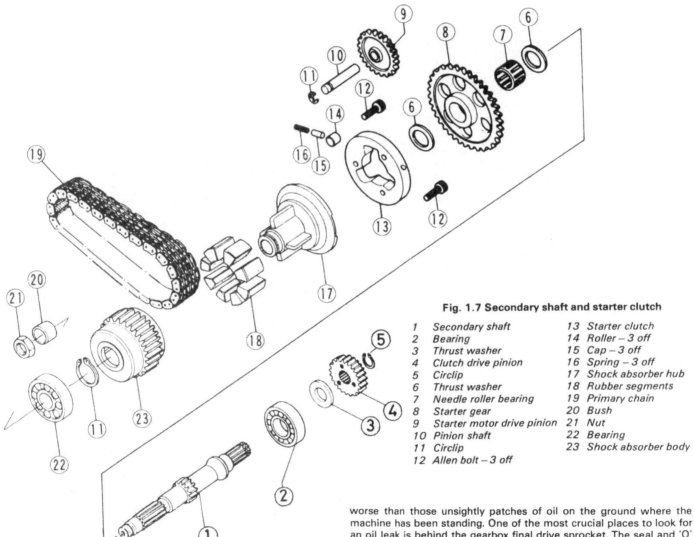

Fig. 1.7 Secondary shaft and starter clutch

1	Secondary shaft	13	Starter clutch
2	Bearing	14	Roller – 3 off
3	Thrust washer	15	Cap – 3 off
4	Clutch drive pinion	16	Spring – 3 off
5	Circlip	17	Shock absorber hub
6	Thrust washer	18	Rubber segments
7	Needle roller bearing	19	Primary chain
8	Starter gear	20	Bush
9	Starter motor drive pinion	21	Nut
10	Pinion shaft	22	Bearing
11	Circlip	23	Shock absorber body
12	Allen bolt – 3 off		

worse than those unsightly patches of oil on the ground where the machine has been standing. One of the most crucial places to look for an oil leak is behind the gearbox final drive sprocket. The seal and 'O' ring that fits on the shaft should be renewed if there is any sign of a leak.

2   Oil seals are relatively inexpensive, and if the unit is being overhauled it is advisable to renew all the seals as a matter of course. This will preclude any risk of an annoying oil leak developing after the unit has been reinstalled in the frame.

23 Examination and replacement: oil seals

1   Oil seal failure is difficult to define precisely. Usually it takes the form of oil showing on the outside of the machine, and there is nothing

24 Examination and renovation: cylinder block

1   The usual indication of badly worn cylinder bores and pistons is excessive smoking from the exhausts. This usually takes the form of blue haze tending to develop into a white haze as the wear becomes more pronounced.

2   The other indication is piston slap, a form of metallic rattle which occurs when there is little load on the engine. If the top of the bore is examined carefully, it will be found that there is a ridge on the thrust side, the depth of which will vary according to the rate of wear which has taken place. This marks the limit of travel of the top piston ring.

3   Measure the bore diameter just below the ridge using an internal micrometer, or a dial gauge. Compare the reading you obtain with the reading at the bottom of the cylinder bore, which has not been subjected to any piston wear. If the difference in readings exceeds 0.05 mm (0.002 in) the cylinder block will have to be bored and honed, and fitted with the required oversize pistons.

4   If a measuring instrument is not available, the amount of cylinder bore wear can be measured by inserting the piston (without rings) so that it is approximately $\frac{3}{4}$ inch from the top of the bore. If it is possible to insert a 0.005 inch feeler gauge between the piston and cylinder wall on the thrust side of the piston, remedial action must be taken.

5   Kawasaki supply pistons in two oversizes: 0.5 mm (0.020 inch) and 1.0 mm (0.040 inch). If boring in excess of 1.0 mm becomes necessary, the cylinder block must be renewed since new liners are not available from Kawasaki.

6   Make sure the external cooling fins of the cylinder block are free from oil and road dirt, as this can prevent the free flow of air over the engine and cause overheating problems.

25 Examination and renovation: pistons and piston rings

1   If a rebore becomes necessary, the existing pistons and piston rings can be disregarded because they will have to be replaced by their new oversizes. If, however, the bores have been checked as described in Section 24 and are to be reused, clean and check the pistons and rings as described below.

2   Remove all traces of carbon from the piston crowns, using a blunt ended scraper to avoid scratching the surface. Finish off by polishing the crowns of each piston with metal polish, so that carbon will not adhere so rapidly in the future. Never use emery cloth on the soft aluminium.

3   Piston wear usually occurs at the skirt or lower end of the piston and takes the form of vertical streaks or score marks on the thrust side of the piston. Damage of this nature will necessitate renewal.

4   After the engine has covered high mileages, it is possible that the ring grooves may have become enlarged. To check this, measure the clearance between the ring and groove with a feeler gauge. A clearance in excess of 0.15 mm (0.006 in) between the top and 2nd ring and their respective grooves will necessitate renewal of the pistons.

5   To measure the end gap, insert each piston ring into its cylinder bore, using the crown of the bare piston to locate it about 1 inch from the top of the bore. Make sure it is square in the bore and insert a feeler gauge in the end gap of the ring. If the end gap exceeds 0.7 mm (0.028 inch) the ring must be renewed. The standard gap is 0.15 – 0.3 mm (0.006 – 0.012 in).

6   When refitting new piston rings, it is also necessary to check the end gap. If there is insufficient clearance, the rings will break up in the bore whilst the engine is running and cause extensive damage. The ring gap may be increased by filing the ends of the rings with a fine file.

7   The ring should be supported on the end as much as possible to avoid breakage when filing, and should be filed square with the end. Remove only a small amount of metal at a time and keep rechecking the clearance in the bore.

8   When fitting new rings to a partly-worn cylinder bore, it is important that any wear ridge in the bore surface is removed to avoid ring breakage. Equally important is the removal of the glazed bore surfaces to allow the new rings to bed in. This job is best entrusted to a Kawasaki dealer or engine reconditioning specialist who will have the necessary equipment to carry out the work quickly and economically.

25.5 Check ring end gap in unworn area of bore

26 Examination and renovation: cylinder head

1   Remove all traces of carbon from the cylinder head using a blunt ended scraper (the round end of an old steel rule will do). Finish by polishing with metal polish to give a smooth shiny surface.

2   Check the condition of the sparking plug hole threads. If the theads are worn or crossed they can be reclaimed by a Helicoil insert. Most motorcycle dealers operate this service which is very simple, cheap and effective.

3   Clean the cylinder head fins with a wire brush, to prevent overheating, through dirt blocking the fins.

4   Lay the cylinder head on a sheet of $\frac{1}{4}$ inch plate glass to check for distortion. Aluminium alloy cylinder heads distort very easily, especially if the cylinder head bolts are tightened down unevenly. If the amount of distortion is only slight, it is permissible to rub the head down until it is flat once again by wrapping a sheet of very fine emery cloth around the plate glass base and rubbing with a rotary motion.

5   If the cylinder head is distorted badly (one way of determining this is if the cylinder head gaskets have a tendency to keep blowing), the head will have to be machined by a competent engineer experienced in this type of work. This will, of course, raise the compression of the engine, and if too much is removed can adversely affect the performance of the engine. If there is risk of this happening, the only remedy is a new replacement cylinder head.

6   Modified exhaust port studs are fitted to the cylinder heads of later models. The studs are 8 mm diameter instead of 6 mm and are of two different lengths. In the event of the later type cylinder head being supplied as a replacement part, note that this will necessitate fitting the modified studs, nuts and exhaust pipe flanged clamps.

7   If the head is serviceable, but the exhaust port studs are stripped or broken, it may be worth considering a modification using the later studs. This is an alternative to having the old stud holes inserted and using the standard 6 mm studs, although it is stressed that this is not an officially-recommended repair.

8   To carry out the repair it will be necessary to drill out the old stud holes for tapping to 8 mm, and also to enlarge the holes in the exhaust pipe holders to suit. If you do not have the facilities or experience in this type of work, it may be safer to have a local engineering specialist carry out the repair. Be warned that breaking off a tap in the soft alloy of the head can be very expensive to rectify.

27 Clean air system: description and renovation – US models only

1   The US models incorporate an air injection system designed to enhance the burning of hydrocarbons in the exhaust gases, thus reducing toxic emissions. The system employs a modified cylinder head and cover, in which air is drawn through a reed valve arrangement into the exhaust ports.

2   The clean air system is automatic in operation and should not normally require attention. The most likely fault is that unfiltered air

may be drawn into the system through a damaged air filter element or leaking hose making the tickover unstable and reducing engine power. Backfiring or other unusual noises may be apparent.

Air suction valve (reed valve)

3   The reed valves may be removed for examination after releasing the covers which house them. Check each valve for signs of deterioration, specifically examining the reeds for signs of cracking, warping or heat damage. Check the bonded rubber valve case for signs of delamination, cracking or scoring. Wash off any contaminants with a suitable solvent, and guard against scraping or scoring the sealing faces. The reeds and stopper plates may be removed if necessary, noting that Loctite or a similar compound must be applied to the screw threads prior to reassembly.

Vacuum switch valve

4   Regular inspection of the vacuum switch valve is unnecessary, and should be avoided unless a fault has been indicated. A vacuum gauge and a syringe-type vacuum pump is required to check that the valve closes at 40 – 46 cmHg. Since few owners are likely to have access to this type of equipment, it is suggested that the check is done by a dealer, or by temporary substitution of a sound valve. Note that if the test is to be performed, no more than 50 cmHg should be applied to the valve. Note also that Kawasaki state that a faulty valve **must** be renewed and that 'adjustment is not permitted' despite the adjuster screw and locknut fitted to the valve.

Hoses

5   Damaged or perished hoses are probably the most likely cause of trouble in the system, and are fortunately the cheapest problem to remedy. Remember that air leaks will cause erratic running, and if located between the vacuum switch valve and reed valves, will allow unfiltered air to enter the reed valves.

---

28 Examination and renovation: valves, valve seats and valve guides

1   Remove the valve tappets and shims, keeping them separate for installation in their original locations. Compress the valve springs with a valve spring compressor, and remove the split valve collets, also the oil seals from the valve guides, as it is best to renew these latter components.
2   Remove the valves and springs, making sure to keep to the locations during assembly. Inspect the valves for wear, overheating or burning, and replace them as necessary. Normally, the exhaust valves will need renewal far more often than the inlet valves, as the latter run at relatively low temperatures. If any of the valve seating faces are badly pitted, do not attempt to cure this by grinding them, as this will invariably cause the valve seats to become pocketed. It is permissible to have the valve(s) refaced by a motorcycle specialist or small engineering works. The valve seating angle is 45°. The valve must be renewed if the head thickness (the area between the edge of the seating surface and the top of the head) is reduced to 0.5 mm/0.020 in (inlet valve), 0.7 mm/0.027 in (exhaust valve).
3   Measure the bore of each valve guide in at least four places using a small bore gauge and micrometer. If the measurement exceeds 7.08 mm (0.2787 in) the guide should be replaced with a new one. If a small bore gauge and micrometer are not available, insert a new valve into the guide, and set a dial gauge against the valve stem. Gently move the valve back and forth in the guide and measure the travel of the valve in each direction. The guide will have to be renewed if the clearance between the valve and guide exceeds the specified figures.
4   It is worthwhile pausing at this juncture to consider the best course of action. It must be borne in mind that valve guide renewal is not easy, and will require that the valve seats be recut after the guide has been fitted and reamed. It is also remarkably easy to damage the cylinder head unless great care is taken during these operations. It may, therefore, be considered better to entrust these jobs to a competent engineering company or to a Kawasaki Service Agent. For the more intrepid, skilled and better equipped owner, the procedure is as follows:
5   Heat the cylinder head slowly and evenly, in an oven to prevent warpage, to 120 – 150°C (248 – 302°F). Using a stepped drift, tap the guide(s) lightly out of the head, taking care not to burn yourself on the hot casting. New guides should be fitted in a similar manner,

ensuring that they seat squarely in the head casting. If a valve guide is loose in the head, it may be possible to have an oversize guide machined and fitted by a competent engineering works, noting that the cylinder head must be bored to suit the new guide. The popular 'dodge' of knurling the outside of the guide is crude and is not recommended.
6   After the guide has been fitted it must be reamed using a Kawasaki reamer (Part Number 57001 – 162). Make sure that the reamer passes squarely though the valve guide bore, taking care not to accidentally gouge out too much material. The valve seat must now be re-cut in the following manner:
7   If a valve guide has been renewed, or a valve seat face is worn or pitted, it must be re-cut to ensure efficient sealing. The process requires the use of three cutters (30°, 45° and 60°). These are normally available as a set. Assemble the tool according to the manufacturer's instructions, with the 45° cutter fitted. Arrange the tool with the spigot inserted in the valve guide, and remove just enough metal to ensure an even, pit-free seating surface. Note that if too much metal is removed, the valve will become pocketed, and the complete cylinder head will have to be renewed. **Kawasaki do not supply valve seat inserts,** so the utmost care must be taken.
8   The 30° and 60° cutters should be used next, and in that order, to leave the raised 45° seating face as an even band between 0.5 and 1.0 mm in width. The valve(s) should now be ground-in, in the normal manner.
9   The valves should be ground in, using ordinary oil-bound grinding paste, to remove any pitting or to finish off a newly cut seat. Note that it is not normally essential to resort to using the coarse grade of paste which is supplied in dual-grade containers. Valve grinding is a simple task. Commence by smearing a trace of fine valve grinding compound (carborundum paste) on the valve seat and apply a suction tool to the head of the valve. Oil the valve stem and insert the valve in the guide so that the two surfaces to be ground in make contact with one another. With a semi-rotary motion, grind in the valve head to the seat, using a backward and forward action. Lift the valve occasionally so that the grinding compound is distributed evenly. Repeat the application until an unbroken ring of light grey matt finish is obtained on both valve and seat. This denotes the grinding operation is now complete. Before passing to the next valve, make sure that all traces of the valve grinding compound have been removed from both the valve and its seat and that none has entered the valve guide. If this precaution is not observed, rapid wear will take place due to the highly abrasive nature of the carborundum paste.
10 Reassemble the valve and valve springs by reversing the dismantling procedure. Fit new oil seals to each valve guide and oil both the valve stem and the valve guide, prior to reassembly. Take special care to ensure the valve guide oil seal is not damaged when the valve is inserted. As a final check after assembly, give the end of each valve stem a light tap with a hammer, to make sure the split collets have located correctly.

28.1a Remove cam followers, or tappets ...

28.1b ... followed by adjustment shims

28.1c Use valve spring compressor and release collet halves

28.1d Remove the spring seat ...

28.1e ... followed by the inner and outer valve springs

28.1f Valves may now be displaced and removed

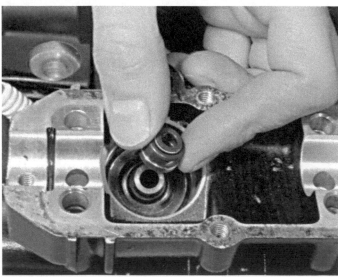
28.10a Fit new valve stem oil seals ...

28.10b ... and lubricate seal and guide

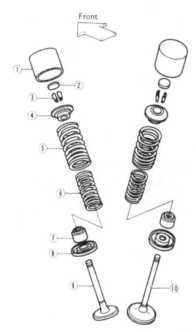

Fig. 1.8 Valves

1 Tappet
2 Shim
3 Split collets
4 Spring upper seat
5 Outer spring
6 Inner spring
7 Oil seal
8 Spring lower seat
9 Exhaust valve
10 Inlet valve

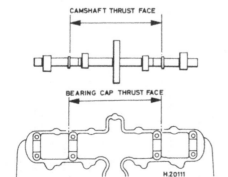

Fig. 1.9 Camshaft thrust face measurement

29 Examination and renovation: camshafts, tappets and camshaft drive mechanism

1 Examine the camshaft lobes for signs of wear or scoring. Wear is normally evident in the form of visual flats worn on the peak of the lobes, and this may be checked by measuring each lobe at its widest point. If worn to 35.65 mm (1.4035 in) or less the camshaft must be renewed. Scoring or similar damage can usually be attributed to a partial failure of the lubrication system, possibly due to the oil filter element not having been renewed at the specified mileage, causing unfiltered oil to be circulated by way of the bypass valve. Before fitting new camshafts, examine the bearing surfaces of the camshafts, and cylinder head, and rectify the cause of the failure.

2 If the camshaft bearing surfaces are marred, it is likely that renewal of both the cylinder head and the camshafts will be the only solution. This is because the camshaft runs directly in the cylinder head casting, using the alloy as a bearing surface. Assemble the bearing caps and measure the internal bore using a bore micrometer. If the diameter is 22.12 mm (0.871 in) or more, it will be necessary to renew the cylinder head and bearing caps. Note that it is not possible to renew the caps alone, because they are machined together with the cylinder head and are thus matched to it.

3 Measure the camshaft bearing journals, using a micrometer. If the journals have worn to 21.93 mm (0.863 in) or less, the camshaft(s) should be renewed. The clearance between the camshafts and their bearing surfaces can be checked using Plastigage or by direct measurement. The clearance must not exceed the wear limit of 0.19 mm (0.0074 in).

4 Camshaft run-out can be checked by supporting each end of the shaft on V-blocks, and measuring any run-out using a dial test indicator running on the camshaft sprocket boss (having first removed the sprocket). This should not normally be more than 0.01 mm (0.004 in). The camshaft must be renewed if run-out exceeds 0.1 mm (0.0039 in).

5 The camshaft endfloat must be checked at this stage, and this requires that the camshafts are temporarily refitted into the cylinder head and the caps refitted to locate them. The check requires a dial gauge to be set up so that the probe rests against the end of the camshaft.

6 Push the camshaft as far as it will go in one direction and zero the dial gauge. Move the camshaft in the opposite direction until it stops, then read off the amount of endfloat indicated. Note the reading, then repeat the check on the remaining camshaft.

7 If the indicated endfloat exceeds the prescribed range of 0.28 – 0.82 mm (0.0110 – 0.0323 in) the camshaft(s), cylinder head or both parts may require renewal. Remove the camshafts and measure across the thrust faces of the camshaft and the bearing cap using a micrometer. The specified limits are as follows:

Camshaft thrust face .... 194.50 – 194.60 mm (7.6574 – 7.6614 in)

Bearing cap thrust face ..... 193.78 – 194.22 mm (7.6291 – 7.6464 in)

8 If the camshaft(s) are below limits they should be renewed to restore endfloat to the correct range. Note that if the bearing caps are below limits, they must be renewed together with the cylinder head, with which they form a matched assembly; the caps cannot be renewed alone.

9 The single camshaft drive chain should be checked for wear, particularly if tensioner adjustment has failed to prevent chain noise, this latter condition being indicative that the chain is probably due for renewal. Lay the chain on a flat surface, and get an assistant to stretch it taut. Using a vernier caliper gauge, measure a 20 link run of the chain. Repeat this check in one or two other places. The maximum length (service limit) is 128.9 mm (5.075 in).

10 Inspect the cam chain tensioner blade, guide blade and the tensioner assembly, renewing any parts which are worn or damaged, especially if a new chain has been fitted. Early KZ/Z750 E and H models, prior to engine number KZ750EE037631 are fitted with a manual type tensioner and all other models have an automatic cross-wedge type tensioner. If the manual tensioner shows signs of premature wear it is recommended that it be replaced with the automatic type. Check both camshaft sprockets for wear or damage. These components can normally be expected to give many miles of service if correctly maintained. The tachometer worm drive on the

exhaust camshaft is unlikely to suffer undue wear.

11 In the case of machines fitted with a mechanically driven tachometer, the worm drive to the tachometer is an integral part of the camshaft which meshes with a pinion attached to the cylinder head. If the worm is damaged or badly worn, it will be necessary to renew the camshaft complete.

12 The tachometer driven worm gear shaft is fitted in a housing which is a press fit in the cylinder head cover. If the worm gear is chipped or broken the gear and integral shaft should be renewed.

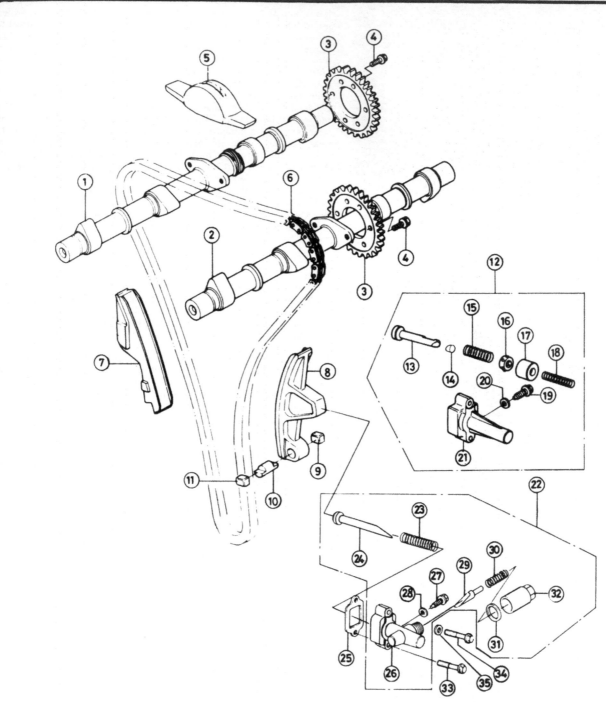

Fig. 1.10 Camshafts, chain and tensioner

1	Exhaust valve camshaft	10	Retaining pin	19	Adjustment bolt
2	Inlet valve camshaft	11	Block	20	Washer
3	Sprocket – 2 off	12	Tensioner – Manual type	21	Tensioner body
4	Bolt – 4 off	13	Plunger	22	Tensioner – Automatic type
5	Cam chain guide	14	Plunger end piece	23	Spring
6	Cam chain	15	Spring	24	Plunger
7	Cam chain guide	16	Nut	25	Gasket
8	Tensioner blade	17	Collar	26	Tensioner body
9	Block	18	Spring	27	Adjustment bolt

28	Washer
29	Cross-wedge
30	Spring
31	Washer
32	Housing bolt
33	Bolt
34	Bolt
35	Washer

30 Examination and renovation: clutch

1   After an extended period of service the clutch linings will wear and promote clutch slip. When the overall width reaches the limit given in Specifications, the inserted plates must be renewed, preferably as a complete set.

2   The plain plates should not show any excess heating (blueing). Check the warpage of each plate using plate glass or surface plate and a feeler gauge. The maximum allowable warpage is 0.40 mm (0.0157 in).

3   Examine the clutch assembly for burrs or indentation on the edges of the protruding tongues of the inserted plates and/or slots worn in the edges of the outer drum with which they engage. Similar wear can occur between the inner tongues of the plain clutch plates and the slots in the clutch inner drum. Wear of this nature will cause clutch drag and slow disengagement during gear changes, since the plates will become trapped and will not free fully when the clutch is withdrawn. A small amount of wear can be corrected by dressing with a fine file; more extensive wear will necessitate renewal of the worn parts. Note that the clearance between the clutch drum slots and the tangs of the clutch plates must not exceed 1.0 mm (0.040 in).

4   The clutch release mechanism attached to the final drive sprocket cover does not normally require attention provided it is greased at regular intervals. It is held to the cover by two crosshead screws and operates on the worm and quick start thread principle.

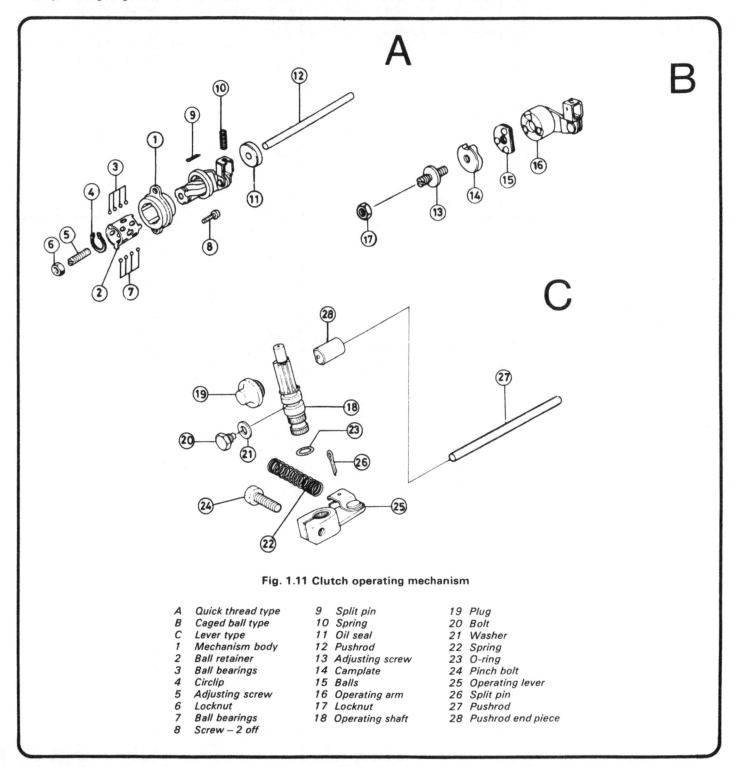

Fig. 1.11 Clutch operating mechanism

A	Quick thread type	9	Split pin	19	Plug
B	Caged ball type	10	Spring	20	Bolt
C	Lever type	11	Oil seal	21	Washer
1	Mechanism body	12	Pushrod	22	Spring
2	Ball retainer	13	Adjusting screw	23	O-ring
3	Ball bearings	14	Camplate	24	Pinch bolt
4	Circlip	15	Balls	25	Operating lever
5	Adjusting screw	16	Operating arm	26	Split pin
6	Locknut	17	Locknut	27	Pushrod
7	Ball bearings	18	Operating shaft	28	Pushrod end piece
8	Screw – 2 off				

31 Examination and renovation: gearbox components

1   Examine each of the gear pinions to ensure that there are no chipped or broken teeth and that the dogs on the end of the pinions are not rounded. Gear pinions with any of these defects must be renewed; there is no satisfactory method of reclaiming them.

2   After thorough washing in petrol, the bearings should be examined for roughness and play. Also check for pitting on the roller tracks.

3   It is advisable to renew the gearbox oil seals irrespective of their condition. Should a reused oil seal fail at a later date, a considerable amount of work is involved to gain access to renew it.

4   Check the gear selector rod for straightness by rolling it on a sheet of plate glass. A bent rod will cause difficulty in selecting gears and will make the gear change particularly heavy.

5   The selector forks should be examined closely, to ensure that they are not bent or badly worn. The case-hardened pegs which engage with the cam channels are easily renewable if they are worn.

6   The tracks in the selector drum, with which the selector forks engage, should not show any undue signs of wear unless neglect has led to under-lubrication of the gearbox. Check the tension of the gearchange pawl, gearchange arm and drum stopper arm springs. Weakness in the springs will lead to imprecise gear selection. Check the condition of the gear stopper arm roller and the pins in the change drum end with which it engages.

7   If it is found necessary to renew any of the gearbox components the accompanying line drawings will give details of the assembly sequence. In addition, gear cluster reassembly is covered in the accompanying photographic sequence.

8   Removing the individual pinions presents no undue problems provided a good quality pair of circlip pliers is used. It is advisable to lay each pinion, thrust washer and circlip out in sequence to prevent mistakes during reassembly. The output shaft incorporates an ingenious neutral finder mechanism which consists of three steel balls running in radial drillings spaced 120° apart within the 4th gear pinion.

9   To remove the pinion from the shaft, hold the shaft vertically with the 4th gear pinion facing downwards over the workbench. If the shaft is now spun briskly, the balls will be thrown outwards, allowing the pinion to drop clear of the shaft. Take care not to lose the three balls.

10   When reassembling the shafts always use new circlips, taking care not to open the clips any more than is essential to ease them into position. The clip ends should be supported by the raised areas of the splines for security. When fitting the output shaft 4th gear pinion do not omit the neutral location balls. Fitting these and retaining them whilst the shaft is inserted can provide hours of entertainment if not done carefully. To avoid undue frustration, use a few drops of engine oil to 'stick' the balls in place, then insert the shaft carefully with the assembly held vertical.

11   In the case of the Z750 P models, note that care must be taken to ensure that the gearchange shaft is well lubricated, both to ensure smooth operation and to prevent corrosion. If the shaft is in good condition and has been operating normally, make sure that it is well greased, using molybdenum disulphide grease, prior to installation.

12   In severe cases, or where the shaft had seized in its bore, the problem can be overcome by fitting a modified outer cover which incorporates a bush and seal. Refer to an authorized Kawasaki dealer for details of the modified parts. Note that the seal and bush used in the later cover cannot be fitted to the earlier type.

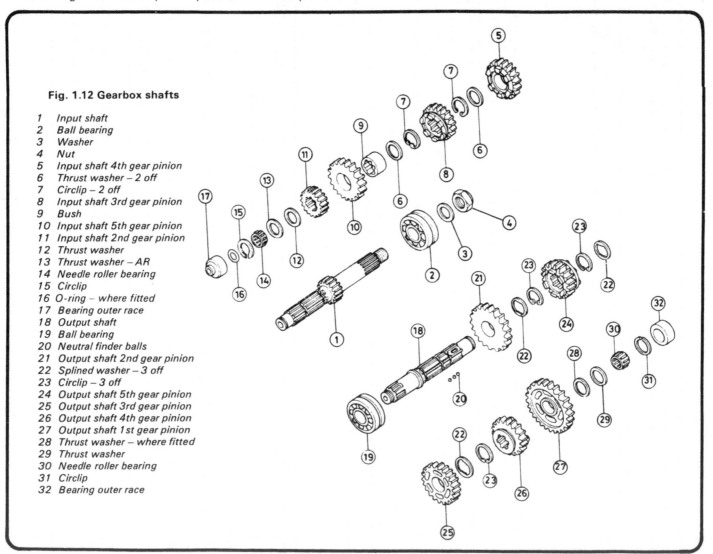

Fig. 1.12 Gearbox shafts

1   Input shaft
2   Ball bearing
3   Washer
4   Nut
5   Input shaft 4th gear pinion
6   Thrust washer – 2 off
7   Circlip – 2 off
8   Input shaft 3rd gear pinion
9   Bush
10   Input shaft 5th gear pinion
11   Input shaft 2nd gear pinion
12   Thrust washer
13   Thrust washer – AR
14   Needle roller bearing
15   Circlip
16   O-ring – where fitted
17   Bearing outer race
18   Output shaft
19   Ball bearing
20   Neutral finder balls
21   Output shaft 2nd gear pinion
22   Splined washer – 3 off
23   Circlip – 3 off
24   Output shaft 5th gear pinion
25   Output shaft 3rd gear pinion
26   Output shaft 4th gear pinion
27   Output shaft 1st gear pinion
28   Thrust washer – where fitted
29   Thrust washer
30   Needle roller bearing
31   Circlip
32   Bearing outer race

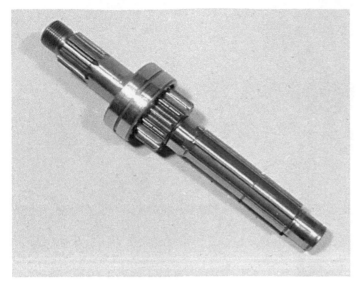

31.7a The input shaft incorporates integral 1st gear pinion

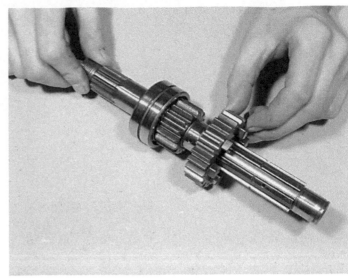

31.7b Slide 4th gear pinion over shaft, noting direction of dogs

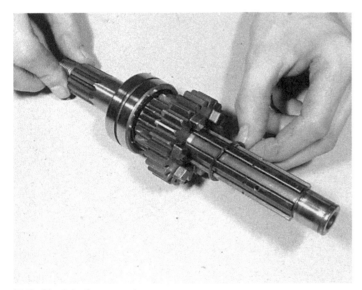

31.7c Fit plain thrust washer ...

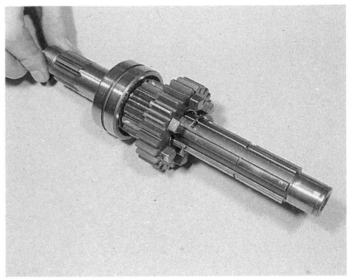

31.7d ... followed by circlip

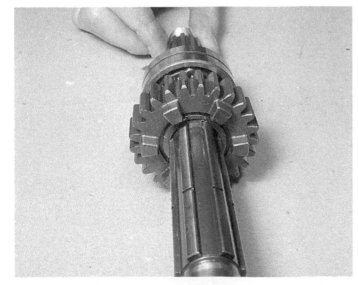

31.7e Note that circlip ends must be positioned as shown

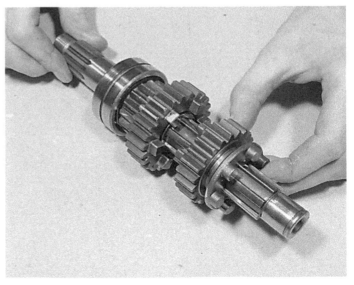

31.7f 3rd gear pinion is fitted with selector groove outwards

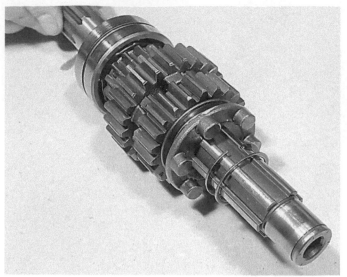

31.7g Fit circlip, followed by thrust washer

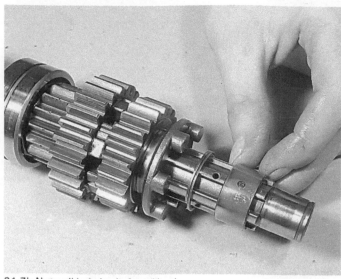

31.7h Note oil hole in shaft and bush

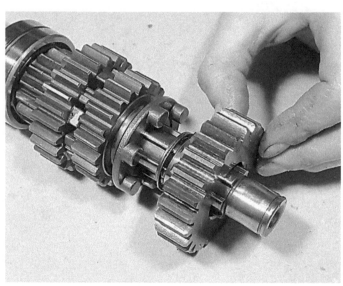

31.7i Fit the 5th gear pinion ...

31.7j ... followed by the 2nd gear pinion

31.7k Place thrust washer over shaft ...

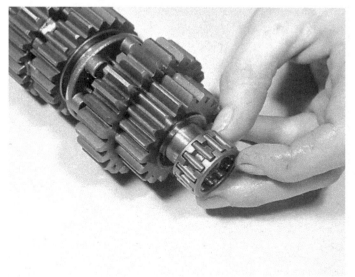

31.7l ... and then fit needle roller bearing

31.7m Secure bearing with circlip ...

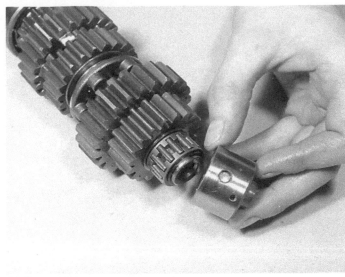

31.7n ... and fit outer race to complete

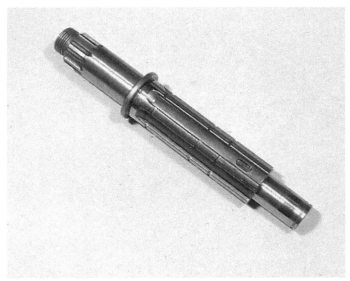

31.8a The bare output shaft

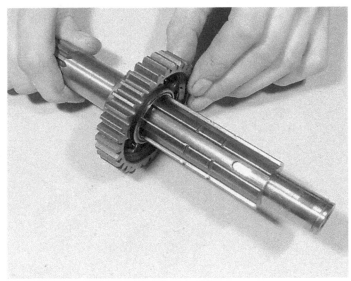

31.8b Slide the 2nd gear pinion into position

31.8c ... and secure with splined washer and circlip

31.8d 5th gear pinion is fitted in direction shown ...

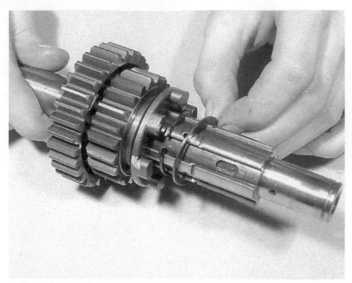

31.8e ... followed by circlip, washer ...

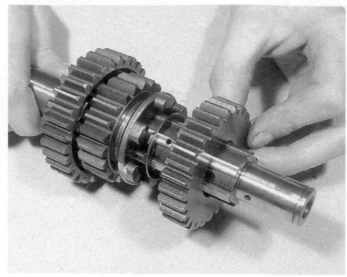

31.8f ... and the 3rd gear pinion

31.8g Fit washer and circlip

31.8h Place neutral finder balls in 4th gear pinion ...

31.8i ... and fit over shaft as shown

31.8j Fit 1st gear pinion and plain washer ...

31.8k ... followed by bearing and circlip

31.8l Place outer bearing over needle roller race

31.8m Fit journal ball race to sprocket end

31.8n Hold output shaft as shown, then spin it ...

31.8o ... until 4th gear pinion drops clear

32 Examination and renovation: front gear case unit – shaft drive models only

1   The models featuring shaft final drive are equipped with a front gear case unit which replaces the final drive sprocket and gear selector mechanism cover of the chain drive machines. The front gear case forms a separate unit which can be detached from the crankcase proper.

2   The unit contains a pair of bevel gears through which the final drive is turned through 90° to meet the drive shaft. A cam-type shock absorber is also included, this being designed to absorb shock loadings which would otherwise be transmitted to the rear wheel.

3   The gear case unit is a robust assembly and should not normally require attention unless it has been set up incorrectly at some previous date, or one or more of the bearings or bevel gears has worn and requires renewal. Normal maintenance should be confined to careful cleaning of the unit and checking that the teeth are not wearing unevenly. If necessary, the manufacturer recommends that a checking compound is applied to the driven bevel gear teeth after cleaning and

degreasing. If the gears are now rotated an impression of the tooth contact points will be visible. If all is well an elliptical contact patch will be evident at the centre of the teeth. If this is offset to either edge the need for adjustment is indicated. The checking compound is a specialist product and should be available from engineering companies which specialise in the reconditioning of car differential units.

4   The bearings are of the tapered roller type and are designed to operate under a carefully adjusted preload. If any discernible play is evident or if there appears to be excessive drag in the bearings, adjustment may be required.

5   Although dismantling and assembling the bevel gears and bearings is not beyond the capabilities of most owners, a number of specialist tools are required because it is vital that mesh depth, gear backlash and bearing preload are set with great precision. For this reason, it is recommended that in the event of a suspected fault the unit is taken to a Kawasaki dealer for checking and overhaul. The dealer will have the necessary tools and test equipment to carry out the work, plus the range of shims and preload collars which are essential during assembly.

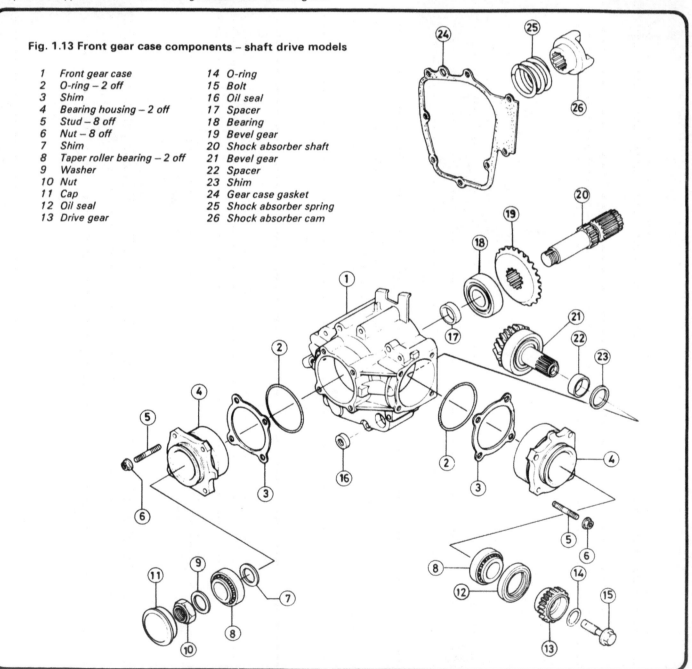

Fig. 1.13 Front gear case components – shaft drive models

1	Front gear case	14	O-ring
2	O-ring – 2 off	15	Bolt
3	Shim	16	Oil seal
4	Bearing housing – 2 off	17	Spacer
5	Stud – 8 off	18	Bearing
6	Nut – 8 off	19	Bevel gear
7	Shim	20	Shock absorber shaft
8	Taper roller bearing – 2 off	21	Bevel gear
9	Washer	22	Spacer
10	Nut	23	Shim
11	Cap	24	Gear case gasket
12	Oil seal	25	Shock absorber spring
13	Drive gear	26	Shock absorber cam

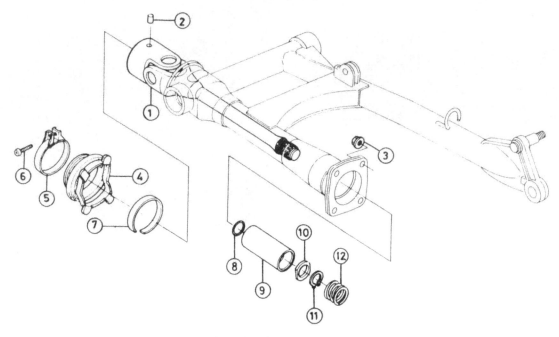

Fig. 1.14 Final driveshaft

1 Driveshaft	4 Rubber gaiter	7 Spring clip	10 Spacer
2 Locking pin	5 Clamp	8 O-ring	11 Circlip
3 Nut – 4 off	6 Screw	9 Drive sleeve	12 Spring

33 Engine/gearbox reassembly: general

1 During assembly ensure that all components are kept quite clean. Make sure all traces of old gaskets have been removed and that the mating surfaces are clean and undamaged. One of the best ways to remove old gasket cement is to apply a rag soaked in methylated spirit.
2 Gather together all the necessary tools and have available an oil can filled with clean engine oil. Make sure all new gaskets and oil seals are to hand, also all replacement parts required.
3 Refer to the torque and clearance settings wherever they are given. Many of the smaller bolts are easily sheared if over-tightened. If the existing screws show evidence of maltreatment in the past, it is advisable to renew them as a complete set.

34 Engine/gearbox reassembly: fitting the upper crankcase components

1 Check that all the bearing shells are laid out in the correct order, then refit them to their respective recesses. Ensure that the locating tang on each shell corresponds with the depression in which it engages. Ensure that each shell is firmly located before proceeding further. Oil each shell liberally.
2 Invert the upper crankcase and place it on the workbench. Commence reassembly by lowering the crankshaft assembly into position, taking care to locate all the holes in the outer tracks of the main bearings with the corresponding dowels located in the crankcase. Make sure the cam drive chain and Morse primary chain are in position on the crankshaft at this stage. Feed the connecting rods through the apertures in the crankcase and once the crankshaft is in position, rotate it to make sure all the main bearings revolve freely.
3 If the crankshaft does not appear to seat correctly, check that the crankshaft seals, which locate in grooves in the crankcase bosses, are seating properly. Note that the seals are handed and are designed to operate in the direction of normal crankshaft rotation, and to this end the seals have arrow marks on their outer faces.
4 Fit the dowel pins in the recesses to the rear of the outer main bearing bosses. Fit the locating half rings to right-hand input shaft boss and the left-hand output shaft boss, noting that with the crankcase half inverted these will appear reversed. The two remaining

gearbox shaft bosses should be fitted with the small, solid, locating pins, if these were removed. Pay particular attention to the small rubber ball which blanks the oil passage in the right-hand input shaft boss. If this is damaged in any way, fit a new one as a precautionary measure. Press the ball into place until it is just flush with the bearing surface. If the starter idler pinion was removed this should be refitted.
5 Place the gearbox input and output shafts in position, making sure that they locate correctly. Lubricate the gears with clean engine oil and check that the shafts turn smoothly and evenly. In particular, check that the output shaft 1st gear turns smoothly. If it appears tight it will be necessary to fit a thinner shim washer. These are available from Kawasaki dealers in 1.0 mm, 0.7 mm and 0.5 mm thicknesses. The clearance between the input shaft 2nd gear and the adjacent copper washer should also be checked using a feeler gauge. If it is not 0.1 – 0.3 mm (0.0039 – 0.0118 in) add or change a shim washer to suit.

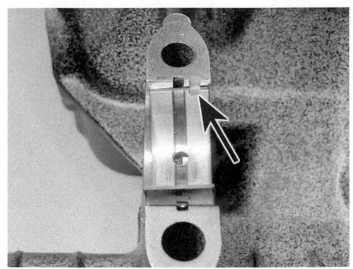

34.1 Fit and lubricate the main bearing shells, noting location tangs

34.2a Fit primary and cam chains, then place crankshaft in position

34.2b Lubricate main bearings and check crankshaft rotation

34.2c Note that crankshaft seals are handed and are marked to show direction of rotation

34.4a Fit half-rings in the grooves of gearbox shaft bosses ...

34.4b ... followed by dowel pins if these were removed

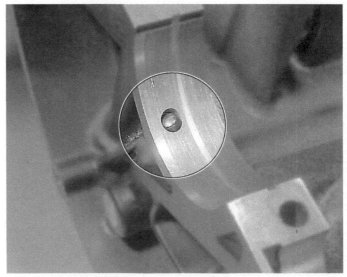

34.4c Renew rubber blanking ball if damaged

34.5a Place gearbox output shaft

34.5b ... and input shaft into casing, check that shafts turn smoothly

35 Engine/gearbox reassembly: fitting the crankcase lower components

1   Check that the oil passages are properly cleaned out. The main oil gallery is closed by a hexagon-headed plug at each end, and these may be removed to allow the oilways to be flushed with a suitable solvent. It is advisable to coat the plug threads with Loctite or a similar compound in the interests of security.

2   Fit the two oil nozzles, one in the right-hand output shaft boss and the other just to the left of the centre main bearing boss. Fit the remaining crankshaft main bearing inserts, checking that they locate properly and lubricating them with engine oil.

3   Fit and lubricate the gear selector drum caged needle roller bearing. Slide the drum half way into position, placing the 4th/5th gear selector fork over the drum. Note that the fork is offset in relation to the boss, the longer end of the boss facing towards the right-hand side of the unit. Push the selector drum fully home. Fit the detent cam over the projecting right-hand end of the drum, ensuring that it locates over the small pin. Fit the circlip which retains the cam.

4   Turn the casing over and fit the guide bolt in its hole near the right-hand end of the selector drum, taking care not to omit the tab washer. Tighten the bolt securely and bend up the locking tab. Moving to the left-hand end of the drum, insert the neutral detent plunger, spring and cap bolt, tightening the latter securely. Rotate the selector drum until the neutral position is found. The detent plunger should be seated in the small cutout located 180° from the cam location pin.

5   Drop the guide pin into its hole in the 4th/5th gear selector fork, so that it engages in the **centre** of the three selector drum tracks. Fit a new split pin, bending the ends over to retain it. The two remaining selector forks are identical, and control the movement of the 2nd/3rd (output) gear, and the 1st (output) gear pinions. Note that the fork section of both is to be fitted with the longer boss section to the left-hand side. Slide the support shaft in from the left-hand side, fitting each fork in turn and checking that the guide pins locate correctly in their tracks. Finally, lubricate the selector components with engine oil prior to crankcase joining.

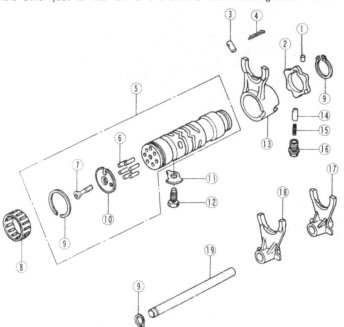

Fig. 1.15 Selector drum assembly

1 Locating pin	11 Tab washer
2 Detent cam	12 Guide bolt
3 Guide pin	13 Selector fork 4th & 5th
4 Split pin	gear
5 Selector drum	14 Neutral detent plunger
6 Change pin – 6 off	15 Spring
7 Screw	16 Cap bolt
8 Needle roller bearing	17 Selector fork
9 Circlip	18 Selector fork
10 Retaining plate	19 Selector fork shaft

35.1a Clean out oilways and fit main passage plugs

35.1b Do not forget blanking plug on machines without oil pressure switches

35.2 Oil nozzles must be staked into crankcase

35.3a Place selector drum needle roller bearing in place ...

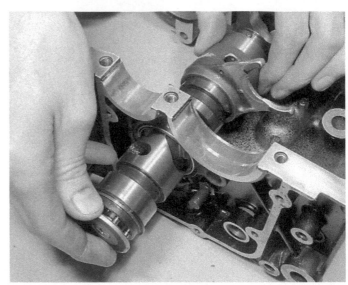

35.3b ... then fit drum and fork as shown

35.3c Fit locating pin, aligning holes

35.3d Pin is secured by fitting split pin

35.3e Slide cam over drum end, noting locating pin

35.3f Secure cam with circlip

35.4a Fit guide bolt and tab washer

35.4b Tighten bolt and bend up tab to secure

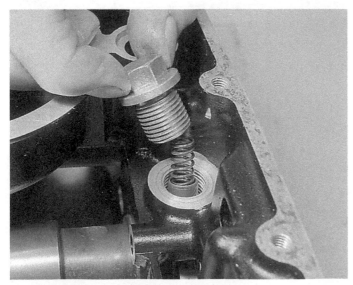

35.4c Fit neutral detent plunger, spring and bolt

35.5a Fit selector fork shaft and remaining forks

35.5b Note circlip which locates shaft end

36 Engine/gearbox reassembly: joining the upper and lower crankcase halves

1   Wipe off the mating surfaces of the crankcase halves to remove any residual oil. Check, and if necessary refit, any of the locating dowels which may have come loose.

2   Apply a thin film of non-hardening gasket compound to the mating surface of the lower crankcase half, taking care not to obstruct the oil passages. When an oilway is encountered, leave a narrow margin around it, so that the gasket compound will not get squeezed into it when the joint is tightened. If the sealing face has been marred or scratched, and oil leakage has been a problem in the past, it is permissible to use one of the Silicon RTV (Room Temperature Vulcanising) liquid gasket compounds now available. The rubbery nature of this substance will take up any small discrepancy in the mating face, but additional care must be taken to avoid obstructing oilways.

3   Lower the lower crankcase half into position, checking to ensure that the selector forks engage in their respective grooves. Push the casing down evenly, ensuring that the locating dowels align correctly. Fit the ten 8 mm and seven 6 mm retaining bolts in place, finger-tight only.

4   The correct tightening sequence is stamped on the lower crankcase half, each of the 8 mm bolts having a number denoting in which order they are to be tightened. Set a torque wrench to about 1.5 kgf m (11 lbf ft) initially, and tighten the 8mm bolts in sequence. Next reset the torque wrench to 2.5 kgf m (18.0 lbf ft) and tighten the bolts to their final torque setting. Tighten progressively the remaining seven 6 mm bolts to 1.0 kgf m (86.8 lbf in), in a diagonal pattern to avoid warpage. Turn the unit over and fit the remaining thirteen 6 mm bolts, securing these in the same way.

5   Check at this stage that the crankshaft will turn freely and that there are no tight spots in evidence. The gearbox assembly should also be checked carefully for signs of resistance or rubbing, and the gear selection checked by turning the camplate to select each ratio. This checking sequence is important, as it is far preferable to spot any problems at this stage, rather than find them when reassembly is complete.

6   It is likely that the crankshaft will be a little stiff in its movement, particularly if new shells have been fitted. However, any undue resistance in either the crankshaft or gear components will necessitate rectification before rebuilding is continued.

37 Engine/gearbox reassembly: refitting the secondary shaft, oil pump and sump

1   Lower the secondary shaft sprocket/starter clutch assembly into the casing, and fit the Morse chain around the sprocket. Slide the secondary shaft into position, guiding the end of the shaft through the sprocket/clutch unit boss. Before driving the bearing home, fit the retainer plate, and hold this is position while the shaft is tapped into position. Refit the countersunk retaining screws and remember to peen these into position after tightening.

2   Place the oil pump in position in the casing, then fit and tighten the mounting bolt and the two long screws which also retain the secondary shaft retaining plate. Note that the latter screws must be peened over after tightening.

3   Refit the secondary shaft bearing cap and tighten the mounting screws. Check that three new oilway 'O'-rings are fitted in their respective positions. Two of the 'O'-rings fit into recesses in the casing and a third is fitted in the base of the oil pump. Make sure that the lower casing and sump mating faces are clean, and that a new 'O'-ring is fitted to the oil filter chamber groove. A new sump gasket should be placed in position on the lower crankcase face, and the sump fitted and secured. Tighten the retaining bolts evenly to 1.0 kgf m (87 lbf in) of torque. A new oil filter can be fitted at this stage, and the cover fitted and tightened. It is worth priming the system by filling the inverted filter chamber with engine oil. This will speed up oil circulation when the engine is first started.

36.3 Lower crankcase half is fitted over inverted upper half

37.1a Engage the starter clutch unit in primary chain

37.1b Check that bearing is in position in crankcase

37.1c Slide secondary shaft and bearing home

37.1d Fit countersunk retainer screws ...

37.1e ... and stake heads to lock them

37.2a Check that pump locating dowels are in position ...

37.2b ... then lower pump into crankcase and secure

37.3a Check that secondary shaft nut is tight, then fit cover and retaining screws

37.3b Fit new O-rings as shown ...

37.3c ... and check that relief valve is in place

37.3d Sump (oil pan) can now be refitted

38 Engine/gearbox reassembly: refitting the clutch

1   Place the special thrust washer over the end of the gearbox input shaft, noting that its chamfered face should be installed towards the bearing. Fit the bearing sleeve and the caged needle roller bearing next, followed by the clutch drum and the plain thrust washer. Note that the bearing sleeve can be mistaken for the output shaft sleeve quite easily. Check prior to installation that the component fitted as the bearing sleeve contains an oil hole. Slide the clutch centre into place, followed by the special thrust washer. Note that this is normally of the Belville type, being conical in section. The word 'OUTSIDE' is stamped on its outer face, and it should be fitted accordingly. On the machine featured in the accompanying photographs, however, a plain thrust washer had been fitted as standard, so it is possible that this arrangement may be encountered. Fit a **new** clutch centre nut, tightening it to 13.5 kgf m (98 lbf ft), holding the clutch centre in the same manner as was used during dismantling.

2   If new clutch plates are to be fitted, apply a coating of oil to their surfaces to prevent seizure. Fit the clutch plates alternately, starting and finishing with a friction plate. Insert the ⅜ in steel ball into the input shaft end, followed by the mushroom-headed pushrod, having applied a coating of molybdenum disulphide grease to both items. Refit the clutch pressure plate, springs, washers and bolts, tightening the latter evenly and progressively to 0.9 kgf m (78 lbf in) of torque.

3   Ensure that the mating surfaces of the clutch cover and crankcase are clean, and that the two locating dowels are in position. Fit a new gasket and offer up the clutch cover. Fit the nine Allen-headed retaining screws, noting that guide clips for the ignition pickup leads are fitted to the outer two of the three lower cover screws. Tighten the screws evenly and securely, leaving loose the two screws which retain the wiring clip if the leads are not yet in position.

39 Engine/gearbox reassembly: refitting the ignition pickup assembly, oil pressure switch and alternator assembly

1   Where an oil pressure switch is fitted, this should be installed where necessary. Use a thread sealant on it, and tighten it to 1.5 kgf m (11.0 lbf ft). On all other models, where the oil passage blanking plug was removed for cleaning, this should be cleaned and a non-hardening thread locking compound applied. Tighten it to 1.3 kgf m (113 lbf in).

2   Apply one or two drops of light machine oil to the moving surfaces of the automatic timing unit. Offer up the ATU, ensuring that it engages correctly over the locating roll pin. Fit the large crankshaft rotation hexagon, again checking that it locates correctly, and then fit

and tighten the central retaining bolt to 2.5 kgf m (18.0 lbf ft).

3 On machines equipped with an oil pressure switch, reconnect the switch lead ensuring that the terminal blade is angled away from the pickup assembly to prevent accidental short circuits. The terminal screw should be tightened securely. On all models, refit the ignition pickup backplate, securing it with its three screws and washers. The output leads should be routed below the clutch cover and secured with the two guide clips. Once the wiring is in place, tighten the two remaining clutch cover screws. Note that the ignition pickup cover should not be fitted at this stage during a full engine rebuild.

4 Clean the crankshaft taper and the internal taper of the alternator rotor. Before refitting the rotor, note that two types may be encountered, and that each has a different torque wrench setting for the rotor retaining bolt. The main differences are shown below, and if there is any doubt as to the type fitted on a particular model, check the dimensions before proceeding further so that the correct torque figure may be applied. Note also that parts cannot be interchanged between the two types; an important point when replacement parts are to be ordered.

5 The early model arrangement uses a 10 mm diameter bolt (14 mm socket size) which should be tightened to a torque setting of 7.0 kgf m (51 lbf ft). The diameter of the crankshaft end on these models is 22 mm. Later models are fitted with a 12 mm diameter bolt (17 mm socket size) and the diameter of the crankshaft end is 25 mm. The 12 mm bolt should be tightened to a torque setting of 13.0 kgf m (94 lbf ft).

6 Once the correct torque setting has been determined, offer up the rotor and fit the retaining bolt. Lock the crankshaft using the method employed during removal, and tighten the bolt to the prescribed figure.

7 Check that the stator is fitted correctly in the outer cover, then fit the cover using a new gasket. Do not omit to fit the locating dowels.

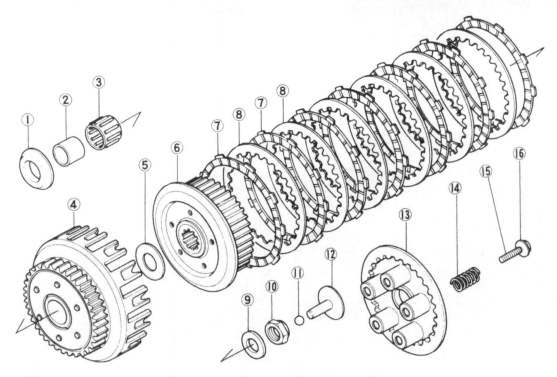

Fig. 1.16 Clutch

1	Thick thrust washer	9	Washer
2	Inner sleeve	10	Nut
3	Needle roller bearing	11	Steel ball
4	Outer drum	12	Pushrod
5	Thrust washer	13	Pressure plate
6	Clutch centre	14	Spring – 5 off
7	Friction plate – 7 off	15	Bolt – 5 off
8	Plain plate – 6 off	16	Washer – 5 off

38.1a Fit thrust washer with chamfered face inwards

38.1b Fit bearing sleeve. **Note oil hole (see text)**

38.1c Needle roller bearing is fitted next ...

38.1d ... followed by clutch drum

38.1e Plain thrust washer is fitted next ...

38.1f ... followed by the clutch centre

38.1g Fit Belville or plain washer and clutch centre nut

38.1h Hold clutch centre as shown and secure nut

38.2a Build up friction and plain plates alternately

38.2b Fit steel ball into hollow shaft end ...

38.2c ... followed by mushroom-headed pushrod

38.2d Refit cover and clutch springs ...

38.2e ... and secure with retaining bolts and washers

39.2 Offer up ATU, noting locating pin

39.3 Place pickup baseplate in position and secure

39.4a Place alternator rotor over crankshaft end

39.4b Tighten bolt to correct torque figure

39.4c Refit cover/stator assembly using new gasket

40 Engine/gearbox reassembly: refitting the selector mechanism – chain drive models

1   Check that the circlip is correctly located in its groove in the end of the selector fork shaft. Offer up the selector claw assembly, having first checked that the locating pin for the centralising spring is secure. If necessary, refit the pin using a thread locking compound. As the assembly is fitted, ensure that the claws are placed on each side of the selector drum end, and make sure that the centring spring engages on the locating pin. It is worthwhile temporarily refitting the gear change pedal to check that gear selection is positive.

2   Examine the seals in the outer cover. These can be renewed as necessary by driving the old seal out using a suitably sized socket as a drift. The casing should be arranged to give as much support as possible to the boss. The new seals may be driven into place in the same manner, taking care not to damage the sealing lip.

3   Slide a new 'O' ring (where fitted) into position on the end of the gearbox output shaft, and place a new gasket in position. Offer up the casing, easing the seals into position. Fit and tighten the retaining screws and bolts. Note that the two bolts are fitted with aluminium washers. Where a separate sprocket guard is used, this should now be fitted. Slide the output shaft spacer into position noting that this is easily mistaken for the clutch centre bearing sleeve. The latter has a small oil hole whilst the output shaft spacer is plain. The spacer should be greased prior to installation, and care must be taken to avoid damaging the seal lip. Grease and insert the clutch pushrod.

40.1 Offer up selector claw assembly. Note position of centralising spring ends

40.2 Check cover oil seals and renew if damaged

40.3a Place a new O-ring over end of output shaft

40.3b Grease and fit output shaft spacer

40.3c Fit dowels, gasket and outer cover

40.3d Pushrod fits as shown (cover removed for clarity)

40.3e The two arrowed bolts have aluminium washers

41 Engine/gearbox reassembly: refitting the selector mechanism and front gear case – shaft drive models

1   Start by refitting the external selector mechanism components as described in paragraph 1 of Section 40. The engine half of the shock absorber cam is fitted next, using a **new** retaining nut. Tighten the nut to 12.0 kgf m (87 lbf ft) and then carefully stake the nut's shouldered section into the slot in the shaft. Take care not to strike the shaft too hard or damage to the gearbox pinions or bearings may result. **Note:** Where the gear case unit is being fitted as part of a full engine rebuild, work should be halted at this point until the engine/gearbox unit has been installed in the frame. Once installed, complete the fitting of the front gear case unit as described below.

2   Offer up the gear case unit, taking great care not to damage or distort the selector shaft oil seal as it is slid over the shaft end. It is preferable to use the correct oil seal guide tool, Part Number 57001-264 if it is available, but it is possible to get by without it if care is taken.

3   Sort the case mounting bolts according to length, and fit them as shown in the accompanying figure. The bolts should be tightened evenly and progressively to a final torque figure of 1.0 kgf m (87 lbf in). Apply a coat of high melting point grease to the output splines and the corresponding internal splines of the drive shaft. Arrange the two components so that the locking pin will align with the hole in the shaft end. If the rubber gaiter (boot) was removed this should be fitted in readiness for the shaft being reconnected. The shaft can now be pulled forward to engage the front gear case splines. The locating pin should click into place, and this should be checked by pulling hard on the end of the shaft. Once the shaft is secure, refit the rubber gaiter to cover the joint.

4   It is now possible to reassemble the rear wheel end of the drive shaft. Lubricate the shaft splines with high melting point grease. Fit the coil spring over the pinion nut and offer up the final drive casing. Refit the four retaining nuts, having first cleaned the stud threads and applied a locking compound such as Loctite or similar. Tighten the four nuts evenly to 2.3 kgf m (16.5 lbf ft).

5   Reconnect the shock absorber(s) noting that the mounting nuts should be tightened to 2.5 kgf m (18.0 lbf ft). Refit the rear wheel in the reverse order of the removal sequence, noting that the splined centre and outer edge of the coupling plate should be greased. The rear wheel spindle nut should be tightened to 14.0 kgf m (101 lbf ft) the brake torque arm nut to 3.1 kgf m (22 lbf ft). Use a **new** split pin to secure the spindle nut. Reconnect and test the rear brake before the machine is ridden.

42 Engine/gearbox reassembly: fitting the pistons and cylinder block

1   To minimise the risk of oil leakage around the cylinder base joint, it is essential that the mating surfaces are completely clean and free from oil prior to reassembly. Pay particular attention to the area around the cylinder liners. There is a small groove at this point, and this must be absolutely clean. Use electrical contact cleaner or a similar residue-free solvent to remove any hint of oil from the grooves, and wipe over the gasket faces after spraying them with the same solvent.

2   After thorough cleaning of the surfaces and grooves, apply a very thin bead of Kawasaki liquid gasket (part number 56019– 120) to the groove around each liner using a fine nozzle. Do not apply excessive amounts of sealant. Kawasaki recommend that the cylinder block is then left for 24 hours before assembly proceeds.

3   Before refitting the pistons, pad the mouths of the crankcase with rag in order to prevent any displaced component from accidentally dropping into the crankcase.

4   Fit the pistons in their original order with the arrow on the piston crown pointing toward the front of the engine.

5   If the gudgeon pins are a tight fit, first warm the pistons to expand the metal. This can be done by immersing each piston in turn in very hot water prior to fitting. Shake off any water droplets before the piston is installed. Oil the gudgeon pins and small end bearing surfaces, also the piston bosses, before fitting the pistons.

6   Always use new circlips, **never** the originals. Always check that the circlips are located properly in their grooves in the piston boss. A displaced circlip will cause severe damage to the cylinder bore, and possibly an engine seizure.

7   Clean the two oil feed orifices and install them with the small hole uppermost. Fit a new O-ring to each orifice. Apply a thin coating of RTV instant gasket to both sides of the cylinder base gasket in the area shown in the accompanying figure. The gasket can now be eased over the holding studs and connecting rods, feeding the cam chain through as it is lowered. Before the gasket is lowered completely, position the cam chain rear guide in its casing recess.

8   Note that the installation of the cylinder block will require an assistant, unless a set of piston ring clamps and piston supports are available.

9   Carefully lower the cylinder block over the holding studs and then slip a block of wood beneath it to act as a support whilst the cylinder bores are lubricated with engine oil and the camshaft chain is threaded through the tunnel between the bores. The chain must engage with the crankshaft drive sprocket.

10   The cylinder bores have a generous lead in for the pistons at the bottom, and although it is an advantage on an engine such as this to use the special Kawasaki ring compressor, in the absence of this, it is possible to gently lead the pistons into the bores, working across from one side. Great care has to be taken NOT to put too much pressure on the fitted piston rings. When the pistons have finally engaged, remove the rag padding from the crankcase mouths and lower the cylinder block still further until it seats firmly on the base gasket.

11   Check that the block is seated correctly and that the crankshaft can be rotated smoothly. Hold the block down during this check, and hold the cam chain taut to prevent it from 'bunching' around the crankshaft sprocket.

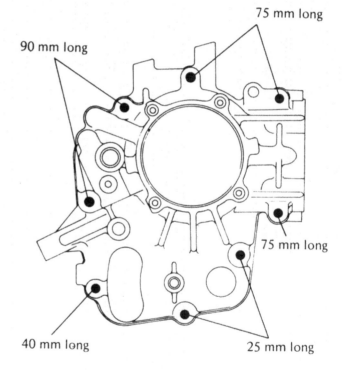

Fig. 1.17 Front gear case mounting bolt locations

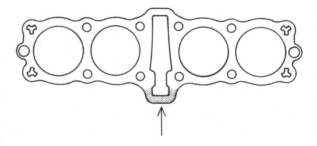

Fig. 1.18 Apply sealant to this area of the cylinder base gasket

42.5 Pack crankcase mouth with rag and fit pistons, noting arrow marks which should face forward

42.6 Use **new** circlips to retain gudgeon pins

42.7a Clean and fit oil feed orifices, using new O-rings

42.7b Position cam chain rear guide, then fit gasket

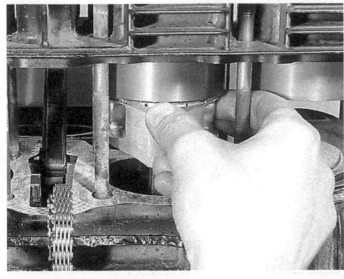

42.8 Lower cylinder block over pistons, feeding rings into lead-in or using piston ring clamps

42.10 Cylinder block can now be pushed down onto gasket

43 Engine/gearbox reassembly: refitting the cylinder head

1   Care must be taken during cylinder head installation to ensure an oil-tight joint between the cylinder head and cylinder block surfaces. Note that oil leakage can also be caused by leakage around the holding studs, and this is discussed in Section 19 of this Chapter. If oil leaks have been noted, check that the studs are sealed correctly before proceeding further.

2   During 1983, a revised cylinder head gasket and oil orifices were made available to reduce the risk of leakage. The new gasket has a film of sealant on its surfaces, while the oil orifices were lengthened and redesigned to give better support to the O-rings which form the seal around them. If oil orifices of the type shown in the accompanying photographs are encountered, purchase and fit the later type.

3   Fit the oval oil orifices and their sealing O-rings into the cylinder block mating surface. Fit the locating dowels over the front outer studs. Place a new cylinder head gasket over the studs, noting that the modified gasket should be used. Place the cam chain front guide in position, ensuring that it locates correctly.

4   Lower the cylinder head over the holding studs, feeding the cam chain through the central tunnel. Check that the cylinder head locates properly, then fit the twelve plain washers and retaining nuts. The nuts should be tightened in two stages, following the tightening sequence shown in the accompanying illustration. Tighten the nuts to 2.5 kgf m (18.0 lbf ft) initially, then to 4.0 kgf m (29.0 lbf ft). Fit the two cylinder head bolts and tighten these to 3.0 kgf m (22 lbf ft).

44 Engine/gearbox reassembly: fitting the camshafts and setting the valve timing

1   If the camshaft sprockets were removed during the course of the engine overhaul, they should be refitted, noting the following points. The exhaust camshaft on models equipped with mechanically driven tachometers has an integral drive gear and thus is easily identified. In the case of models fitted with an electronic tachometer, the gear is not machined, but a raised boss will be found in its place. The sprockets are identical, differing only in which of the three pairs of fixing holes is used. These are illustrated in the accompanying illustration. When fitting the sprockets, use a non-permanent thread locking compound on the bolt threads and tighten them to 1.5 kgf m (11.0 lbf ft). When fitted correctly, the identifying letters and timing marks on the sprockets should face the notched end of the camshafts.

2   Fit the sixteen hollow dowel pins which locate the camshaft bearing caps. On mechanical tachometer engines, check that the tachometer driven gear has been removed. It is quite easy to chip the gear teeth during camshaft installation if this precaution is not taken. On all models, remove each cam follower or 'tappet' in turn and make absolutely certain that the adjustment shim is located correctly. These are very easily displaced, and can cause a lot of unnecessary work later if not checked now.

3   Hold the cam chain taut to prevent it from jamming under the crankshaft and turn the crankshaft via the large hexagon until the '1-4' T mark is aligned exactly with the fixed mark. Lubricate the exhaust camshaft and pass it through the cam chain loop, noting that its notched end faces towards the right-hand side of the engine. Arrange the camshaft so that the 'Z7EX' mark faces forward and lies parallel to the cylinder head gasket face. Fit the inlet camshaft, following the same procedure, leaving the 'IN' mark facing rearwards and lying parallel to the gasket face.

4   Locate the chain pin which lies next to the 'Z7EX' mark. Starting with this pin as zero (0), count along the chain towards the inlet camshaft until the 45th pin is reached. Check that the scribed line of the 'IN' mark lies between the 45th and 46th pin, and adjust as necessary until it does. Refer to the accompanying valve timing illustration for details.

5   Once the camshafts are set correctly, lubricate the camshaft bearing surfaces, and fit the camshaft caps. Each cap is numbered and must be fitted so that it corresponds with the adjacent number cast into the cylinder head, and with the arrow mark facing forward. Fit the retaining bolts finger tight, then tighten them evenly and progressively until the camshafts are pulled down fully against valve spring pressure. The final torque figure is 1.2 kgf m (104 lbf in) and the bolts must be tightened in the order shown in the accompanying figure.

6   Check that the various timing marks are aligned correctly, noting

42.11 Do not omit to fit rubber plugs in holes next to holding studs

43.3a Clean gasket face and fit oil feed orifices (early type shown)

43.3b Position **new** O-rings around orifices ...

43.3c ... then place new cylinder head gasket over studs

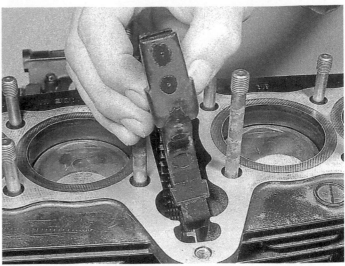

43.3d Fit cam chain front guide, ensuring that it locates properly

43.4 Lower cylinder head over holding studs

that if even slightly mistimed serious engine damage can result when the engine is next started. If a new camshaft, cylinder head, valve or cam follower was fitted it is **essential** that all valve clearances are checked and adjusted as required. The procedure is described in detail in the Routine Maintenance Chapter of this manual. Where a mechanical tachometer is fitted, apply a small film of molybdenum disulphide grease to the driven gear shaft, and install it and its holder.

7   Prime the valve pockets with clean engine oil. Fit the half-moon rubber end plugs at both ends of the camshafts, sticking them in place with RTV instant gasket. Fit a new cylinder head cover gasket noting that it is handed and can only be fitted one way. Offer up the cover noting that the arrow mark faces forward. Fit the retaining bolts and tighten them evenly to 0.8 kgf m (69 lbf in).

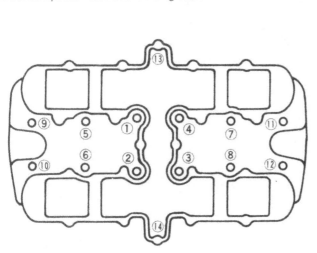

Fig. 1.19 Cylinder head nut tightening sequence

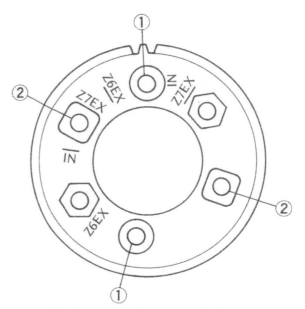

Fig. 1.20 Camshaft sprocket fixing holes

1   Inlet camshaft   2   Exhaust camshaft

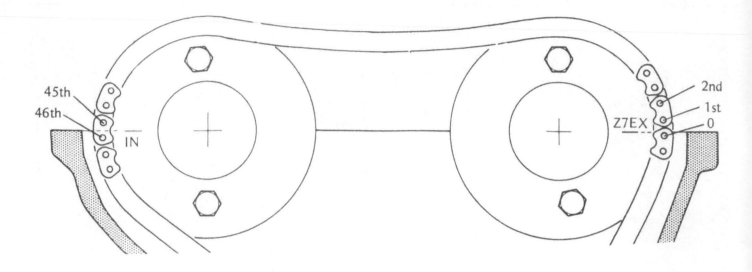

Fig. 1.21 Valve timing cam chain position

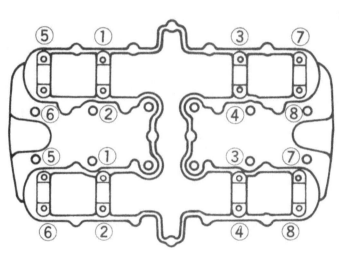

Fig. 1.22 Camshaft cap bolt tightening sequence

44.3 'Z7EX' mark should be positioned as shown

44.4 'IN' mark of inlet camshaft should appear as shown

44.5a Note camshaft cap numbers cast into cylinder head

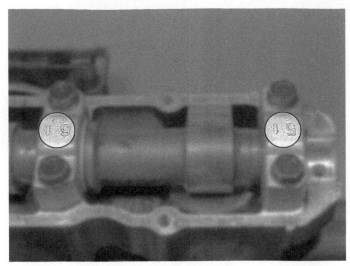

44.5b Fit caps with numbers matching those of head and with arrow marks facing forward

44.5c Arrow shows raised tachometer gear blank which identifies exhaust cam on electronic tachometer models

44.7a Fit half-moon plugs and new cover gasket

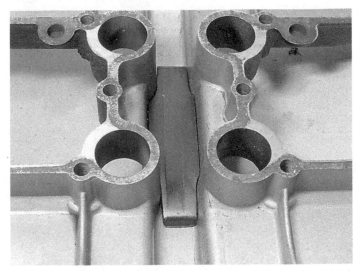

44.7b Check that chain guide is in position before fitting cylinder head cover

45 Engine/gearbox reassembly: refitting the camshaft chain tensioner – manual type

Note: In cases of premature tensioner wear, a revised tensioner mechanism of the cross-wedge type should be fitted in place of the manual version described below. This modification is applicable to KZ/Z750 E and H models from engine number KZ750EE000001 to 037630. It is likely that the cross-wedge tensioner will have already been fitted to all affected machines, but if the old type manual tensioner is still in place, refer to an authorized Kawasaki dealer for details of the new version. The procedure for fitting and adjusting the new tensioner mechanism is described in Section 46 of this Chapter.

1   Remove the lock bolt to free the pushrod and remove the pushrod stop and the light spring which fits behind the pushrod head. Compress the spring against the pushrod head until a small cross-drilling is revealed. Insert a piece of wire through this hole to retain the spring. Check that the heavy spring is in place in the tensioner body.

2   Lightly grease the pushrod and position the stop on the end of the pushrod so that the stop is installed correctly as shown in the accompanying illustration.

3   Assemble the pushrod and stop, holding the open end of the body downwards so that the retainer balls will drop clear, allowing the pushrod to enter the body. Check that the flat face of the tensioner is towards the lock bolt hole. Holding the pushrod into the body so that the wire rests across the gasket face, fit the original lock bolt and plain washer, tightening it securely to retain the pushrod. The wire can now

be pulled out.

4   Fit the tensioner, using a new gasket where necessary, and install the retaining bolts. Note that the upper bolt is longer and is fitted with an aluminium washer. Check that one pair of pistons are at TDC (the '1-4' or '2-3' T marks should be aligned) and then loosen the lock bolt to allow the tensioner to take up any chain slack. The lock bolt can now be tightened to retain the tensioner setting.

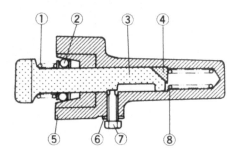

Fig. 1.23 Cam chain tensioner – manual type (see note at beginning of Section 45)

1	Spring	4	Stop	7	Lockbolt
2	Ball retainer	5	Tensioner body	8	Heavy spring
3	Pushrod	6	Washer		

46 Engine/gearbox reassembly: refitting the camshaft chain tensioner – automatic type

1   This tensioner can be identified by the secondary plunger housing and its cylindrical cap, running at 90° to the main plunger bore.
2   Remove the stop bolt and washer and withdraw the tensioner pushrod and spring from the tensioner body. Unscrew and remove the large cap and cross-wedge plunger and spring. Clean all tensioner components in solvent, and then lubricate each part with a thin film of molybdenum disulphide grease. Note that absolute cleanliness is vital.
3   Assemble the pushrod and spring in the tensioner body. Compress the spring and turn the pushrod until the location groove is visible through the stop bolt hole. The bolt and plain washer should now be fitted and tightened securely.
4   Install the tensioner using a new gasket. Note that the upper bolt is the larger of the two and is fitted with an aluminium washer. Tighten the two bolts evenly. With the engine set with either pair of pistons at TDC (check that one of the 'T' marks is aligned) fit the cross-wedge plunger so that its tapered face is at 90° to that of the pushrod. If fitted correctly, the end of the cross-wedge plunger will protrude by about 10 mm. If this is not the case, it means that full chain slack is not on the tensioner side of the chain. Turn the crankshaft slowly in the direction of normal rotation until the cross-wedge plunger is set up as described above. The cap and spring can now be fitted, using a new sealing washer where necessary. Tighten the cap to 2.5 kgf m (18.0 lbf ft).

47 Refitting the engine/gearbox unit into the frame

1   As mentioned during the engine removal sequence, the engine/gearbox unit is unwieldy, requiring at least two, or preferably three, people to coax it back into position. This is even more important during reassembly, as the unit must be offered up at the right angle, and then manoeuvred into position. Care must be taken not to damage the finish on the frame tubes, and it is worthwhile protecting these with rag or masking tape.

All models except R1, P1 and N1 models
2   Fit the rear upper and front upper engine mounting brackets, leaving the bolts slightly loose. Note that lock washers are fitted to the rear bolts and to the front nuts. Insert the six engine mounting bolts. The bolts should be fitted loosely at first, and then tightened to the appropriate torque figure.

R1 model
3   This model employs rubber-bushed anti-vibration mountings in the front engine brackets. Prior to assembly, check the condition of the rubber inserts. The short inner bushes are bonded in place and can be pulled off if required, whilst the longer main bushes are a push fit in the engine mounting brackets. If the bushes are cracked or perished they should be renewed. Fit the longer main bushes using soapy water as a lubricant. Dry the bracket assembly, and then stick the short inner bushes in place with RTV instant gasket.
4   The two front mounting brackets are handed, and will only fit with the rubber bushed boss offset to the outside and down (see photograph). Assemble the mounting bolts finger tight. Once all of the bolts are in place, tighten to the appropriate torque figure.

Shaft drive models
5   The shaft drive versions employ anti-vibration rubber mountings front and rear. These should be checked and renovated as described in paragraph 3 above. The two front brackets are identical, and are marked '1194'. The lettering should face **outwards** when installed. The rear upper left and rear upper right brackets are marked '1195' and '1196' respectively, and are fitted with the numbers facing **inwards**. The lower rear mountings have rubber bushes inserted directly into the frame lugs.
6   To assist in identification, the overall length of the rubber bush inserts is given below.

Rubber insert overall lengths

Front (right and left)	59 mm (2.32 in)
Rear upper (right and left)	50 mm (1.97 in)
Rear lower (right and left)	42 mm (1.65 in)

7   After fitting the mounting bolts loosely, tighten the nuts to the recommended torque settings.

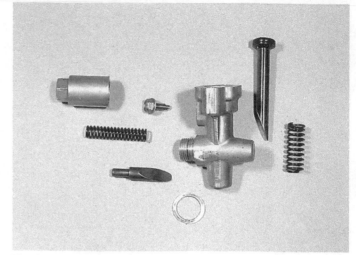
46.1 The 'cross-wedge' tensioner components

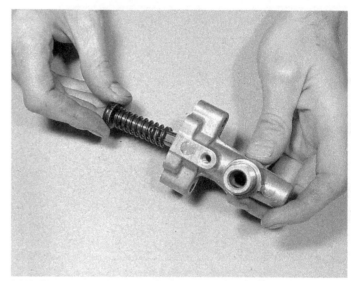
46.3a Fit plunger and spring with groove facing as shown

46.3b Hold spring compressed, then secure with stop bolt

46.4a Offer up tensioner, noting longer bolt with aluminium washer goes uppermost

46.4b Fit and secure cross-wedge, spring and cap with its sealing washer

47.1 Place engine unit in frame as shown

47.3a Check condition of engine mounting rubbers (where fitted)

47.3b Longer bush can be displaced as shown

47.4a Front mounting fit as shown

47.4b Assemble rear upper mounting ..

47.4c ... noting spacer at left-hand end

48 Engine/gearbox installation: final assembly and adjustments

1   Coat the clutch pushrod with molybdenum disulphide grease and slide it into the centre of the gearbox input shaft. On shaft drive models, complete the installation of the front gear case as described in Section 41. If the clutch release arm was removed, refit it so that the angle between the arm and the clutch cable is 70°-80°. and secure the pinch bolt. On all models, reconnect the neutral switch lead and the alternator leads. Where an oil level switch is fitted, this too should be reconnected.

2   If the starter motor was removed during the course of an engine overhaul, check that the Õ-ring is in good condition, and apply a thin film of engine oil to make fitting easier. Offer up the motor, taking care not to damage the O-ring. Fit and tighten the two retaining bolts. Check that the terminal nut is secure and then slide the rubber boot over the terminal to protect it.

3   On machines equipped with an oil cooler, offer up the oil cooler assembly and retain the radiator to the frame loosely with the four rubber-bushed mounting bolts. Clean the crankcase mounting flanges and the hose unions, fitting a new O-ring to each. Fit the pipe unions and tighten the retaining bolts securely. The radiator mounting bolts can now be secured. Tighten the flexible hose gland nuts to 2.2 kgf m (16.0 lbf ft).

4   On chain drive machines, place the final drive sprocket inside the drive chain loop and slide the sprocket over the splined end of the gearbox output shaft. Fit the splined locking washer, then fit the retaining nut and tighten it to 8.0 kgf m (58 lbf ft), whilst holding the rear wheel or applying the rear brake to prevent rotation. Secure the nut by bending the locking washer against its flats.

5   Refit the crankcase oil breather cap and tighten securely its central retaining bolt. Refit the battery tray assembly and place the air cleaner casing in position. Do not bolt it into place at this stage. Check that the intake adaptors and the air cleaner adaptors are in position. In the case of the latter, make sure that they are located properly and that the spring retainers are rolled back towards the air cleaning casing. Fit the hose clips to the intake adaptors with the screws at the bottom and facing outwards. Check that they are slackened off.

6   Manoeuvre the carburettor bank about half way into position, and then reconnect the throttle cable. Ease the carburettor bank between the air cleaner and intake adaptors, and work the carburettor stubs into their respective rubber adaptors. This stage is rather awkward. Tighten the intake adaptor clamps. Check that the air cleaner casing is correctly positioned, and then fit and tighten its two retaining bolts. Reconnect the breather hose between the air cleaner casing and the

breather cover, and secure it with the spring wire clip. Drop the air filter element into the casing. Check that it is located properly, and refit the lid.

7   Reconnect the ignition pickup leads at the four pin connector block. Where fitted, reconnect the oil pressure switch lead. Connect the engine earth lead to the crankcase noting that it is retained by a hexagon headed screw behind the oil filler cap. Refit the right-hand footrest assembly and refit the brake pedal, ensuring that it is set at the correct height. Reconnect the brake light switch operating spring. On machines fitted with a mechanical tachometer, reconnect the drive cable at the cylinder head.

8   Refit the left-hand footrest assembly, remembering to reconnect the side stand interlock switch operating spring, where fitted. Re-route and install the clutch cable and check that the clutch release mechanism operates smoothly. On chain drive machines, offer up the sprocket cover and secure it. In the case of shaft drive models, refit the rear section of the left-hand engine cover. Refit the gearchange pedal or linkage, as appropriate.

9   Use a dab of grease to stick a new sealing ring into each exhaust port. Reassemble the exhaust system by reversing the dismantling sequence, but do not tighten any of the mounting bolts yet. With the system fitted loosely into position, install the exhaust port split collets and retainers. Fit the two flanged nuts to each and tighten evenly and progressively. Once all the exhaust pipes are secure, tighten the remaining system mountings.

10   Refit the electrical panel to the left-hand side of the battery tray assembly. Refit the ignition coils below the frame top tubes, and on US models only, reconnect the vacuum switch valve and its associated hoses. Fit the fuel tank, remembering to connect up the fuel pipe, vacuum pipe and fuel sender lead as appropriate. Make a final close visual check around the machine and check all fasteners for security. Pay close attention to the routing of all electrical leads, hoses and cables, and re-route these where necessary. Refit the side panels, reconnect the battery and check that the electrical system functions normally. Check, and where necessary adjust, the final drive chain adjustment (chain drive models), the rear brake pedal height, brake switch and the clutch cable. Refit the dual seat.

11   Remove the oil filler cap and add sufficient oil to bring the level to the upper mark in the inspection window. This should take about $3\frac{1}{2}$ litre (6.2 Imp pint, 3.7 US qt). Note that the oil level will be full somewhat after the oil has been circulated around the rebuilt engine, so remember to re-check the oil level after the machine has been run for a few minutes. Note that an SE or SF grade oil having a viscosity of 10W/40, 10W/50, 20W/40 or 20W/50 should be used.

48.1 Reconnect neutral switch and alternator wiring

48.2a Check condition of O-ring, then refit starter motor ...

48.2b ... securing it with two bolts at rear

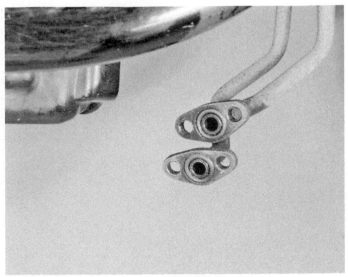
48.3 Fit new O-rings to oil cooler unions

48.5 Refit crankcase breather cover and secure with single bolt

48.6a Connect throttle cable, then offer up carburettor assembly

48.6b Tighten the intake adaptor clamps ...

48.6c ... then roll spring retainers into place

48.7 Note brake pedal alignment marks

48.8a Clean clutch release mechanism recess ...

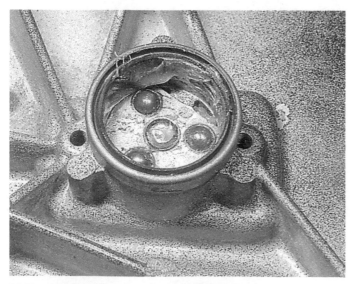
48.8b ... and drop plate and balls into position

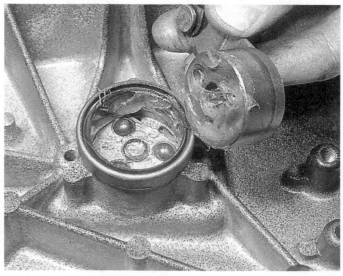
48.8c Grease and install clutch operating arm

48.8d Engage cable in arm and retain with a new split pin

48.8e Slide operating link over splined shaft ...

48.8f ... and fit pedal as shown (rear – set linkage)

48.9a Stick exhaust port gaskets in place with grease

48.9b Fit split collets and secure retainers

49 Starting and running the rebuilt engine unit

1　Make sure that all the components are connected correctly. The electrical connectors can only be fitted one way, as the wires are coloured individually. Make sure all the control cables are adjusted correctly. Check that the fuse is in the fuse holder, try all the light switches and turn on the ignition switch. Close the choke lever to start.

2　Switch on the ignition and start the engine by turning it over a few times with the kickstart or the electric starter, bearing in mind that the fuel has to work through the four carburettors. Once the engine starts, run at a fairly brisk tick-over speed to enable the oil to work up to the camshafts and valves.

3　Check the exterior of the engine for signs of oil leaks or blowing gaskets. Before taking the machine on the road for the first time, check that all nuts and bolts are tight and nothing has been omitted during the reassembling sequence.

50 Taking the rebuilt machine on the road

1　Any rebuilt engine will take time to settle down, even if the parts have been replaced in their original order. For this reason it is highly advisable to treat the machine gently for the first few miles, so that the oil circulates properly and any new parts have a reasonable chance to bed down.

2   Even greater care is needed if the engine has been rebored or if a new crankshaft and main bearings have been fitted. In the case of a rebore the engine will have to be run-in again as if the machine were new. This means much more use of the gearbox and a restraining hand on the throttle until at least 500 miles have been covered. There is not much point in keeping to a set speed limit; the main consideration is to keep a light load on the engine and to gradually work up the performance until the 500 mile mark is reached. As a general guide, it is inadvisable to exceed 4000 rpm during the first 500 miles and 5000 rpm for the next 500 miles. These periods are the same as for a rebored engine or one fitted with a new crankshaft. Experience is the best guide since it is easy to tell when the engine is running freely.

3   If at any time the oil feed shows signs of failure, stop the engine immediately and investigate the cause. If the engine is run without oil, even for a short period, irreparable engine damage is inevitable.

Chapter 2 Fuel system and lubrication

Refer to Chapter 7 for information relating to the 1983 on models

Contents

Specifications

Fuel tank capacity

	Litre	Imp gal	US gal
E1, E2, E3:			
Overall	17.3	3.8	4.6
Reserve	1.7	0.37	0.45
H1, H2, H3:			
Overall	12.4	2.7	3.3
Reserve	1.8	0.4	0.48
L1, L2, R1:			
Overall	21.7	4.8	5.7
Reserve	2.0	0.44	0.53
N1:			
Overall	14.8	3.3	3.9
Reserve	2.2	0.48	0.58
P1:			
Overall	23.4	5.2	6.2
Reserve	1.8	0.4	0.48

Fuel grade .... Unleaded or leaded. Minimum octane rating 91 (Research method/RON)

Carburettors

	All E, H and L models	N1
Make	Keihin	Keihin
Model:		
UK	CV34	N/App
US	CV34-30	CV34
Jet needle	NO1A	NO1A
Needle jet:		
UK	N/Av	N/App
US	N/Av	N426-01B36
Primary main jet	62	65
Secondary main jet	125	90
Pilot air jet	110	110
Primary main air jet	130	100
Secondary main air jet	60	60
Pilot jet	35	35
Pilot screw turns out:		
UK	2	N?App
US	Fixed	Fixed
Fuel level	4.0 ± 1 mm (0.157 ± 0.04 in)	

Carburettors

	P1	R1
Make	Mikuni	Mikuni
Model	BS34	BS34
Main jet	110R	110
Needle jet	Z-2	Y-9
Jet needle/clip position*:		
UK	4BE3/3	4BE3/3
US	N/App	4BE0/4
Pilot jet	37.5	37.5
Starter jet	N/Av	50
Main air jet	250	250
Pilot air jet	300	300
Pilot screw turns out	2.0	N/Av
Fuel level	3.0 ± 1 mm (0.118 ± 0.04 in)	
Float height	18.6 mm (0.7322 in)	

Number of needle clip groove, counting from the top

Transmission oil

Grade	SAE 10W/40, 10W/50, 20W/40 or 20W/50, SE or SF class
Capacity:	
Oil change only	3.0 litre (5.3 Imp pint/3.2 US qt)
Oil and filter change	3.5 litre (6.2 Imp pint/3.7 US qt)

Oil pump

Type	Trochoidal
Inner to outer rotor clearance service limit	0.30 mm (0.0118 in)
Outer rotor to body clearance service limit	0.30 mm (0.0118 in)
Rotor side clearance service limit	0.12 mm (0.0047 in)
Oil pressure @ 4000 rpm/90°C (194°F)	2.0 - 2.5 kg/cm² (28 - 36 psi)

Torque settings

Component	kgf m	lbf ft	lbf in
Carburettor adaptor Allen bolts - R1 model only	1.40	10.0	–
Oil passage plug - R1 model	1.30	–	113.0
Oil cooler gland nuts - R1 model	2.20	16.0	–
Oil filter bolt	2.00	14.5	–
Oil pressure switch	1.00	–	87.0
Oil pressure relief valve	1.50	11.0	–
Sump retaining bolts	1.00	–	87.0
Sump drain plug	3.80	27.0	–

1 General description

The fuel system comprises a fuel tank from which fuel is fed by gravity to the bank of four carburettors via a vacuum operated fuel tap. All models except the GT750 (KZ750 P1) and the GPz 750 (KZ/Z750 R1) are fitted with a bank of four Keihin CV34 constant depression carburettors. The two excepted models make use of Mikuni BS34 instruments of generally similar design. Air is drawn to the carburettors from a moulded plastic air cleaner casing containing a pleated paper type air filter element.

Engine lubrication is of the wet sump type, the oil being contained in a sump at the bottom of the crankcase.

The gearbox is also lubricated from the same source, the whole engine unit being pressure fed by a mechanical oil pump that is driven off the crankshaft.

2 Fuel tank: removal and replacement

1   Lift the dualseat to gain access to the single fixing bolt which secures the rear of the tank to the frame via a rubber-bushed mounting lug. Check that the fuel tap is turned to the 'On' or 'Res' positions (**not** 'Pri'), then lift the rear of the tank, after removing the fixing bolt. Pull off the fuel and vacuum hoses at the top stubs, prising them off with a screwdriver to avoid straining the hoses. On machines equipped with a fuel level sensor, trace and disconnect the lead.

2   Grasp the tank firmly, and, lifting the rear slightly, pull it backwards to disengage it from the mounting rubbers at the steering head. The tank can be refitted by reversing the removal sequence. Where the rubber buffers that hold the front of the tank prove reluctant to engage, try lubricating them with soapy water.

3 Fuel tap: removal and overhaul

1   The fuel tap is of the vacuum type and is automatic in operation. The tap lever has three positions, marked 'On', 'Res' (reserve) and 'Pri' (prime). In the first two of these settings, fuel flow is controlled by a diaphragm and plunger, held closed by a light spring. When the engine is started, the low pressure in the intake tract opens the plunger, allowing fuel to flow through the tap to the carburettors. When the tap is set to the 'Pri' position, the diaphragm and plunger are bypassed.

2   In the event of failure, the most likely culprits are the vacuum hose or the diaphragm. If a leak develops in either of these the tap will not operate in anything other than the 'Pri' position. Check the vacuum hose for obvious splits or cracks, and renew it if necessary. If the diaphragm itself is suspect, set the tap lever to 'On' or 'Res' and disconnect the fuel and vacuum hoses at the carburettor. Suck gently on the vacuum hose. If fuel does not flow, remove the tap for inspection as described below.

3   Remove the fuel tank as described in Section 2. If the tank is full or nearly full, drain it into a clean metal container taking great care to avoid any risk of fire. Place the tank on its side on some soft cloth, arranging it so that the tap is near the top. Slacken and remove the two tap mounting bolts and lift the tap away, taking care not to damage the O-ring which seals it.

4   From the front of the tap, remove the two small cross-head screws which secure the tap lever assembly. Withdraw the retainer plate, wave washer, tap lever, O-ring and the tap seal. Examine the tap seal and tap lever O-ring, especially if there has been evidence of leakage. Check that the tap seal has not become damaged and caused the blockage of the outlet hole. Fit a new O-ring and tap seal as required, and reassemble the tap lever assembly by reversing the above sequence.

5   Working from the rear of the tap, remove the four countersunk screws which retain the diaphragm cover, noting the direction in which the vacuum stub faces, and lift it away taking great care not to damage the rather delicate diaphragm. Remove the small return spring. Very carefully dislodge the diaphragm assembly and remove it from the tap body. The diaphragm assembly comprises a plastic diaphragm plate sandwiched between two thin diaphragm membranes. Carried through the centre of the assembly is the fuel plunger which supports a sealing O-ring.

6   Examine closely the diaphragm for signs of splitting or other damage. Carefully remove any dust or grit which may have found its way into the assembly. Check the condition of the O-ring on the end of the plunger. If wear or damage of the above components is discovered, it will be necessary to renew the diaphragm assembly complete. Note that one side of the diaphragm plate has a groove in it, and this must face towards the O-ring on the plunger. When fitting the diaphragm assembly and cover, check that the diaphragm lies absolutely flat, with no creases or folds. Fit the cover with the vacuum stub facing in the correct direction (this varies according to the model). Tighten the securing screws evenly and firmly.

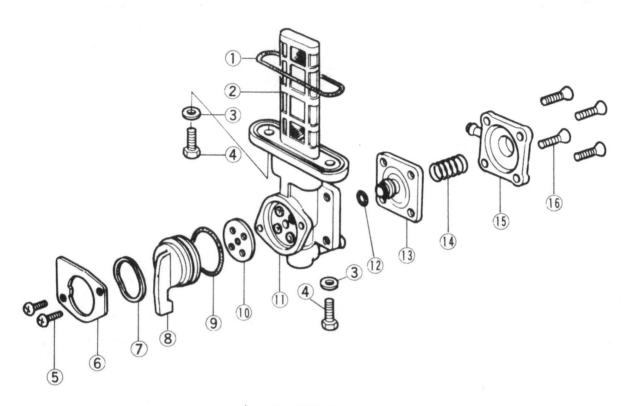

Fig. 2.1 Fuel tap

1 O-ring	5 Screw - 2 off	9 O-ring	13 Diaphragm
2 Filter stack	6 Lever retaining plate	10 Seal	14 Spring
3 Washer - 2 off	7 Wave washer	11 Tap body	15 Diaphragm cover
4 Bolt - 2 off	8 Tap lever	12 O-ring	16 Screw - 4 off

3.3 Fuel tap is retained by two bolts to underside of tank

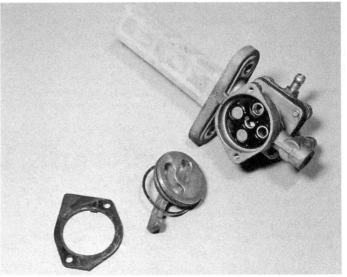

3.4 Remove screws and plate to allow tap lever removal

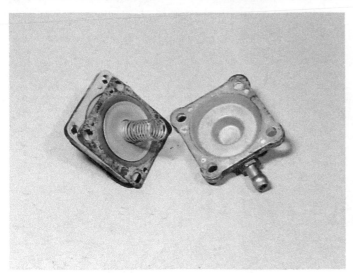

3.5 Tap diaphragm cover and diaphragm plate assembly

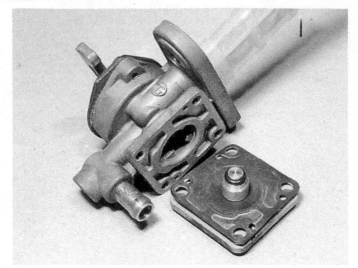

3.6 Examine tap diaphragm for splits or holes

4 Carburettors: general description and removal

1 The majority of the Kawasaki 750 models are equipped with four Keihin CV34 carburettors, the R1 and P1 models being fitted with Mikuni BS34 instruments. In the event of carburettor problems, access to the jets and other internal components is restricted, and it is recommended that the instruments are removed as an assembly.

2 Remove the fuel tank (Section 2) and pull off the left-hand and right-hand side panels. Slacken the retaining clips which secure the carburettors to the intake adaptors and roll back the spring retainer bands on the air cleaner rubbers. Displace the spring wire clips, and remove the vacuum hoses, and pull off the overflow pipes. Loosen the throttle cable adjuster locknuts and free the adjuster from its bracket. In the case of Mikuni instruments, free the cable by lifting it out of its recess in the bracket.

3 Pull the assembly back and twist it slightly to disengage the carburettor stubs from the mounting rubbers. Withdraw the assembly to the right and free the throttle cable inner to complete removal.

5 Carburettors: dismantling, overhaul and reassembly

1 Dismantle, check and assemble each carburettor separately to avoid interchanging components between instruments. Remove the four diaphragm cover screws and lift away the cover and spring, taking care not to tear the diaphragm. Peel off the edge of the diaphragm and remove it together with the throttle valve assembly. Unscrew the needle holder plug from the inside of the valve and displace the needle.

2 Unscrew and remove the pilot screw, spring, plain washer and O-ring. On US models this assembly is covered by a small blanking plug. Use a punch or scriber to deform the plug, then prise it out. Screw the pilot screw inwards (clockwise) until it seats, counting the number of turns to allow it to be set at the same position during reassembly, then remove as described above.

3 Unscrew the four screws and lockwashers which retain the float bowl and lift it away. Refer to the accompanying illustration, and remove the main jet, main jet bleed pipe, secondary main jet and the needle jet holder, followed by the needle jet. Prise off the plastic plug which covers the pilot jet and remove the pilot jet.

4 Displace the float pivot pin and lift away the float assembly, complete with the float valve. The latter can be disengaged from its retaining wire clip once the assembly has been lifted clear. Note that the Mikuni carburettors are generally similar to the Keihin type described above, and can be dismantled in much the same fashion. Refer to the accompanying photographic sequence and line drawing for details.

5 Wash the carburettor components in a degreasing solution, or use a proprietary carburettor cleaner to remove gum deposits. Blow the carburettor body drillings out with compressed air. Check the end of the pilot screw for wear or damage, renewing it if marked or deformed. Clean the jet orifices by blowing them clear. If a stubborn blockage is encountered, use a fine nylon bristle to clear it, **never** use wire which may scratch or enlarge the jet. Examine the jet needle, renewing it if scored or bent, together with the needle jet. Check the throttle valve assembly for wear and the diaphragm for holes or splits. Check the float assembly for signs of leakage, renewing it if fuel can be heard inside the floats when shaken. Check the float valve needle and seat, renewing the needle if the seating area has become grooved. Before commencing reassembly, check that each component is absolutely clean and that all sediment has been removed from the instrument.

6 Fit the float bowl components first, in the reverse order of their removal. As each part is fitted, check that it corresponds with the details shown in the specifications. Renew any O-ring or plastic plug which has worn or was damaged during removal. Fit the assembled throttle valve assembly, ensuring that the diaphragm lies flat and is located by the small tab. Offer up the spring and cover, holding the valve open to avoid any creases in the diaphragm. Fit and secure the cover retaining screws. On US models, press in a new blanking plug to cover the pilot screw, which should have been set to its original position. Apply a trace of locking compound to the outside of the plug to secure it.

Carburettor separation – Keihin type

7 It is not necessary to separate the individual instruments unless attention to the bodies or linkage is required. Note that a choke shaft kit should be obtained before starting work. Unhook the end of the choke shaft spring. Remove the eight retaining screws and lift away the upper mounting bracket. Remove the lower mounting bracket in the same way and separate the right-hand and left-hand pairs of carburettors.

8 To separate the left-hand pair of instruments, flatten the locking washer and remove the nut which secures the choke lever. Remove the bolt, wave washer, plain washer and plastic washer to free the choke linkage shaft, noting the small spring and detent ball which will be displaced during removal. Remove the two screws which retain each choke valve to the shaft, noting that it may prove necessary to file off the staked-over screw ends. Remove the choke valves and shaft and separate the carburettors.

9 When reassembling the carburettor bank, refer to the appropriate line drawing for guidance. Check and renew the O-rings as required. Use new screws to retain the choke valves, staking the ends using the adaptor supplied with the kit in conjunction with pliers. Before tightening the mounting bracket screws, place the carburettors on a flat surface to ensure alignment.

Carburettor separation – Mikuni type

10 Proceed as described in paragraphs 7 to 9, noting the following. The 'choke' mechanism takes the form of a cold-start fuel circuit

operated by a plunger rod running across the carburettor bank. This can be removed after pulling off the circlips which hold the four plunger operating links to the rod. Note the relative positions of each link, circlip, spring seat and spring. When the rod is withdrawn, two detent balls and springs will be released from the centre carburettors. Take care not to lose them and grease them during installation. Check the operation of the choke mechanism after reassembly, ensuring that each of the four plungers is fully closed in the off position.

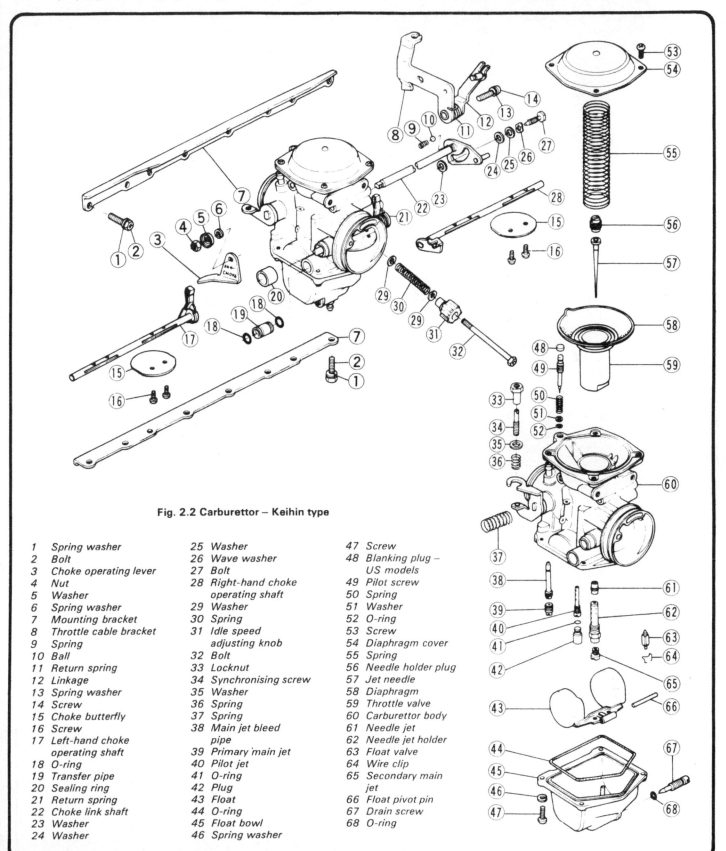

Fig. 2.2 Carburettor – Keihin type

1 Spring washer	25 Washer	47 Screw
2 Bolt	26 Wave washer	48 Blanking plug – US models
3 Choke operating lever	27 Bolt	49 Pilot screw
4 Nut	28 Right-hand choke operating shaft	50 Spring
5 Washer	29 Washer	51 Washer
6 Spring washer	30 Spring	52 O-ring
7 Mounting bracket	31 Idle speed adjusting knob	53 Screw
8 Throttle cable bracket	32 Bolt	54 Diaphragm cover
9 Spring	33 Locknut	55 Spring
10 Ball	34 Synchronising screw	56 Needle holder plug
11 Return spring	35 Washer	57 Jet needle
12 Linkage	36 Spring	58 Diaphragm
13 Spring washer	37 Spring	59 Throttle valve
14 Screw	38 Main jet bleed pipe	60 Carburettor body
15 Choke butterfly	39 Primary main jet	61 Needle jet
16 Screw	40 Pilot jet	62 Needle jet holder
17 Left-hand choke operating shaft	41 O-ring	63 Float valve
18 O-ring	42 Plug	64 Wire clip
19 Transfer pipe	43 Float	65 Secondary main jet
20 Sealing ring	44 O-ring	66 Float pivot pin
21 Return spring	45 Float bowl	67 Drain screw
22 Choke link shaft	46 Spring washer	68 O-ring
23 Washer		
24 Washer		

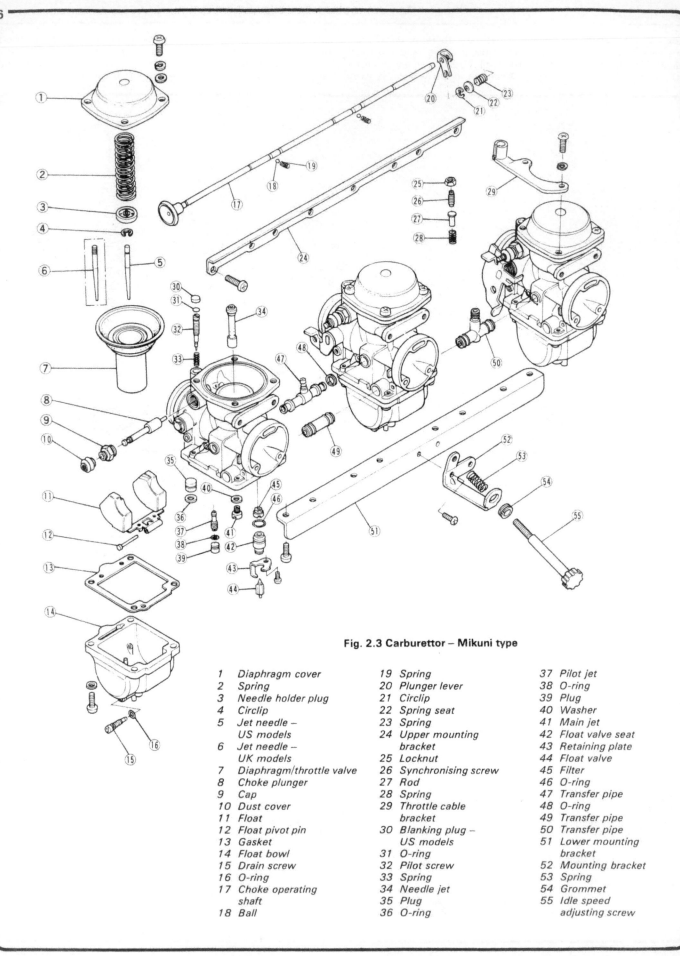

Fig. 2.3 Carburettor – Mikuni type

1 Diaphragm cover	19 Spring	37 Pilot jet
2 Spring	20 Plunger lever	38 O-ring
3 Needle holder plug	21 Circlip	39 Plug
4 Circlip	22 Spring seat	40 Washer
5 Jet needle – US models	23 Spring	41 Main jet
6 Jet needle – UK models	24 Upper mounting bracket	42 Float valve seat
7 Diaphragm/throttle valve	25 Locknut	43 Retaining plate
8 Choke plunger	26 Synchronising screw	44 Float valve
9 Cap	27 Rod	45 Filter
10 Dust cover	28 Spring	46 O-ring
11 Float	29 Throttle cable bracket	47 Transfer pipe
12 Float pivot pin	30 Blanking plug – US models	48 O-ring
13 Gasket	31 O-ring	49 Transfer pipe
14 Float bowl	32 Pilot screw	50 Transfer pipe
15 Drain screw	33 Spring	51 Lower mounting bracket
16 O-ring	34 Needle jet	52 Mounting bracket
17 Choke operating shaft	35 Plug	53 Spring
18 Ball	36 O-ring	54 Grommet
		55 Idle speed adjusting screw

5.1a Remove throttle valve diaphragm cover and return spring

5.1b Peel edge of diaphragm away from carburettor

5.1c Throttle valve assembly and diaphragm may now be removed

5.1d Note white plastic needle retainer (Mikuni type)

5.1e Needle can be displaced from valve for inspection

5.2a Pilot screw (arrowed) is covered by plug on US models

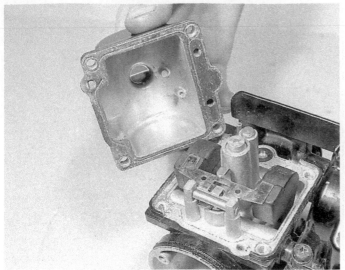

5.2b Float bowl is secured by four screws

5.2c Remove main jet and washer

5.2d Needle jet assembly can now be displaced . . .

5.2e . . . and removed from top of carburettor

5.2f Remove plug which closes pilot jet passage . . .

5.2g . . . and unscrew jet

5.4a Displace pin and remove float assembly

5.4b Float valve can now be removed

5.4c Release retainer to remove float seat

5.4d Note fuel strainer incorporated in seat

5.5a Check needle for straightness

5.5b Note needle number and clip locating grooves

6 Carburettor adjustment

1   Carburettor adjustment should be carried out with the engine at normal operating temperature and with the valve clearances and spark plug gaps set correctly. Check also that throttle cable free play is set to the recommended 2 - 3 mm (0.079 - 0.118 in).

2   Set the idle speed to 1000 - 1100 rpm, making any necessary adjustment via the large throttle stop knob. The individual synchronisation screws should not be disturbed. Check that the idle speed remains stable after opening and closing the throttle a few times. Moving the handlebar should not affect idle speed. If it does,, check the throttle cable routing.

3   The pilot screws should not require adjustment once set up as described in the specifications. Note that on US models, the pilot screws are set at the factory to comply with exhaust emission laws. The screws are covered by anti-tamper plugs and should not be adjusted. Note that **no standard setting** is supplied for these models. If it is necessary to disturb the screws, always check their settings first.

4   If continual over-richness or over-weakness occurs the fuel level should be checked as shown in Fig. 2.4. If the level is outside the specification, adjustment may be made, after removal of the float bowl, by judicious bending of the float needle operating tang on the float unit.

7 Carburettors: synchronisation

1   Carburettor synchronisation must be checked at the interval specified in Routine Maintenance, and whenever the carburettors have been disturbed or if the engine is running roughly. A set of accurate vacuum gauges is essential for the synchronisation, and if these are not available the job should be entrusted to a Kawasaki dealer. On no account attempt to adjust synchronisation by 'feel'. It will almost always make things worse.

2   Remove the fuel tank and arrange a temporary fuel supply, either by using a small temporary tank or by using extra long fuel pipes to the now remote fuel tank on a nearby workbench. Note that if the vacuum hose is bypassed it is important to plug its open end before attempting the check. Connect the vacuum gauge hoses to the four vacuum take-off points, having first disconnected the relevant hose(s) and cap(s). Start the engine and run it until it has reached normal operating temperature. Where the gauges have clamping adjustment, set this to give a needle flutter of 3 cmHg or less.

3   Run the engine at idle speed and check that a vacuum reading of about 22 cmHg is shown for each cylinder. More importantly, there must be less than 2 cmHg difference in the readings on each cylinder. To adjust the synchronisation, slacken the synchronising screw locknuts, these being located between each pair of carburettors on the engine side. Turn the screws until synchronisation is obtained, then open and close the throttle a few times before re-checking. Once synchronised, hold the screws and secure the locknuts. The Kawasaki adjusting tool, part number 57001-351, is very useful but not essential.

8 Air cleaner: location, removal and maintenance

The procedure for maintenance of the air filter is given in Routine Maintenance.

9 Lubrication system: checking the oil pressure

1   The efficiency of the lubrication system is dependent on the oil pump delivering oil at the correct pressure. This can be checked by fitting an oil pressure gauge to the right-hand oil passage plug, which is located immediately below the ignition pickup housing. Note that the correct threaded adaptor must be obtained or fabricated for this purpose. The best course of action is to obtain the correct Kawasaki pressure gauge and adaptor, Part Numbers 57001-164 and 57001-403 respectively.

2   Remove the end plug and fit the adaptor and gauge into position. Start by checking the pressure with the engine cold. Note that if the

engine is warm, **hot oil may be expelled when the end plug is removed**. Start the engine and note the pressure reading at various engine speeds. If the system is working normally, a reading of 4.4 - 6.0 kg/cm^2 (63 - 85 psi) should be maintained. If it exceeds the higher figure by a significant amount it is likely that the relief valve is stuck closed. Conversely, an abnormally low reading indicates that the valve is stuck open or the engine very badly worn. The test should now be repeated after the engine has warmed up. The correct pressure at 4000 rpm and 90°C (194°F) should be 2.0 - 2.5 kg/cm^2 (28 - 36 psi). If the oil pressure is significantly below this figure, and no obvious oil leakage is apparent, the oil pump should be removed for examination. On no account should the machine be used with low oil pressure, as plain bearing engines in particular rely on oil pressure as much as volume for effective lubrication.

3   It is likely that the normal oil pressure will be slightly above the specified pressure, but if it proves to be abnormally high, it is likely to be due to the oil pressure relief valve being jammed or damaged. The latter component is fitted to the inside of the sump. Refer to Section 12 of this Chapter.

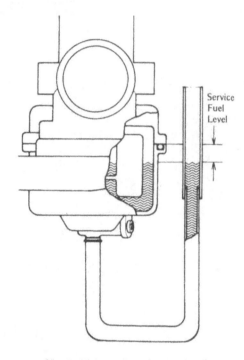

Fig. 2.4 Measuring the fuel level

7.3 Synchronising screws and locknuts are located between carburettors

10 Oil pump: removal, examination and renovation

1 It is possible to remove the oil pump for examination after the clutch cover, clutch assembly, and sump have been removed, having, of course, drained the sump beforehand. Refer to the relevant Sections of Chapter 1 for details. Examination of the pressure relief valve and renewal of the oil filter element should be undertaken before re-assembling these components. The oil pump can be removed from the underside of the crankcase after releasing the mounting bolts.

2 Remove the oil pump cover screws and the circlip on the pump spindle. Lift the cover away. The inner and outer rotors can be shaken out of the pump body, the driving pin displaced, and the pump spindle withdrawn.

3 Wash all the pump components with petrol and allow them to dry. Check the pump casing casting for breakage or fracture, or scoring on the inside perimeter.

4 Reassemble the pump rotors and measure the clearance between the outer rotor and the pump body, using a feeler gauge. If the measurement exceeds the service limit of 0.30 mm (0.012 in) the rotor or the body must be renewed, whichever is worn. Measure the clearance between the outer rotor and the inner rotor, using a feeler gauge. If the clearance exceeds 0.30 mm (0.012 in) the rotors must be renewed as a set. With the pump rotors installed in the pump body lay a straight edge across the mating surface of the pump body. Again with a feeler gauge measure the clearance between the rotor faces and the straight edge. If the clearance exceeds 0.12 mm (0.005 in) the rotors should be replaced as a set.

5 Examine the rotors and the pump body for signs of scoring, chipping or other surface damage which will occur if metallic particles find their way into the oil pump assembly. Renewal of the affected parts is the only remedy under these circumstances, bearing in mind that the rotors must always be replaced as a matched pair.

6 Reassemble the pump components by reversing the dismantling procedure. The component parts must be ABSOLUTELY clean or damage to the pump will result. Replace the rotors and lubricate them thoroughly before refitting the cover.

7 Check that the pump turns smoothly, then refit it to the casing. Before refitting the sump, remove and examine the pressure relief valve as described in Section 13. Do not omit the large 'O' ring which must be fitted to the oil pump before the sump is refitted.

10.2 Remove circlip and screws to free pump cover

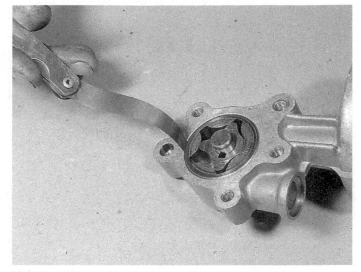

10.4a Measure clearance between outer rotor and pump body

10.4b Clearance between inner and outer rotors can be checked as shown

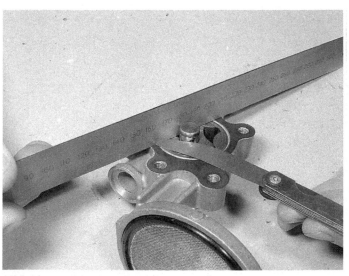

10.4c Use straightedge to check rotor endfloat

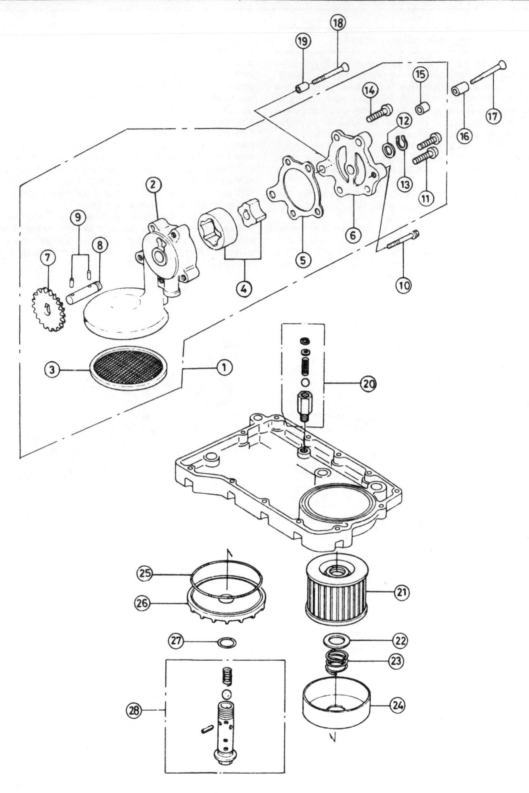

Fig. 2.5 Oil pump and filter

1	Oil pump	8	Shaft	16	Dowel pin	22 Washer
2	Pump body	9	Pin - 2 off	17	Screw	23 Spring
3	Filter gauze	10	Bolt	18	Screw	24 Cover
4	Inner and outer	11	Screw - 2 off	19	Dowel pin	25 O-ring
	rotors	12	Washer	20	Oil pressure	26 Filter cover
5	Gasket	13	Circlip		relief valve	27 O-ring
6	End cover	14	Screw	21	Oil filter	28 Centre bolt
7	Drive gear	15	Dowel pin		element	

11 Oil pressure relief valve: dismantling, examination and renovation

1   If problems with the lubrication system have been experienced, it is advisable to check the operation of the pressure relief valve, whilst the sump is removed. The valve can be unscrewed from the inside of the sump, using a spanner on the hexagonal body.

2   Kawasaki caution against dismantling the valve, because it is felt that doing so may itself upset the valve assembly and cause inaccuracy. Using a wooden dowel or plastic rod, push the ball off its seat against spring pressure, noting that the ball should move smoothly, with no rough spots. If any hesitation is noted, wash the valve assembly thoroughly in a high flash point solvent, and blow it dry with compressed air. If no improvement is noted, renew the valve assembly as a unit, there being no provision for obtaining the parts individually.

12 Oil filter: renewing the element

1   The procedure for renewing the oil filter element is given in Routine Maintenance.

2   Never run the engine without the filter element or increase the period between the recommended oil changes or oil filter changes.

13 Oil pressure warning switch – all E, H and L models

1   An oil pressure warning switch is incorporated in the lubrication system to give warning of impending disaster in the event of oil pressure failure. The switch is located inside the ignition pickup housing and normally gives very little trouble. In the event that the oil warning light does not come on when the ignition is just switched on, it is imperative that the fault is isolated and rectified before the machine is ridden.

2   Remove the ignition pickup cover, and disconnect the pressure switch lead. With the ignition switched on, earth the lead from the warning lamp against the crankcase. If the warning lamp comes on, the switch should be renewed. If, however, the warning lamp still does not work, attention should be turned to the bulb and wiring.

3   The switch may be unscrewed from the casing after the ignition pickup base plate has been removed. Note that the terminal which screws onto the top of the switch should be positioned away from the contact breaker to avoid any possibility of it earthing across.

If the light comes on suddenly whilst riding, stop the machine immediately, and investigate the cause, noting the above comments. On no account ride the machine with the warning lamp on. When refitting the switch, tighten it to 1.5 kgf m (11.0 lb ft).

14 Oil level switch: R1 and all shaft drive models

The GPz750 (KZ/Z750-R1) model and the shaft drive Spectre (KZ750-N1) and GT750 (KZ750-P1) models are equipped with an oil level switch. The switch unit is mounted on the underside of the sump and may be removed for testing after the engine oil has been drained, by removing the two retaining bolts. A multimeter or battery and bulb arrangement should be used to make a continuity check on the unit, and the general test procedure is described in Chapter 6, Section 20, paragraph 8.

Chapter 3 Ignition system

Refer to Chapter 7 for information relating to the 1983 on models

Contents

Specifications

Ignition system
Type	Electronic
Advance	10° BTDC @ 1050 rpm
	40° BTDC @ 3650 rpm

Spark plugs
	NGK	ND
Make	NGK	ND
Type:		
US	B8ES	W24ES-U
UK, Canada	BR8ES	W24ESR-U
Cold weather type:*		
US	B7ES	W22ES-U
UK, Canada	BR7ES	W22ESR-U

For use below 10°C (50°F) or at constant low speed only

Gap	0.7 – 0.8 mm (0.028 – 0.031 in)

Torque settings
Component	kgf m	lbf ft	lbf in
Spark plugs	2.80	20.00	–
Automatic timing unit	2.50	18.00	–
IC ignitor unit – 1982 models	0.65	–	56.0

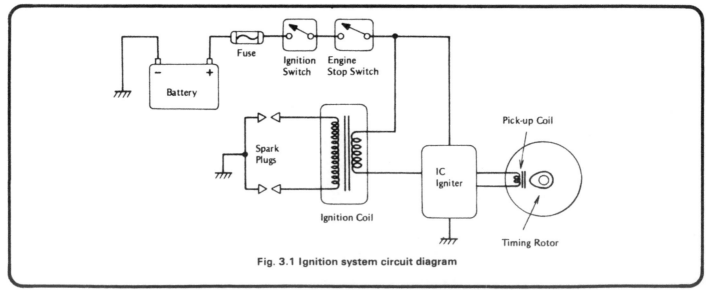

Fig. 3.1 Ignition system circuit diagram

1 General description

The Kawasaki KZ/Z750 models are equipped with a magnetically-triggered battery and coil ignition system of proven design. The system comprises two identical circuits, each circuit firing two of the four spark plugs. Cylinders 1 and 4 are grouped together, as are cylinders 2 and 3. For any given cylinder, the plug is fired twice for every engine cycle, but one of the sparks occurs during the exhaust stroke and thus performs no useful function. The arrangement is usually known as a 'spare spark' or 'wasted spark' system for obvious reasons.

2 Ignition system: locating and identifying faults

1 If problems occur it is essential that any fault-finding is approached in a logical sequence.

2 The accompanying flow chart shows the sequence in which the system should be checked in order to locate the source of a problem quickly and efficiently. Follow the flow chart until the likely cause is located, then refer to the following Sections for details of the various test procedures.

3 Before attempting any work on the ignition system components, note that accidental short circuits will almost certainly damage the ignitor unit. Even a momentary reversed connection or short circuit can cause irreparable damage to the internal circuitry. When disconnecting the battery **always** separate the negative (–) lead from its terminal first, and reconnect it last to avoid any risk of shorting if a spanner touches the frame.

4 Note that on some later machines, a revised ignitor unit was fitted. This had insulated mounting bushes, whilst the earlier type was bolted directly to the frame. If the later type unit is fitted to an early machine, note that the mounting bolts should be tightened to no more than 0.5 – 0.8 kgf m (3.6 – 5.8 lbf ft) or the bushes may be damaged.

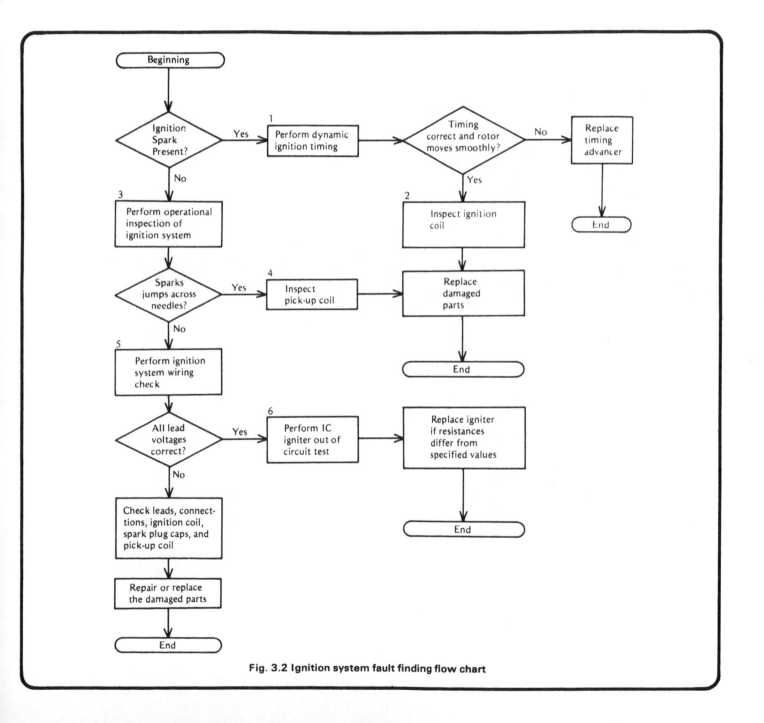

Fig. 3.2 Ignition system fault finding flow chart

3 Checking the ignition timing

1 The ignition timing should be checked dynamically using a stroboscopic timing lamp, preferably of the bright xenon tube type rather than the cheaper neon types which produce a much dimmer light. Connect the strobe according to the manufacturer's instructions, to the spark plug lead of cylinder number 1 or 4. Remove the circular inspection cover from the right-hand end of the crankcase so that the timing marks can be viewed through the oval hole in the ignition pickup backplate.

2 Start the engine, and direct the strobe at the timing marks, which will appear to be 'frozen' by the flashes from the lamp. At low engine speeds, up to about 1900 rpm, the 'F' mark should align with the fixed timing mark. Gradually increase the engine speed, noting that at about 3500 rpm the automatic timing unit should begin to advance the ignition. Maximum advance is indicated by a pair of parallel lines just to the right of the '4' mark. This should occur at about 3800 rpm.

3 If the timing does not correspond to the above conditions it will be necessary to remove the automatic timing unit for inspection and overhaul as described in the following Section. If this fails to effect a cure then the automatic timing unit must be deemed faulty and renewed. Note that there is no provision for timing adjustment as such, and none should be required.

4 Automatc timing unit (ATU): removal and overhaul

1 If a timing check has indicated erratic or inaccurate ignition (see Section 3), it will be necessary to remove the automatic timing unit (ATU) for examination. It is worth bearing in mind that the ATU, being the only purely mechanical part of the ignition system, is the only possible culprit with this type of ignition problem.

2 Remove the circular ignition pickup cover, if this is still in place, to reveal the ignition pickup baseplate. Remove the three large cross-head screws at the outer edge of the baseplate and lift it away from the crankcase. **Note:** On no account disturb the smaller pickup securing screws.

3 Before removing the ATU, check its operation by twisting the rotor. Note carefully any sign of stiffness or jerkiness as the mechanism operates. Both of the above symptoms are often attributable to inadequate lubrication, and will cause erratic or limited ignition advance.

4 Slacken the smaller of the two hexagons and remove the retaining bolt, the large crankshaft rotation hexagon and the ATU. Dismantle the unit on a clean surface, laying each part out carefully to avoid incorrect assembly. Unhook the bobweight springs, and free the weights by

prising off the clips which secure them to their pivot posts. The rotor can now be slid off the ATU centre.

5 Wash each part in petrol or a similar solvent, taking care that all residual grease is removed from the groove inside the rotor. Check the pivot pins and the corresponding holes in the bobweights. If these are badly worn, inaccurate timing will be unavoidable and a new ATU must be fitted.

6 Fill the internal groove in the rotor with grease. Slide the rotor into position, noting that the rotor projection should face towards the raised stop which has 'TEC' stamped near it. Refit the weights and springs, lubricating both pivot posts with a drop of engine oil before fitting the retaining clips. Check that the unit operates smoothly, then refit it by reversing the removal sequence. Check that the crankshaft rotation hexagon locates properly, then fit the retaining bolt which should be tightened to 2.5 kgf m (18.0 lbf ft).

5 Operational test of the ignition system

1 This test requires the use of a separate 6 or 12 volt motorcycle battery in addition to that fitted to the machine. Kawasaki recommend the use of an electrotester to provide a pre-set 5 – 8 mm electrode gap. In view of the fact that this device will probably not be available it is recommended that a similar piece of equipment be fabricated using an insulated stand of wood, and four electrodes of stiff wire or fashioned from nails. The opposing ends of the electrodes, across which the spark will jump, should be sharpened, and the outer ends of the two electrodes to which the HT voltage is to be applied should be ground down so that the sparking plug caps are a push fit. The accompanying figure shows the general arrangement.

2 Open the dual seat, then trace and separate the four-pin connector between the IC ignitor and the pick-up coil leads. The second battery must be connected to the IC ignitor as shown in Fig. 3.4. Note that the blue lead is connected to the battery negative (-) terminal and the black lead to the positive (+) terminal. The switch shown in the circuit is not vital; in practice the test circuit can be made and broken by joining and separating the black lead to the battery positive terminal.

3 Connect the test apparatus so that the plug caps are connected to one side of the electrodes, and connect the other side of the electrodes to a good earth point on the engine. Check that the electrode gaps do not exceed 8 mm. If intermittent earthing occurs or if the spark gap is too great to allow the spark to jump (in normal operation) damage to the electronic ignition may result. Turn the ignition switch on.

4 The test is conducted by applying voltage from the second battery to the IC ignitor for a few seconds, then disconnecting the supply. This energises the low tension windings in the ignition coil, and a spark should be produced at both plugs when the supply is disconnected. If

4.2 Do not disturb screws which retain pickup to baseplate

4.3 Check operation of ATU. If stiff, dismantle and clean

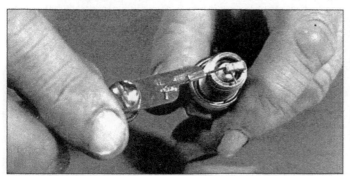

Electrode gap check - use a wire type gauge for best results

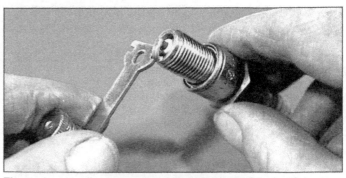

Electrode gap adjustment - bend the side electrode using the correct tool

Normal condition - A brown, tan or grey firing end indicates that the engine is in good condition and that the plug type is correct

Ash deposits - Light brown deposits encrusted on the electrodes and insulator, leading to misfire and hesitation. Caused by excessive amounts of oil in the combustion chamber or poor quality fuel/oil

Carbon fouling - Dry, black sooty deposits leading to misfire and weak spark. Caused by an over-rich fuel/air mixture, faulty choke operation or blocked air filter

Oil fouling - Wet oily deposits leading to misfire and weak spark. Caused by oil leakage past piston rings or valve guides (4-stroke engine), or excess lubricant (2-stroke engine)

Overheating - A blistered white insulator and glazed electrodes. Caused by ignition system fault, incorrect fuel, or cooling system fault

Worn plug - Worn electrodes will cause poor starting in damp or cold weather and will also waste fuel

no spark occurs, investigate the system wiring as described in Section 6. If the plugs spark properly, the fault must lie in the pick-up coil, which can be checked as described in Section 8. **Important note:** do not connect the second battery for more than 30 seconds, otherwise the ignition coil and ignitor may overheat and be damaged. If the test shows the circuit for cylinders 1 and 4 to be operating normally, repeat the test on the cylinders 2 and 3 circuit, connecting the appropriate high tension leads to the test apparatus, and noting that the auxiliary battery should be connected to the yellow lead (positive terminal) and the red lead (negative terminal) at the 4-pin connector.

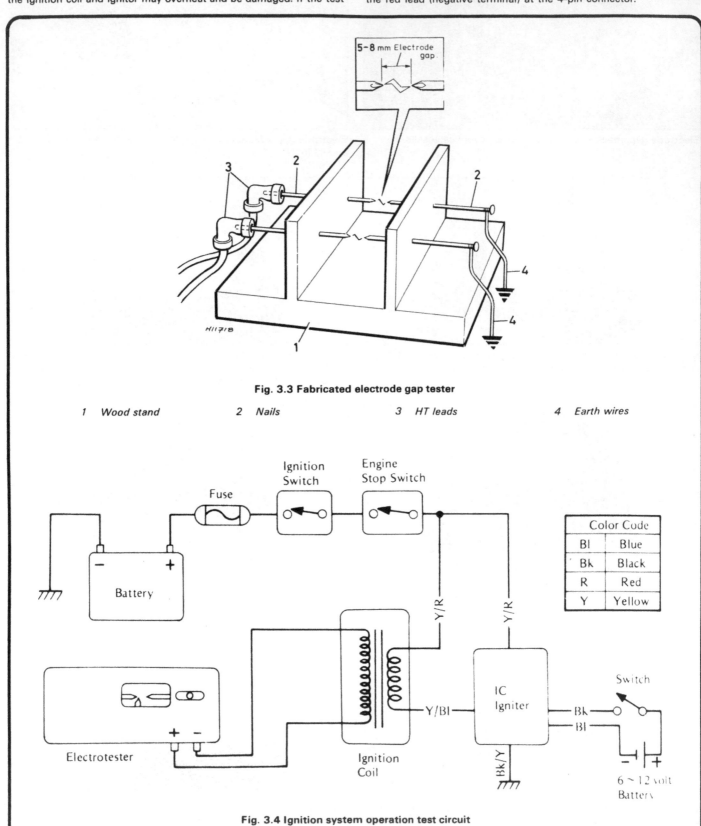

Fig. 3.3 Fabricated electrode gap tester

| 1 | Wood stand | 2 | Nails | 3 | HT leads | 4 | Earth wires |

Fig. 3.4 Ignition system operation test circuit

6 Ignition system wiring and connection check

1 In the event of a fault occurring check the various connector blocks for corrosion or moisture. Should these be discovered a water dispersing fluid such as WD40 can be used as a temporary remedy in most instances. For a more permanent cure, clean off as much of the corrosion as possible to ensure a sound connection, and pack the connector halves with silicone grease to prevent further trouble. The engine 'kill' switch can suffer from similar problems and should be dealt with in a similar manner.

2 A general wiring test can be performed using a multimeter set on the 0 – 20 volts dc range. The system should be intact, all connectors attached normally, and the meter probes should be pushed into the back of the IC ignitor terminals as shown in the table below. The test is performed with the engine stopped but with the ignition switched on.

7 IC ignitor: out-of-circuit test

1 If the system wiring test described in Section 6 indicates that all voltages are correct, the IC ignitor unit should be tested out of circuit by measuring its internal resistances. Check that the ignition switch is off, then pull off the connector at the ignitor. Using the table shown in Fig. 3.6, connect the meter probes to the appropriate terminals and note the resistance reading obtained. If the figures obtained differ markedly from those shown the unit may require renewal. Note that different meters may give slightly different readings than the Kawasaki tester, so confirm the diagnosis by taking the suspect unit to a Kawasaki dealer before buying a new one.

8 Ignition pickup coils: resistance test

1 If the operational test of the system (Section 5) has indicated that the fault lies in the ignition pickup assembly, the resistances of the pickup coils should be checked. Trace and disconnect the pickup coil leads at the 4-pin connector at the IC ignitor. Set the multimeter to the ohm x 100 scale and connect one probe to the Black lead and the other to the Blue lead, noting the reading. Repeat the test on the remaining coil by connecting the probes to the Yellow and Red leads.

2 In each case, a reading of 360 – 540 ohm should be obtained. If a reading which differs significantly is found, the pickup coil should be considered faulty and renewed.

9 Ignition coils: checking

1 The ignition coils are a sealed unit designed to give long life, and are mounted on the frame tubes in the upper cradle behind the steering head. The most accurate test of an ignition coil is with a three point coil and condenser tester (electrotester).

2 Connect the coil to the tester when the unit is switched on, and open out the adjusting screw on the tester to 6 mm (0.24 inch). The spark at this point should bridge the gap continuously. If the spark starts to break down or is intermittent, the coil is faulty and should be renewed.

3 In the absence of a coil tester, the winding may be checked for broken or shorted windings using a multimeter, noting that the test will not reveal insulation breakdown which may only be evident under high voltage.

4 The primary winding resistance can be measured by connecting

Meter Range	Connections*	Location	Reading
20V DC	Meter (+) → Yellow/Red, Black, or Green	At the 4-pin connector for the ignition coils	Battery voltage
	Meter (+) → Black, Blue, Yellow, or Red	At the 4-pin connector for the pick-up coils	0.5~1.0 V

*Connect the meter (−) lead to ground.

Fig. 3.5 General wiring test table

Meter Range	Connections	Location	Reading*
x 1 kΩ	Meter (+) → Black/Yellow Meter (−) → Black, Green	At the 4-pin connector for the ignition coils	∞
x 100 Ω	Meter (+) → Black, Green Meter (−) → Black/Yellow	,,	200~500 Ω
	Meter (+) → Yellow/Red Meter (−) → Black/Yellow	,,	200~600 Ω
	Meter (+) → Black/Yellow Meter (−) → Yellow/Red	,,	300~700 Ω
x 1 kΩ	Meter (+) → Blue (Red) Meter (−) → Black (Yellow)	At the 4-pin connector for the pick-up coils	25~45 kΩ
	Meter (+) → Black (Yellow) Meter (−) → Blue (Red)	,,	20~40 kΩ

*Measured with the Kawasaki Hand Tester (57001-983)
A tester other than the Kawasaki Hand Tester may show slightly different readings.

Fig. 3.6 IC ignitor test table

the meter probes between the two low tension terminals. Set the meter on the ohm x 1 range and note the reading, which should be 1.8 – 2.8 ohm. To check the secondary windings, remove the plug cap from the high tension (plug) lead, and connect the meter probes to the high tension leads. With the meter set on the ohm x 1000 (kohms) scale, a reading of 10 – 16 kohm is to be expected.

5   Finally check for continuity between each low tension lead and the coil core, followed by the high tension lead, repeating the test with the second low tension lead. If anything other than infinite resistance (insulation) is shown, the coil must be renewed.

9.1 Coils are mounted below fuel tank

10 Spark plugs: checking and resetting the electrode gaps

1   The recommended spark plug types are shown below. Note that where the machine is used in an ambient temperature that remains at or below 10°C (50°F) for long periods, or where the machine is confined to low speed use or both, a change to the 'cold weather' plug grade may be advised. In normal circumstances however, the standard grade of plug should be used.

2   The spark plug electrode gaps should be checked in accordance with the intervals specified in the Routine Maintenance Section, or in the event of ignition problems. Note that if the spark plugs in 1 and 4 or 2 and 3 cylinders appear to malfunction simultaneously, the relevant ignition system components should be checked as described earlier in this Chapter.

3   With some experience, the condition of the spark plug electrodes and insulator can be used as a reliable guide to engine operating conditions. See the accompanying diagram.

4   Always carry a spare pair of spark plugs of the recommended grade. In the rare event of plug failure, they will enable the engine to be restarted.

5   Beware of over-tightening the spark plugs, otherwise there is risk of stripping the threads from the aluminium alloy cylinder heads. The plugs shuld be sufficiently tight to seat firmly on their copper sealing washers, and no more. Use a spanner which is a good fit to prevent the spanner from slipping and breaking the insulator.

6   If the threads in the cylinder head strip as a result of over-tightening the spark plugs, it is possible to reclaim the head by the use of a Helicoil thread insert. This is a cheap and convenient method of replacing the threads; most motorcycle dealers operate a service of this nature at an economic price.

7   Make sure the plug insulating caps are a good fit and that the plugs, caps and high tension leads are clean and dry. Accumulated dirt or moisture can cause tracking and thus misfiring.

Chapter 4 Frame and forks

Refer to Chapter 7 for information relating to the 1983 on models

Contents

Specifications

Frame

Type .... Welded tubular steel

Front forks

Type .... Hydraulically damped telescopic, air assisted, linked on R1, P1 and N1

Air pressure range:
    E1, E2, E3, L1, L2, R1 .... 0.6 – 0.9 kg/cm² (8.5 to 13.0 psi)
    H1, H2, H3 .... 0.5 to 1.0 kg/cm² (7.1 – 14.0 psi)
    N1, P1 .... 0.5 – 0.7 kg/cm² (7.1 – 10.0 psi)

Standard air pressure:
    E1, E2, E3, L1, L2, R1 .... 0.7 kg/cm² (10 psi)
    H1, H2, H3, N1, P1 .... 0.6 kg/cm² (8.5 psi)

Spring free length (service limit)
    E1, E2, E3, L1, L2 .... 497 mm (19.57 in) US, 477 mm (18.78 in) UK
    H1, H2, H3 .... 483 mm (19.02 in)
    R1 .... 496 mm (19.53 in)
    P1, N1 .... N/av

Oil capacity per leg:

	At oil change	dry	level
E1, E2, E3, L1, L2 (US)	230 cc	248 cc	355 ± 4 mm
E1, E2, E3, L1, L2 (UK)	215 cc	232 cc	382 ± 4 mm
H1, H2, H3	260 cc	280 cc	436 ± 4 mm
R1	240 cc	256 ± 4 cc	168 ± 4 mm
N1	250 cc	293 ± 2.5 cc	457 ± 2 mm
P1	260 cc	306 ± 2.5 cc	408.5 ± 2 mm

Oil grade:
    N1 and P1 .... SAE 5W/20
    All others .... SAE 10W

Rear suspension

Type .... Swinging arm, incorporating final drive shaft on N1 and P1 models

Rear suspension units:

E, H and L .... Oil damped coil spring, 5 spring preload settings, 4 damper settings
R1 .... Oil damped coil spring, 7 spring preload settings, 5 damper settings
N1 and P1 .... Interconnected air suspension, 4 damper settings

Swinging arm unit

Chain drive models:

Sleeve diameter service limit .... 21.96 mm (0.8646 in)
Pivot shaft runout service limit .... 0.14 mm (0.0055 in)
Pivot shaft repair limit .... 0.70 mm (0.2756 in)

Shaft drive models:

Side clearance .... 1.4 – 1.6 mm (0.055 – 0.063 in)

Torque settings

Component	kgf m	lbf ft	lbf in
Front wheel spindle nut(s):			
E and H models	8.00	58.0	–
Front wheel spindle clamp:			
E model (nuts)	1.80	13.0	–
H model (bolt)	2.00	14.5	–
Front fork air valves	1.20	–	104.0
Fork damper rod Allen bolts:			
UK models	3.00	21.5	–
US models	2.30	16.5	–
Fork pinch bolts:			
Upper	2.00	14.5	–
Lower	3.80	27.0	–
Fork top plugs	2.30	16.5	–
Handlebar clamp bolts	1.80	13.0	–
Rear wheel spindle nut	12.0	87.0	–
Rear suspension unit mounting	3.00	22.0	–
Steering stem top bolt	4.00	29.0	–
Steering stem clamp bolt nut	1.80	13.0	–
Steering stem adjuster locknut	2.00	14.5	–
Swinging arm pivot nut	10.0	72.0	–
Brake torque arm nut	3.00	22.0	–

Additional torque settings – R1 model

	kgf m	lbf ft	lbf in
Front fork air valve	0.80	–	69.0
Handlebar clamp bolts	3.00	22.0	–
Handlebar alloy section to yoke bolts	10.0	72.0	–
Steering stem top bolt	3.00	22.0	–

Additional torque settings – N1 and P1 models

	kgf m	lbf ft	lbf in
Front wheel spindle nut	6.50	47.0	–
Front wheel spindle pinch bolt	1.40	10.0	–
Rear wheel spindle nut	14.0	101.0	–
Final drive case components:			
Case mounting nuts	2.3	16.5	–
Cover bolts	2.3	16.5	–
Drain plug	2.0	14.5	–
Pinion gear nut	10.0	72.0	
Handlebar clamp bolts:			
N1 model	1.9	13.5	–
P1 model	2.5	18.0	–
Handlebar alloy section to yoke bolts – P1	10.0	72.0	–
Steering stem top bolt	4.30	31.0	–
Front fork damper rod Allen bolts – US models	2.00	14.5	–
Front fork clamp bolts:			
Upper	1.70	12.5	–
Lower	2.50	18.0	–
Front fork air valve	0.55	–	48.0
Front fork drain bolts	0.75	–	65.0
Rear suspension units:			
Air valve	0.55	–	48.0
Mounting	2.50	18.0	–
ar suspension unit hose:			
Fittings	1.00	–	87.0
Male pipes	1.20	–	104.0
Swivel nuts	1.30	–	113.0
Swinging arm pivot stubs	1.30	9.5	113.0

1 General description

The KZ/Z 750 models employ a conventional welded tubular steel frame of the full cradle type.

Front forks are of the hydraulic alloy-damped telescopic variety and employ air pressure as a supplementary springing medium.

Rear suspension is by pivoted rear fork, or swinging arm, on all models. In the case of the P1 and N1 versions the final drive shaft forms an integral part of the swinging arm assembly, whilst a plain welded tubular steel arrangement is fitted to all other models. The swinging arm is supported by a pair of oil-damped coil spring suspension units which have adjustable pre-load and damping. The N1 and P1 shaft drive machines are fitted with air suspension units.

2 Front fork legs: removal and refitting – all models except R1 and shaft drive models

1 Place the machine securely on its centre stand leaving adequate working space around the front wheel area. Remove the brake caliper mounting bolts and lift the calipers clear of the wheel. Each caliper should be tied to the frame to hold it away from the working area and to prevent strain on the hydraulic hoses. Remove the front wheel as described in Chapter 5 Section 3.

2 Remove the four bolts and lock washers which retain the front mudguard to the fork lower legs. Lift the mudguard clear, taking care to avoid damage. If the fork legs are to be dismantled after removal, unscrew the air valve dust caps and depress the valve cores to release air pressure. Slacken the top plugs to one or two turns to make subsequent removal easier.

3 Slacken the top and bottom yoke pinch bolts, then remove each fork leg in turn by pulling it downward. It will prove easier to do this if the stanchion is twisted and pulled downward simultaneously.

4 When reassembling the forks, reverse the removal sequence, noting the following points. Prior to installation, check the fork oil levels. Refer to Section 12 of this Chapter for details.

5 Slide each leg back into the fork yokes noting that on E and L models the top edge of the stanchion should lie flush with that of the top yoke. On H models, leave the stanchion just below this level so that the flanged edge of the top plug lies level with the yoke. Tighten the yoke pinch bolts and fit and tighten the top plug to the specified torque settings. Refit the air valves using a non-hardening thread locking compound on their threads. Tighten the valves to 1.2 kgf m (104 lbf in). Set up the fork air pressures as described in Section 22 of this Chapter. Refit the front mudguard, wheel and brake calipers.

3 Front fork legs: removal and refitting – R1

1 The general procedure for fork removal and installation is as described in Section 2, noting the following points.

2 The handlebar fairing should be detached to give better access around the steering head. It is retained by two rubber-bushed bolts to the underside of the bottom yoke and by a bracket on each side of the top yoke.

3 Remove the air valve cap and depress the valve core to release the pressurised air. Remove the two screws which retain the top yoke cover, lifting it away to reveal the handlebar mountings. The handlebar assembly consists of two tubular sections which carry the controls and switches, these are clamped to the two cast alloy sections which mount in turn on the top yoke. Each of the alloy sections is retained by a large hexagon headed bolt which screws into the top of each stanchion. Remove the bolts and slide each handlebar assembly clear of the top yoke.

4 Loosen the yoke pinch bolts and work the stanchions clear of the yokes as described in Section 2. It will not normally be necessary to disturb the air unions and headlamp brackets, but should this be desired, the assembly can be pulled from between the fork yokes. It will be noted that the top of the fork shroud/headlamp bracket is located by a rubber seal, whilst at the lower end a combination of three rings seat the shroud against the air union. (See the accompanying illustration for details). Make sure that these are fitted correctly during assembly.

5 When installing the fork legs, apply oil to the inside of the air

unions to help the stanchion pass through the O-rings without damage. Check that the various rings and collars which locate the fork shroud/headlamp bracket are positioned correctly, then slide the fork leg into position. Note that the stanchion should be positioned so that it lies just below the top face of the upper yoke; about 1 mm will suffice.

6 Tighten the upper pinch bolts sufficiently to hold the stanchions in position, but leave the lower pinch bolts loose at this stage. If the forks were dismantled, the recessed top plugs should be tightened to 2.3 kgf m (16.5 lbf ft). Fit the handlebar locating plates, fitting the Allen screws loosely in position. Note that the arrow marks on the plates should face towards the rear of the machine. Offer up each handlebar assembly and fit the large retaining bolts finger tight.

7 Tighten, to the specified torque settings and in the order given, the large handlebar retaining bolts, noting that the stanchions should be drawn hard against the handlebar alloy sections, the upper pinch bolts, the lower pinch bolts to 3.8 kgf m (27 lbf ft) and the locating plate Allen screws.

8 Refit the air valve using a non-hardening thread locking compound on its threads, and tighten it to 0.8 kgf m (69 lbf in). Complete reassembly by refitting the front mudguard, wheel and brake calipers, tightening all bolts to the specified torque settings. Refit the fairing and check the front fork air pressure, brake operation and handlebar mirror adjustments.

3.3a Depress or remove valve to release fork air pressure

3.3b Remove screws to free cover

3.3c Unscrew large flanged bolt ...

3.3d ... and disengage handlebar from yoke

3.4a Slacken upper pinch bolt (note fairing bracket) ...

3.4b ... followed by the lower yoke pinch bolt

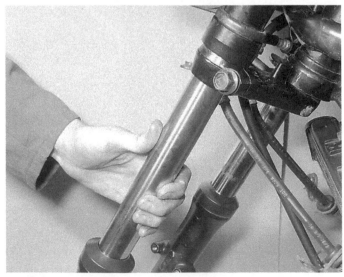

3.4c Fork leg can now be removed from yokes

3.4d Note ring which locates top of headlamp bracket

3.4e Bottom of bracket is located by three rings ...

3.4f ... which seat against air union

3.5 Lubricate union O-rings during assembly

4 Front fork legs: removal and refitting – P1 model

1 The fork arrangement on the P1 model is broadly similar to that of the R1 described in Section 3. It will be noted that no handlebar fairing is fitted, and that the air unions are fitted directly below the top yoke. Rubber gaiters (boots) are fitted to protect the fork stanchions.

2 The procedure described in Section 3 may also be applied to the P1, noting the following exceptions. The handlebar alloy sections are located by small plates of the same type as used on the P1. It should be noted, however, that the arrow mark should, in this instance, face forward.

5 Front fork legs: removal and refitting N1 model

1 The forks on the shaft drive N1 model are broadly similar to those fitted to the R1 (GPz 750) described in Section 3. They differ for removal purposes in that the air unions are fitted above the fork shroud/headlamp brackets, and conventional handlebars are fitted. The stanchions are protected by rubber gaiters, or boots, fitted below the bottom yoke.

2 The fork legs can be removed after the front wheel, mudguard and brake calipers have been detached. Remove the air valve dust cap and depress the valve core to release air pressure before removing the legs. Slacken the yoke pinch bolts and work the legs downwards, twisting the stanchions from side to side to assist this operation.

3 When refitting the legs, apply oil to the inside of the air unions to lubricate the O-rings which seal against the stanchions. Care should be taken to avoid damage to the O-rings during installation. The stanchions should be positioned so that the top edge is level with the top face of the yoke. Tighten the yoke pinch bolts to the specified torque settings and refit the dust caps which close the stanchion tops. Check that the breather holes in the fork gaiters face rearwards.

6 Steering head assembly: removal and refitting – all models except R1 and shaft drive machines

1 The steering head assembly, comprising the top yoke, bottom yoke and steering stem and the steering head bearings, can be removed with or without the fork legs. It is, however, considerably easier to deal with this assembly after the forks have been removed, and for this reason it is recommended that the relevant procedure described in Section 2 is followed as a preliminary step.

2 Remove the fuel tank as described in Chapter 2 Section 2, followed by the headlamp unit as described in Chapter 6 Section 15. Separate the various connectors inside the headlamp shell to permit its removal.

3 The handlebar assembly is removed next noting that two approaches are possible. The official method is to systematically remove the clutch cable, left-hand switch cluster and wiring clips, right-hand switch cluster and wiring clips and the starter lockout switch. The four handlebar clamp bolts are now removed and the handlebar lifted away. An easier method is to remove the handlebar clamps and lift the entire handlebar assembly rearwards, resting across the frame with the various control cables and electrical leads attached. This avoids a great deal of dismantling work, but does mean that it will be necessary to work around the cables and leads during steering head removal.

4 Loosen the top yoke left-hand pinch bolt and remove the top yoke right-hand pinch bolt together with its cable guide. Release the speedometer and tachometer drive cables by unscrewing the knurled ring at the base of each instrument. Remove the two screws which retain the cover at the front of the bottom yoke, lifting the cover away to expose the brake hydraulic union. Remove the two bolts which secure the union to the bottom yoke.

5 Release the three bolts which secure the headlamp shell and subframe to the steering head yokes, and lift the assembly clear of the steering head area. Slacken the steering stem pinch bolt, then remove the large top bolt, plain washer and lock washer. Using a soft-faced mallet, tap the underside of the top yoke until it comes free of the fork stanchions and steering stem. Lift the yoke clear, together with the instruments, and place the assembly to one side, noting that the instruments should be kept upright.

6 Arrange any remaining cables, leads, etc, so that they do not foul the lower yoke. Obtain a small box or tin in which the steering head

balls can be placed safely. Note that as the lower yoke and stem are released, the lower race balls will drop free, and some provision must be made to catch them. As a precaution place a large piece of rag below the headstock to catch any displaced balls. Slacken the lockring, using a C spanner, and lower the steering head stem and lower yoke assembly clear of the frame. Note that there are 20 balls in the lower race and 19 in the upper one, all being of the same size, namely $\frac{1}{4}$ in diameter.

7   Before reassembly, examine and clean the bearing races, then stick the 20 lower race balls into position using high melting point grease. Reassemble the fork unit, following the dismantling sequence in reverse. Note that the steering head should be tightened **just** sufficiently to remove free play. On no account overtighten the head races. It is surprisingly easy to inadvertently apply a loading of several tons to the head bearings, which will quickly break up as a result. When set correctly, there should be no discernible play in the forks when they are shaken. The fork assembly should, however, move easily and without any sign of resistance from lock to lock.

8   The various fasteners should be tightened to the specified torque figures.

7   Steering head assembly: removal and refitting – R1 model

1   The GPz model differs from those described in Section 6 in that tapered roller steering head bearings are used.

2   Remove the front wheel, brake calipers, front mudguard, front fork legs and the handlebar fairing as described in Section 3 of this Chapter. Remove the fuel tank (see Chapter 2, Section 2). Release the two screws which secure the headlamp unit to the shell. Disengage the unit and disconnect the headlamp bulb connector and also the lead to the parking lamp, where fitted. Separate the connector blocks inside the headlamp shell and push the wiring clear of the shell via the cutouts at the back.

3   Remove the two bolts which secure the headlamp shell to the fork shrouds and the single bolt which passes through the adjustment bracket below the shell. Lift the unit clear of the steering head. Disconnect the leads from the indicator lamps and displace the fork shrouds, removing them together with the lamps. Remove the various sealing washers and also the air unions and hose. The handlebar sections should be detached as described in Section 3 and tied clear of the working area. Release the hydraulic union from the lower yoke and manoeuvre the hydraulic hoses clear of the yokes.

4   Slacken and remove the steering stem top bolt and plain washer. Remove the top yoke and instrument panel. If necessary, tap the underside of the yoke to free it from the steering stem.

5   Support the lower yoke, then, using a C-spanner, slacken and remove the two slotted nuts which retain the steering stem. Lower clear the lower yoke and steering stem and displace the upper bearing inner race.

6   When refitting the yokes and bearings, grease the latter prior to installation, and coat the steering stem with grease. Offer up the lower yoke and fit the slotted adjuster nut finger tight. The raised collar should face downward. It is important to bed the bearings in on assembly as described below. If the old bearings were undisturbed proceed as described in paragraph 9 onwards.

7   To set up the bearings initially, a C-spanner, preferably the correct Kawasaki item, Part Number 57001-1100, will be required. If using any other spanner, note that it will be necessary to make some provision for a spring balance to be attached at a point 180 mm (7.1 in) from the centre of the steering stem. To this end, extend the spanner as required and drill a hole in the handle at the correct distance. A spring balance capable of reading above 22.2 kg (48.94 lb) will also be required.

8   Fit the C-spanner and apply 4.0 kgf m (29 lbf ft) to the adjuster nut. This is achieved by hooking the spring balance to the hole in the spanner and pulling on it until a reading of 22.2 kg (48.94 lb) is shown. Check that the steering head assembly turns smoothly with no evidence of play or tightness.

9   Slacken the nut slightly, then turn it slowly clockwise until resistance is **just** evident. Take great care not to apply excessive pressure because this will cause premature failure of the bearings. The object is to set the adjuster so that the bearings are under a **very light** loading, just enough to remove any free play. Once set correctly, run the slotted locknut into place and tighten it firmly whilst holding the adjuster nut in the correct position.

10   Continue assembly by reversing the dismantling sequence. Check that all electrical cables and control cables are routed so that they do not impede steering movement. When refitting the wiring connectors in the headlamp shell, check that the wiring colour codes match up. When assembly has been completed check that the steering turns smoothly and evenly with the front wheel raised clear of the ground. Check the operation of the front brake, throttle and clutch and adjust the headlamp alignment and rear view mirror setting.

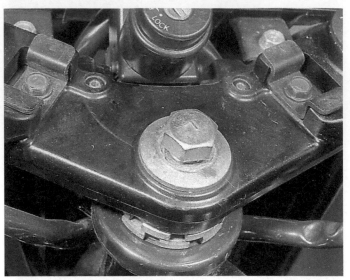

7.4 Remove top bolt to free top yoke

8   Steering head assembly: removal and refitting – shaft drive models

1   The N1 and P1 models are equipped with tapered roller steering head bearings. In the case of the P1 the procedure is almost identical to that described for the R1 model, with the exception of the handlebar fairing and the modified position of the fork air unions. Refer to Section 7 for full details.

2   The N1 model employs the same steering head assembly as the R1 and P1 models, but is equipped with conventional handlebars. Carry out preliminary dismantling as described in Section 6, paragraphs 1 to 5, then refer to Section 7, paragraph 4 onwards for details on removing, fitting and adjusting the yokes and bearings.

9   Steering head bearings: examination and renovation – all models except R1 and shaft drive models

1   Before commencing reassembly of the forks, examine the steering head races. The ball bearing tracks of the respective cup and cone bearings should be polished and free from indentations, cracks or pitting. If signs of wear are evident, the cups and cones must be renewed. In order for the straight line steering on any motorcycle to be consistently good, the steering head bearings must be absolutely perfect. Even the smallest amount of wear on the cups and cones may cause steering wobble at high speeds and judder during heavy front wheel braking. The cups and cones are an interference fit on their respective seatings and can be tapped from position with a suitable drift.

2   Ball bearings are relatively cheap. If the originals are marked or discoloured they **must** be renewed. To hold the steel balls in place during reassembly of the fork yokes, pack the bearings with grease. The upper and lower races contain 19 and 20 $\frac{1}{4}$ in steel balls respectively. Although a small gap will remain when the balls have been fitted, on no account must an extra ball be inserted, as the gap is intended to prevent the balls from skidding against each other and wearing quickly.

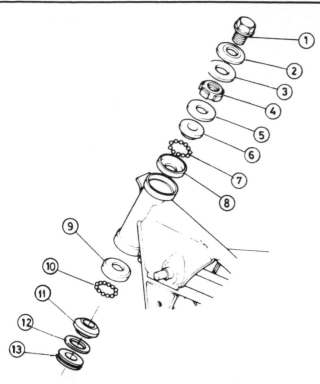

Fig. 4.1 Steering head bearings – KZ/Z750 E, H and L

1	Top bolt	8	Upper bearing cup
2	Washer	9	Lower bearing cup
3	Lockwasher	10	Balls
4	Adjusting nut	11	Lower bearing cone
5	Washer	12	Washer
6	Upper bearing cone	13	Dust seal
7	Balls		

10 Steering head bearings: removal, examination and refitting – R1 and shaft drive models

1 The steering head bearings should be checked for wear or damage after the steering head assembly has been dismantled. The inner races are easily checked after all traces of old grease have been removed by washing in a suitable solvent. Turn each race slowly, checking for marks or discolouration of the roller faces.

2 Clean the outer races, and examine the bearing surface for wear or damage. If any wear is discovered, Kawasaki recommend that **both** bearings, including outer races, should be renewed. They also advise that the lower yoke and steering stem be renewed, but there appears to be little to be gained from this unless this component is obviously worn and a sloppy fit in the bearings.

3 If renewal is necessary, removal of the old bearing outer races and installation of the new outer races may be accomplished using the correct Kawasaki service tools. Failing this proceed as described below.

4 The outer races are a fairly tight fit in the steering head tube. Most universal slide-hammer type bearing extractors will work here, and these can often be hired from tool shops. Alternatively, a long drift can be passed through one race and used to drive out the opposite item. Tap firmly and evenly around the race to ensure that it drives out squarely. It may prove advantageous to curve the end of the drift slightly to improve access. Note that with this method there is a real risk of damage unless care is taken. If the race refuses to move, **stop**. Leave the job until a proper extractor can be obtained.

6 The lower inner race can be levered off the steering stem, using screwdrivers on opposite sides to work it free. To fit the new item, find a length of tubing slightly larger in its internal diameter than the steering stem. This will suffice as a tubular drift. Grease the bearing thoroughly and wipe a trace of grease around the steering stem. Drive the bearing home evenly and fully.

7 The new outer races can be installed using a home-made version of the drawbolt arrangement shown in the accompanying illustration.

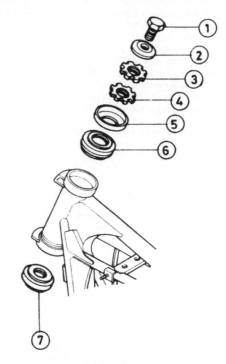

Fig. 4.2 Steering head bearings – KZ/Z750 R1, Z750 P1 and KZ750 N1

1	Top bolt	5	Dust cover
2	Washer	6	Upper bearing race
3	Slotted nut	7	Lower bearing race
4	Slotted nut		

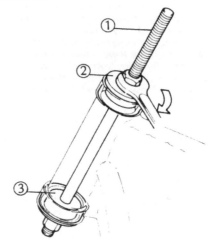

Fig. 4.3 Drawbolt arrangement for fitting bearing outer races

1	High tensile bolt	3	Guide
2	Heavy washer		

11 Front fork legs: dismantling, overhaul and reassembly

1 Having removed the fork legs as described earlier in this Chapter they may be dismantled for further examination. Always deal with one leg at a time and on no account interchange components. On machines with separate air valves, these should be removed together with their sealing O-rings and the fork top bolts removed. On R1 and P1 models, unscrew the recessed top bolt using an Allen key. In the case of the N1 model, depress the top plug against fork spring pressure using a large screwdriver, then displace and remove the circlip which retains it. Invert the fork leg over a drain tray or bowl and pump the fork to expel the damping oil.

2   Referring to the appropriate line drawing, commence dismantling. To separate the fork stanchion (tube) and lower leg (slider) it is necessary to unscrew the recessed damper rod securing Allen bolt from the bottom of the lower leg. The damper rod is free to rotate inside the stanchion, and will almost invariably do so before the bolt will unscrew. The answer is to hold the head of the damper rod whilst the bolt is removed. A Kawasaki tool, Part Numbers 57001-193 and 57001-1011, is ideal for this, but unlikely to be available to most owners. The accompanying photograph shows a home-made alternative which we used during the workshop project. A scrap bolt was found, having a hexagon size of 19 mm measured across the flats. This was inserted in the end of a length of steel tubing and the two were pinned together. The tool was then dropped into the stanchion and held with a self-locking wrench while the damper bolt was freed.

3   On machines with fork gaiters (boots) slacken the retaining clips so that the stanchion can be pulled free. Separate the stanchion assembly by pulling it clear of the lower leg. Invert the stanchion and tip out the damper rod assembly and rebound spring. The R1 (GPz 750) model has two rebound springs which should be fitted with the heavier spring uppermost. Note also that the latter model and the N1 and P1 shaft drive machines have bushed forks. This means that the top bush, washer and oil seal must be displaced as the stanchion is withdrawn. To this end, pull back the dust seal to expose the wire retaining clip. Prise out the clip, then knock the bush and seal out by pulling the stanchion sharply outwards. The remaining models have unbushed forks, the material of the lower leg providing the bearing surface, and the stanchion and lower leg may be separated once the damper rod bolt is freed.

4   The parts most liable to wear over an extended period of service are the internal surfaces of the lower leg and the outer surfaces of the fork stanchion or tube. If there is excessive play between these two parts they must be replaced as a complete unit. Check the fork tube for scoring over the length which enters the oil seal. Bad scoring here will damage the oil seal and lead to fluid leakage.

5   Owners of the N1, P1 and R1 have the theoretical advantage of bushed forks. The top bush is pressed into the top of the lower leg and provides a bearing surface for the stanchion. The lower bush is fitted to a waisted section of the stanchion. It can be removed by spreading the bush slightly by easing a screwdriver blade into the vertical split. Whilst removal of the bushes is straightforward, renewal poses something of a problem. The manufacturer lists only the top bush as a separate part and thus if the lower bush is worn (which is unlikely because it runs on the softer alloy lower leg) it will be necessary to fit a new stanchion complete. The top bush, on the other hand, should be renewed if play is evident between it and the stanchion when the fork is assembled.

6   It is advisable to renew the oil seals when the forks are dismantled even if they appear to be in good condition. This will save a strip-down of the forks at a later date if oil leakage occurs. The oil seal in the top of each lower fork leg is retained by an internal C-ring which can be prised out of position with a small screwdriver. When ordering new seals, specify the later double-lipped type for the E, H and L models. Check that the dust excluder rubbers are not split or worn where they bear on the fork tube. A worn excluder will allow the ingress of dust and water which will damage the oil seal and eventually cause wear of the fork tube.

7   It is not generally possible to straighten forks which have been badly damaged in an accident, particularly when the correct jigs are not available. It is always best to err on the side of safety and fit new ones, especially since there is no easy means to detect whether the forks have been over stressed or metal fatigued. Fork stanchions can be checked, after removal from the lower legs, by rolling them on a dead flat surface. Any misalignment will be immediately obvious.

8   The fork spring wil take a permanent set after considerable usage and will need renewal if the fork action becomes spongy. The service limit for the total free length of each spring is given in Specifications.

9   Reassemble the forks by reversing the dismantling sequence. In the case of machines fitted with a renewable top bush, drive this home after the stanchion has been fitted using a length of tubing as a tubular drift. Press the seal home in a similar manner, not forgetting the washer or spacer fitted between the two. Secure the bush and seal with the retaining clip.

10 The damper bolt should be carefully degreased and a new sealing washer fitted. On US models, apply instant gasket to the bolt head and washer, and a non-hardening locking compound to the bolt threads. Tighten to the specified torque figure. Before refitting the fork legs,

refer to Section 12 for details of oil grades, quantities and levels for the various models.

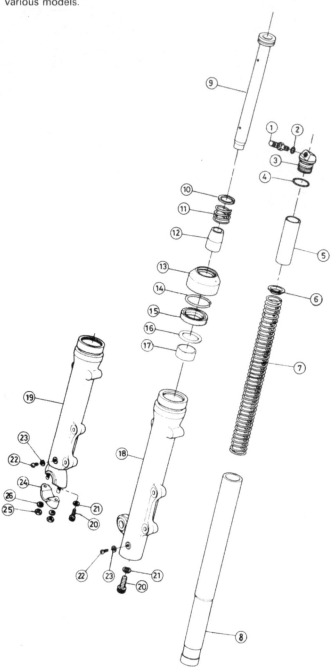

Fig. 4.4 Front forks – KZ/Z750 H, E and L

1	Air valve	15	Oil seal
2	O-ring	16	Washer - 1982 models only
3	Top plug		
4	O-ring	17	Bush - 1982 models only
5	Spacer - H model only		
6	Spring seat	18	Lower leg - H model
7	Spring	19	Lower leg - L and E model
8	Stanchion		
9	Damper rod	20	Allen bolt
10	Damper rod piston ring	21	Sealing washer
11	Rebound spring	22	Drain screw
12	Damper rod seat	23	Sealing washer
13	Dust cover	24	Spindle clamp
14	Circlip	25	Nut - 2 off
		26	Spring washer - 2 off

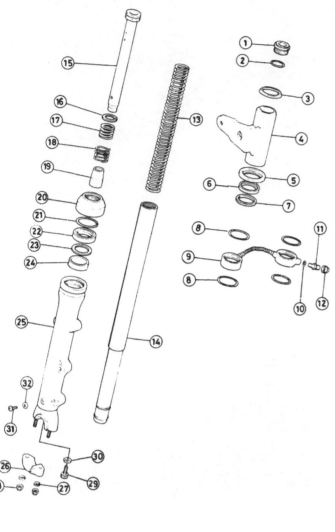

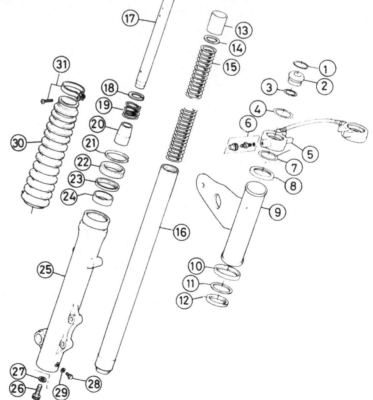

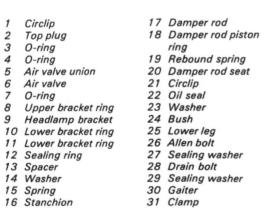

Fig. 4.5 Front forks – KZ/Z750 R1

1	Top plug	17	Rebound spring
2	O-ring	18	Rebound spring
3	Upper bracket ring	19	Damper rod seat
4	Headlamp bracket	20	Dust seal
5	Lower bracket ring	21	Wire retaining clip
6	Lower bracket ring	22	Oil seal
7	Sealing ring	23	Washer
8	O-ring - 4 off	24	Bush
9	Air valve union	25	Lower leg
10	O-ring	26	Spindle clamp
11	Air valve	27	Spring washer - 2 off
12	Air valve cap	28	Nut - 2 off
13	Spring	29	Allen bolt
14	Stanchion	30	Sealing washer
15	Damper rod	31	Drain screw
16	Damper rod piston ring	32	Sealing washer

Fig. 4.6 Front forks – KZ750 N1 and Z750 P1

1	Circlip	17	Damper rod
2	Top plug	18	Damper rod piston ring
3	O-ring	19	Rebound spring
4	O-ring	20	Damper rod seat
5	Air valve union	21	Circlip
6	Air valve	22	Oil seal
7	O-ring	23	Washer
8	Upper bracket ring	24	Bush
9	Headlamp bracket	25	Lower leg
10	Lower bracket ring	26	Allen bolt
11	Lower bracket ring	27	Sealing washer
12	Sealing ring	28	Drain bolt
13	Spacer	29	Sealing washer
14	Washer	30	Gaiter
15	Spring	31	Clamp
16	Stanchion		

11.1 Recessed top plug is fitted to R1 and P1 models

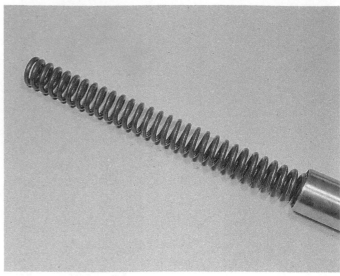

11.2a Remove fork spring, noting direction of fitting

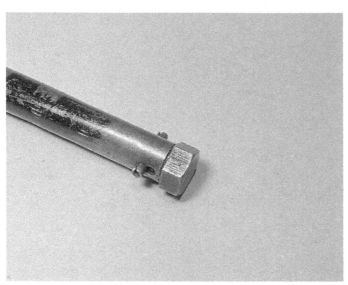

11.2b Home-made damper rod holding tool

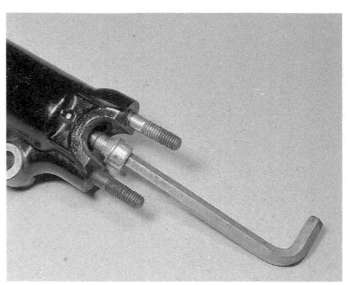

11.2c Slacken damper retaining bolt ...

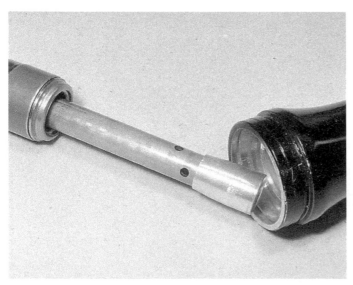

11.2d ... and withdraw stanchion assembly

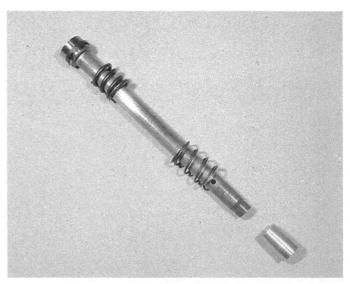

11.3a R1 model has two rebound springs

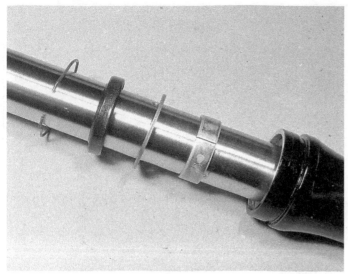

11.3b Release clip to free oil seal, washer and top bush (R1, N1 and P1)

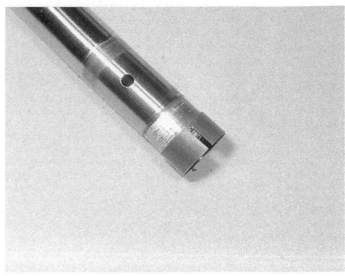

11.5 Bottom bush is easily removed (R1, N1 and P1)

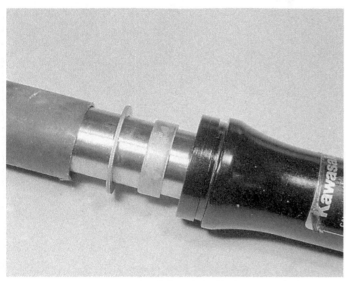

11.9a Use tubing to press in top bush and washer ...

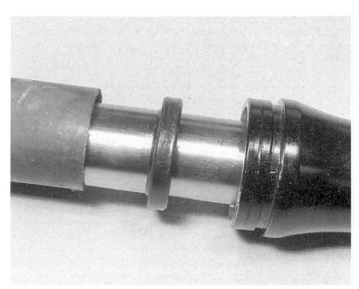

11.9b ... followed by the oil seal

11.9c Retain assembly with wire circlip ...

11.9d ... then slide dust seal into place

12 Changing the front fork oil

1   Fork oil plays an important role in controlling front wheel movement. In use, it tends to heat up, cooling off again when the machine is parked. This in turn tends to allow condensation to form, diluting and degrading the oil. It is important to change the damping oil at the intervals specified in Routine Maintenance, and whenever the forks are overhauled.

2   When changing the fork oil, remember to release fork air pressure to prevent unpleasant surprises when the drain screws are removed. Place a drain tray beneath each leg in turn and remove the drain screw. Draining is speeded up considerably by 'pumping' the forks to expel the oil. Remove the fork top bolts or plugs after placing a jack or blocks beneath the engine to take the weight off the forks. For further details of fork bolt and plug removal refer to the earlier Sections of this Chapter. Remove the fork spring spacer (where fitted) and the springs and place them to one side.

3   Add the prescribed amount of oil according to the table shown below. Whilst this will normally give the correct oil level it is preferable to check this by measuring the oil level below the top of the stanchion. It is important, especially with air forks, that this level is correct and equal in both fork legs.

4   Measure the fork oil level with the fork spring removed and the fork fully extended (front wheel off the ground) using a rod or ruler as a dipstick. Add oil as necessary to bring it to the correct level. If necessary, remove excess oil by draining or by using a syringe or plastic 'squeeze pack'.

13 Steering head lock: maintenance

1   The ignition switch incorporates a locking mechanism which allows the steering to be secured when the machine is parked. The lock and switch mechanism form a sealed unit, and no maintenance is possible or required. In the event of a fault, it will be necessary to renew the complete assembly. It is retained by two bolts to the underside of the top yoke.

14 Frame: examination and renovation

1   The frame is unlikely to require attention unless accident damage has occurred. In some cases, renewal of the frame is the only satisfactory remedy if the frame is badly out of alignment. Only a few frame specialists have the jigs and mandrels necessary for resetting the frame to the required standard of accuracy, and even then there is no easy means of assessing to what extent the frame may have been overstressed.

2   After the machine has covered a considerable mileage, it is advisable to examine the frame closely for signs of cracking or splitting at the welded joints. Rust corrosion can also cause weakness at these joints. Minor damage can be repaired by welding or brazing, depending on the extent and nature of the damage.

3   Remember that a frame which is out of alignment will cause handling problems. If misalignment is suspected, as a result of an accident, it will be necessary to strip the machine completely so that the frame can be checked, and if necessary, renewed.

15 Swinging arm fork: removal and renovation – chain drive models

1   Wear in the swinging arm bearings is characterised by a tendency for the rear of the machine to twitch when ridden hard through a series of bends. This can be checked by placing the machine on the centre stand, and pushing the swinging arm from side to side. Any discernible free play will necessitate the removal of the swinging arm for further examination.

12.3a Remove top bolt or plug and fork spring . . .

12.3b . . . drain old oil, then top up to correct level

2   Place the machine squarely on its centre stand and remove the rear wheel and brake caliper as described in Section 4 Chapter 5. Remove the chain guard mounting bolts or screws and lift it clear of the swinging arm. Remove the rear suspension unit lower mounting bolts and lock washers. Remove the swinging arm pivot shaft nut. Displace the shaft and withdraw the swinging arm clear of the frame.

3   Displace the swinging arm pivot sleeve and clean off all old grease from it and the needle roller bearings. Examine the sleeve for signs of wear, corrosion or indentation. If worn or damaged, the sleeve must be renewed together with the bearings. Check the diameter of the sleeve using a micrometer. If worn to or below the service limit of 21.96 mm (0.865 in) the sleeve and the bearing should be renewed.

4   Wear of the bearings is less easy to establish, since the difference between a serviceable bearing and a badly worn item cannot be checked by direct measurement. Inspect each bearing closely for signs of discolouration or abrasion. If less than perfect it is best to assume that renewal will be required.

5   Make sure that the new bearings are available before attempting to remove the old ones, because the latter will be destroyed during removal. Support the end of the pivot tube on a block of wood, then pass a long drift down through it. Tap around the far bearings until they are displaced. Invert the assembly and repeat the operation to remove the remaining pair. Fit the new bearings using a press or a drawbolt arrangement to pull them into position. On no account try to drive the new bearings home – they will be damaged in the attempt. Place the grease sealing caps over the ends of the pivot tube.

6 In the case of 1982 models, the single pair of bearings should be fitted so that they are 5 mm (0.197 in) below the ends of the pivot tube. Carefully draw the new grease seals into place, taking care not to distort them.

7 On all models, grease the bearings thoroughly, then grease and fit the pivot sleeve. Refit the swinging arm assembly by reversing the removal sequence. Ensure all fasteners are tightened to the specified torque figures.

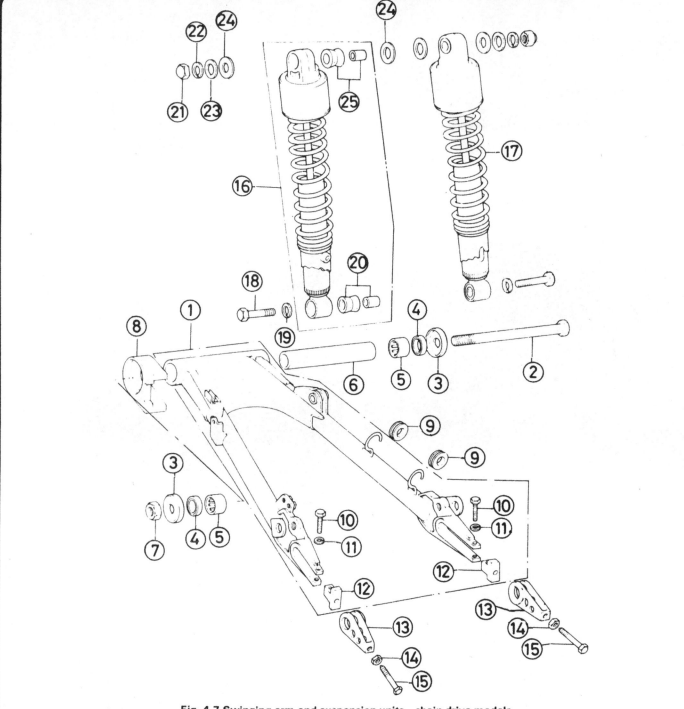

Fig. 4.7 Swinging arm and suspension units – chain drive models

1	Swinging arm	7	Nut	14	Locknut - 2 off	20	Bush - 2 off
2	Pivot shaft	8	Chain guide - R1	15	Bolt - 2 off	21	Nut - 2 off
3	Cap - 2 off		model	16	Left-hand suspension	22	Spring washer
4	Grease seal - 2 off	9	Grommet - 2 off		unit		- 2 off
5	Needle roller bearing	10	Bolt - 2 off	17	Right-hand suspension	23	Washer - 2 off
	- 4 off (1980/81),	11	Spring washer - 2 off		unit	24	Washer - 4 off
	2 off (1982)	12	Adjustment stop	18	Bolt - 2 off		
6	Centre sleeve	13	Chain adjusting	19	Spring washer		
			bracket - 2 off		- 2 off		

15.2a Remove suspension unit lower mounting bolts

15.2b Slacken and remove the pivot shaft nut

15.2c Withdraw the pivot shaft to free the swinging arm assembly

15.3a Remove the drive chain guide block ...

15.3b ... and remove pivot shaft end caps

15.3c Grease seals may be prised out if worn or damaged

15.3d Remove pivot sleeve, clean, and check for wear

15.4 Check needle roller bearings for wear or damage

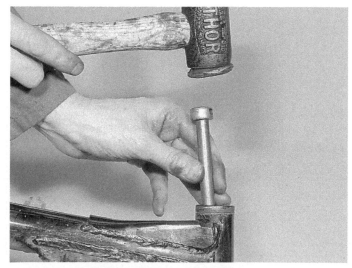

15.5 Old bearings will be destroyed during removal

16 Swinging arm fork: removal and renovation – shaft drive models

1   Free the front end of the drive shaft from the front gear case as described in Chapter 1, Section 5, paragraph 16 onwards, noting that the rear wheel, brake and final drive casing should be detached as part of this procedure. The front gear case need not be removed.

2   Slacken the pivot stub locknuts and unscrew the stubs to allow the swinging arm to be manoeuvred clear of the frame.

3   Prise out and discard the old oil seals and remove the bearing inner races for cleaning and examination. The bearing rollers should be smooth and unmarked, with no signs of pitting or discolouration. Clean the outer races and check these in the same way. If any indication of wear or damage is discovered, it will be necessary to renew the bearings complete with outer races.

4   The outer races are pressed into the pivot tube and should be removed using a slide hammer with a bearing extractor attachment. Whilst it may be possible to drive the old races out using a long drift passed through the opposing bearing, there is little of the bearing available for purchase. If this method is chosen, great care must be taken to avoid damage. The new outer races can be fitted using a large socket as a drift, taking care to ensure that the races enter the bores squarely.

5   Grease and install the inner races, then tap the seals into place using the large socket employed to fit the bearings. Refit the swinging arm unit, screwing the adjuster stubs home finger tight. Set up the clearance between the frame boss and swinging arm by screwing the stubs inwards. The gap, which should be checked using feeler gauges, should be 1.4 – 1.6 mm (0.0551 – 0.0630 in).

6   Continue assembly by reversing the removal sequence. Before fitting the final gear case make sure that the locating pin at the front coupling has engaged properly by pulling hard on the rear of the shaft.

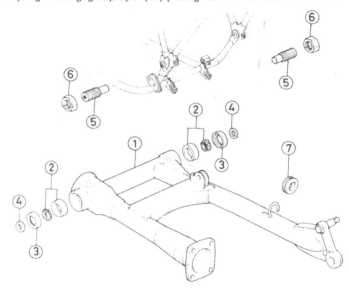

Fig. 4.8 Swinging arm – shaft drive models

1	Swinging arm	4	Thrust washer - 2 off
2	Taper roller	5	Pivot stub - 2 off
	bearing - 2 off	6	Locknut - 2 off
3	Grease seal - 2 off	7	Grommet

17 Rear suspension units: examination

Oil-damped coil spring type

1   The units can be removed from the machine after releasing the top and bottom mountings. If one unit is dealt with at a time it will not be necessary to support the rear of the machine. Note that where a grab rail is fitted it will be necessary to detach it to allow the top mounts to be displaced.

2   The action of the units can be checked by depressing and releasing them. If all is well they should compress fairly easily, with stiffer rebound damping being evident. In the event of failure it will be

necessary to fit new units as a matched pair, the sealed construction obviating any attempt to rebuild the units. If, however, the damper rate adjuster requires renewal, note that a repair kit is available, comprising the plastic pinion and spacer. This can be ordered through an authorized Kawasaki dealer, who will also be able to advise on fitting.

Air suspension units

3   The units can be removed as described in the preceding paragraphs, noting that it will also be necessary to free the air valve union which is secured by two bolts beneath the seat. Ensure that **both** units are kept vertical to avoid the accidental transfer of damping oil between the two units via the air hoses. As with the standard units, the air suspension units are sealed and cannot be rebuilt if defective. Note that when refitting the units, the air hose connections on the 'P' models must face towards the rear of the machine, whilst on the 'N' model they face to the front.

4   The damping oil may be changed or its level checked as follows. You will need a syringe and a length of small-bore tubing to inject the oil into the unit through the air hose union hole, and a measuring cylinder of at least 200 cc capacity, calibrated in ccs.

5   Remove the units from the machine and disconnect the air hose after depressing the air valve to release pressure from the units. Invert the units and pump out the old damping oil. Leave the units a while for the oil to drain.

6   Fill the syringe with SAE 5W fork oil and inject this into the unit until it is completely full. Disconnect the syringe occasionally during filling and move the unit from side to side to distribute the oil and displace any air bubbles.

7   When all air is removed, hold the unit inverted over the measuring cylinder and allow the oil to drain until the volume of oil in the cylinder matches the specified air chamber volume shown below. The remaining oil will then correspond with the specified oil quantity.

8   Repeat the above process on the remaining unit, then refit the units, taking care not to spill any oil or to allow dirt to enter the union hole. Reconnect the air hose unions and adjust the air pressure to within the specified range.

Air volume	Oil volume
105 ± 2.5 cc	396 ± 2.5 cc

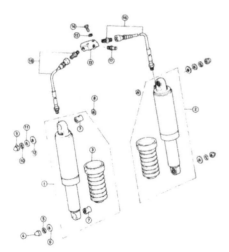

Fig. 4.9 Rear suspension units – KZ/Z750 P1 and N1

1	Left-hand unit	10	Spring washer - 2 off
2	Right-hand unit	11	Washer - 2 off
3	Gaiter	12	Washer - 2 off
4	Nut - 2 off	13	Air valve union
5	Spring washer - 2 off	14	Bolt - 2 off
6	Washer - 2 off	15	Spring washer - 2 off
7	Bush - 4 off	16	Right-hand air hose
8	Washer - 2 off	17	Air valve
9	Nut - 2 off	18	Left-hand air hose

18 Centre stand: examination

1   The centre stand is retained on a pivot tube between two frame bosses and is retracted by a coil spring. The stand and spring should be checked regularly for signs of wear, noting that any failure of the stand while the machine is being ridden can have disastrous consequences. The stand can be removed for examination and lubrication by withdrawing the split pin and washer which secures the pivot shaft. The shaft can now be displaced and the stand assembly removed.

19 Prop stand: examination

1   The prop stand is secured to a plate on the frame with a bolt and nut, and is retracted by a tension spring. Make sure the bolt is tight and the spring is not overstretched, otherwise an accident can occur if the stand drops during cornering.

20 Footrests and rear brake pedal: examination

1   The footrests are of the swivel type and are retained by a clevis pin secured by a split pin. The advantage of this type of footrest is that if the machine should fall over the footrest will fold up instead of bending.

2   The brake pedal is clamped to its splined shaft by a pinch bolt and can be removed once this has been released.

21 Instrument panel: general

1   The instrument panel arrangement varies widely according to the model. On some versions a fairly sophisticated microprocessor – controlled display is incorporated, and for this reason the various arrangements will be covered in detail in Chapter 6.

22 Suspension adjustment

1   All models feature some form of air assistance on the front forks, and this can be varied within prescribed limits to alter the spring rate. The various pressure ranges and standard pressure settings will be found below. On machines with separate air valves for each fork leg, ensure that each leg is under the same pressure. Any difference must not exceed 0.1 kg/cm^2 (1.4 psi).

Front fork air pressure

Model	Standard pressure	Pressure range
E1, E2, E3, L1, L2, R1	0.7 kg/cm^2 (10 psi)	0.6 – 0.9 kg/cm^2 (8.5 – 13 psi)
H1, H2, H3	0.6 kg/cm^2 (8.5 psi)	0.5 – 1.0 kg/cm^2 (7.1 – 14.0 psi)
N1, P1	0.6 kg/cm^2 (8.5 psi)	0.5 – 0.7 kg/cm^2 (7.1 – 10.0 psi)

2   The chain drive models are equipped with hydraulically damped coil spring rear suspension units. On all but the R1 model, five spring preload settings are provided together with four damper settings. In the case of the latter model seven spring preload settings and five damper settings are provided. The spring and damper rates can be varied at the owner's discretion to provide the best compromise between comfort and load carrying capability. It is important that both units are set up identically.

3   The shaft drive models employ interconnected air suspension units at the rear. These can be adjusted to suit load or road surface conditions by increasing or decreasing the air pressure in the units via the common air valve below the seat. Kawasaki recommend the use of a suspension air pressure gauge, Part Number 52005-1003 to ensure accurate readings without excessive air loss during measurement.

Rear suspension unit air pressure

	N1	P1
Maximum safe pressure	5 kg/cm^2 (71 psi)	5 kg/cm^2 (71 psi)
Usable pressure range	1.5 – 3.0 kg/cm^2 (21 – 43 psi)	1.5 – 4.0 kg/cm^2 (21 – 57 psi)

The units also incorporate a four position damping adjuster which can be set to the rider's preference. Again, check that both units are set at the same damping rate.

Chapter 5 Wheels, brakes and tyres

Refer to Chapter 7 for information relating to the 1983 on models

Contents

Specifications

Wheels

Type ....   Seven-spoke cast alloy

Brakes

Front ....   Twin 226 mm hydraulic disc brake
Rear ....   Single 226 mm hydraulic disc brake

Master cylinder – service limits

	Front	Rear
Cylinder bore diameter	15.95 mm (0.6280 in)	14.08 mm (0.5543 in)
Piston outside diameter	15.80 mm (0.6220 in)	13.77 mm (0.5421 in)
Primary cup diameter	16.00 mm (0.6299 in)	14.10 mm (0.5551 in)
Secondary cup diameter	16.40 mm (0.6457 in)	14.50 mm (0.5709 in)
Spring free length	37.40 mm (1.3661 in)	37.20 mm (1.4646 in)

Brake caliper – service limits

Caliper bore diameter	42.95 mm (1.6898 in)
Piston outside diameter	42.72 mm (1.6819 in)

Brake pads

Minimum friction material thickness ....   1.00 mm (0.039 in)

Brake discs – service limits

Runout (warpage)	0.3 mm (0.0118 in)
Thickness	
Front	4.5 mm (0.1772 in)
Rear	6.0 mm (0.2362 in)

Tyres

	Front	Rear
Type	Tubeless	Tubeless
Sizes:		
E1, E2, E3, L1, L2	3.25H-19 4PR	4.00H-18 4PR
H1, H2, H3	3.25H-19 4PR	130/90-16 67H
R1 (US) and P1	100/90-19 57H	120/90-18 65H
R1 (UK)	100/90 V-19	120/90 V-18
N1	100/90-19 57H	120/90-16 67H

Tyre pressures

	R1 (US), E1, E2, E3, L1, L2, P1	H1 (US)
Front	28 psi	25 psi
Rear:		
Up to 97.5 kg, 215 lb loading	32 psi	22 psi
97.5-165 kg, 215-364 lb loading	36 psi	25 psi

H1 (UK)

	Up to 180 kph (110 mph)	Above 180 kph (110 mph)
Front	25 psi	28 psi
Rear:		
Up to 95 kg, 210 lb loading	25 psi	28 psi
95-136 kg, 210-300 lb loading	28 psi	32 psi
136-180 kg, 300-397 lb loading	32 psi	36 psi

R1 (UK)

	Up to 210 kph (130 mph)	Above 210 kph (130 mph)
Front	28 psi	32 psi
Rear:		
Up to 97.5 kg, 215 lb loading	32 psi	41 psi
97.5-180 kg, 215-397 lb loading	36 psi	41 psi

N1

Front	25 psi
Rear:	
Up to 97.5 kg, 215 lb loading	25 psi
97.5-180 kg, 215-397 lb loading	32 psi

Torque settings

Component	kgf m	lbf ft	lbf in
Front wheel spindle nut	8.00	58.0	–
Rear wheel spindle nut	6.50	47.0	–
Brake torque arm nut	3.00	22.0	–
Disc brake components:			
Bleed valves	0.80	–	69.0
Hose union bolts	3.00	22.0	–
Lever pivot bolt	0.30	–	26.0
Lever pivot bolt locknut	0.60	–	52.0
Caliper holder shaft bolts	1.80	13.0	–
Disc mounting bolts	2.30	16.5	–
Front caliper mounting bolts	4.00	29.0	–
Front master cylinder mounting bolts	0.90	–	78.0
Rear caliper torque arm nut	3.00	22.0	–

Additional torque settings – N1 and P1 models

	kgf m	lbf ft	lbf in
Front wheel spindle nut	6.50	47.0	–
Front wheel spindle clamp bolt	1.40	10.0	–
Rear wheel spindle nut	14.00	101.0	–
Caliper mounting bolts	3.00	22.0	–

1 General description

The KZ/Z 750 models are equipped with cast alloy wheels of various sizes according to the model, and carry tubeless tyres front and rear. The front brake is a twin hydraulic disc unit, whilst rear brāking is by a single hydraulic unit.

2 Front wheel: examination and renovation

1 Carefully check the complete wheel for cracks and chipping, particularly at the spoke roots and the edge of the rim. As a general rule a damaged wheel must be renewed as cracks will cause stress points which may lead to sudden failure under heavy load. Small nicks may be radiused carefully with a fine file and emery paper (No 600 – No 1000) to relieve the stress. If there is any doubt as to the condition of a wheel, advice should be sought from a Kawasaki repair specialist.
2 Each wheel is covered with a coating of lacquer, to prevent corrosion. If damage occurs to the wheel and the lacquer finish is penetrated, the bared aluminium alloy will soon start to corrode. This deposit however, should be removed carefully as soon as possible and a new protective coating of laquer applied.
3 Check the lateral run out at the rim by spinning the wheel and placing a fixed pointer close to the rim edge. If the maximum run out is greater than 0.5 mm (0.020 in) axially, or 0.8 mm (0.031 in) radially, Kawasaki recommend that the wheel should be renewed. This is, however, a counsel of perfection; a run out somewhat greater than this

can probably be accommodated without noticeable effect on steering. No means is available for straightening a warped wheel without resorting to the expense of having the wheel skimmed on all faces. If warpage was caused by impact during an accident, the safest measure is to renew the wheel complete. Worn wheel bearings may cause rim run out. These should be renewed.
4 Note that impact damage or serious corrosion on models fitted with tubeless tyres has wider implications in that it could lead to a loss of pressure. If in any doubt as to the wheel's condition, seek professional advice.

3 Front wheel: removal and refitting

1 Place the machine on its centre stand, leaving adequate working space around the wheel area. Slacken the knurled ring which retains the speedometer drive cable to the drive gearbox and pull the cable clear of the wheel. Remove the two bolts which secure one of the brake caliper mounting brackets to the fork leg. Lift the caliper away from the fork and tie it to the frame to avoid straining the hydraulic hose. Place a wood wedge between the brake pads to prevent their being displaced should the brake be operated accidentally.
2 In the case of machines fitted with leading axle type forks, remove the wheel spindle nut, then slacken the spindle clamp bolt. Place blocks beneath the crankcase so that the wheel is raised clear of the ground. Take the weight of the wheel and withdraw the spindle. The wheel can now be lifted away.

3   Where centre axle forks are fitted, loosen, but do not remove, the clamp nuts on the bottom of the forks. Slacken the wheel spindle nuts by two or three turns, then support the engine so that the wheel is raised clear of the ground as described above. Remove the clamp nuts and clamp halves and lift the wheel clear of the forks.

4   Note that on all models the wheel should not be placed on its side with weight resting on one of the brake discs. This can distort the disc. Place wooden blocks beneath the wheel rim or store the wheel against a wall to avoid this.

5   Refit the wheel by reversing the removal sequence. On leading axle fork machines, slide the wheel spindle into place and fit the wheel spindle nut finger tight. Check that the speedometer gearbox is located correctly. Note that in the case of the KZ/Z750 H1 model prior to frame number KZ750 H-006527, the locating lugs on the speedometer drive gearbox must be disregarded because if they are engaged on the fork leg it will be impossible to route the speedometer cable correctly. Instead, position the gearbox so that the cable entry point lies 10°

below horizontal, facing rearwards. Check that this position is maintained while securing the wheel spindle nut. Tighten the wheel spindle nut then the clamp bolt to the specified torque settings.

6   In the case of machines fitted with centre axle forks, check that the wheel is assembled correctly with the speedometer gearbox properly engaged on the left-hand side and that the spacer is fitted on the right. Lift the wheel into position and fit the clamp halves and nuts. Note that each clamp half has an arrow mark which should face forward. Tighten the wheel spindle nuts first, then the clamp nuts. The front clamp nuts should be secured first, then the rear nuts, leaving a small gap towards the rear. Secure all clamp and spindle nuts to the specified torque settings.

7   Remove the wooden wedge from between the caliper pads, and fit the caliper to the fork leg, ensuring that the mounting bolts are tightened to the correct torque figure. Refit the speedometer cable, ensuring that the inner cable engages correctly, and tighten the knurled ring firmly to secure it.

3.3 Release speedometer cable and remove one caliper, then release clamps to free wheel

3.5 Check that locating tang (arrowed) engages in fork

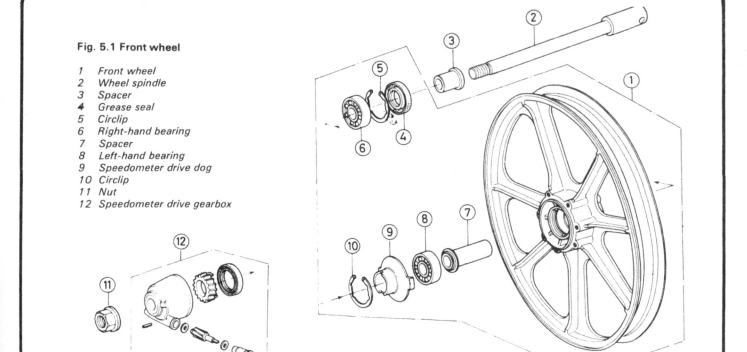

Fig. 5.1 Front wheel

1   Front wheel
2   Wheel spindle
3   Spacer
4   Grease seal
5   Circlip
6   Right-hand bearing
7   Spacer
8   Left-hand bearing
9   Speedometer drive dog
10 Circlip
11 Nut
12 Speedometer drive gearbox

4   Rear wheel: removal and refitting – chain drive models

1   Place the machine on its centre stand, leaving adequate working space to the rear. Loosen the self-locking nut which secures the brake torque arm to the caliper, and free the caliper hose from its guides on the swinging arm. Slacken the two chain adjuster locknuts and unscrew fully the adjuster bolts. Straighten and remove the split pin and unscrew the wheel spindle nut. The adjusters can now be pushed down, clear of the ends of the swinging arm.

2   Slide the wheel forward to allow the chain to be lifted clear of the rear wheel sprocket and rested around the wheel spindle to the outside of the sprocket. Remove the single bolt and lock washer which retains each swinging arm end stop, and remove the stops. Pull the wheel rearwards and remove the spindle to free the adjusters and caliper. Manoeuvre the wheel clear, then refit the spindle through the caliper and swinging arm to support the former until the wheel is refitted. Place a wood wedge between the brake pads to prevent them from being expelled if the brake is operated accidentally. **Note:** When storing the wheel, prop it against a wall or support it on wooden strips to avoid any risk of warping the brake disc.

3   To install the wheel, remove the wedge from between the pads and remove the wheel spindle. Use a screwdriver to act as a temporary support for the caliper. Position the wheel close to the swinging arm ends and fit the spindle through the adjuster, caliper, collar, hub, sleeve coupling, collar and left-hand adjuster, then run the nut into place by a few threads. Check that the alignment marks on the adjusters face outwards.

4   Offer up the wheel and fit the end stops, securing these with their single bolts and lock washers. Loop the chain around the sprocket and then complete the operation by setting the chain free play as described in Routine Maintenance.

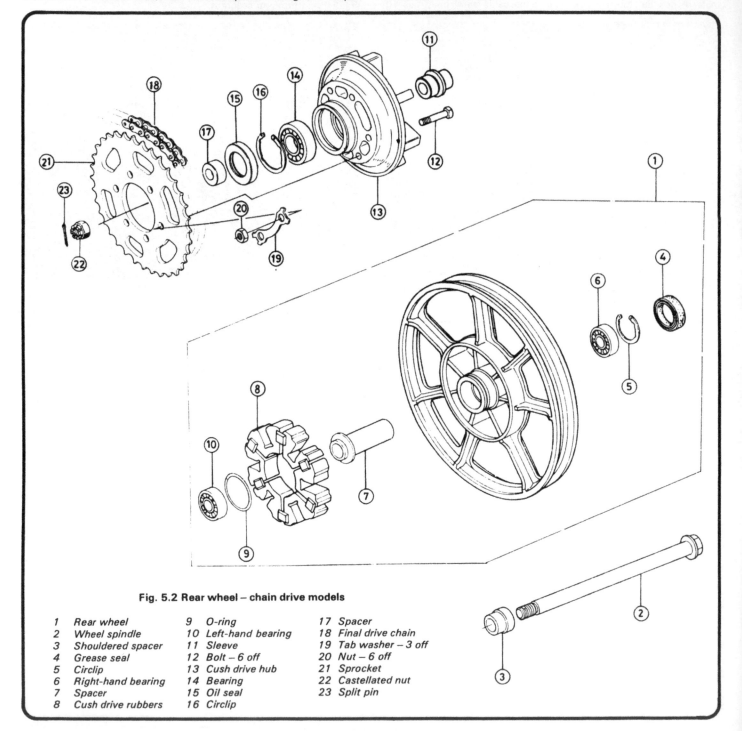

Fig. 5.2 Rear wheel – chain drive models

1	Rear wheel	9	O-ring	17	Spacer
2	Wheel spindle	10	Left-hand bearing	18	Final drive chain
3	Shouldered spacer	11	Sleeve	19	Tab washer – 3 off
4	Grease seal	12	Bolt – 6 off	20	Nut – 6 off
5	Circlip	13	Cush drive hub	21	Sprocket
6	Right-hand bearing	14	Bearing	22	Castellated nut
7	Spacer	15	Oil seal	23	Split pin
8	Cush drive rubbers	16	Circlip		

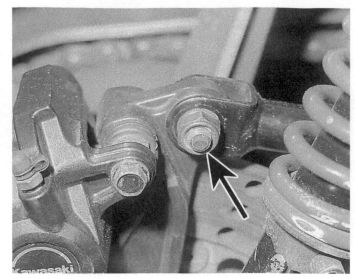
4.1a Remove self-locking nut to free brake torque arm

4.1b Slacken spindle nut and adjusters

4.2a Withdraw wheel, taking care not to lose spacers

4.2b Support caliper assembly after wheel has been freed

4.3a Fit swinging arm end stops ...

4.3b ... and adjusters, noting indent for adjuster bolt

4.4a Fit spindle, and adjust chain noting alignment marks

4.4b Tighten wheel spindle nut and secure with split pin

5   Rear wheel: removal and refitting – shaft drive models

1   Place the machine on its centre stand. Remove the wheel spindle nut split pin and unscrew the nut, then withdraw the spindle to allow the caliper and spacer to be displaced, noting that it will be necessary to slacken the caliper torque arm nut to permit its movement.
2   Slide the wheel to the right and remove the spindle completely to allow the wheel to be disengaged from the cush drive rubbers and removed. Do not place the wheel so that it rests on the brake disc.
3   To install the wheel, grease the cush drive outer periphery and its internal splines, then fit the wheel by reversing the above sequence. Tighten the torque arm nut and the spindle nut to the specified torque settings. Use a new split pin to secure the spindle nut.

6   Front wheel bearings: removal and installation

1   Remove the front wheel as described in Section 3. Remove the speedometer drive gearbox and the circlip which retains the drive dog and its counterpart on the right-hand side. Displace and remove the dog to expose the bearing. Prise out and discard the grease seal.
2   Arrange the wheel with the left-hand side uppermost and supported so that the hub is clear of the bench surface. Pass a long drift through the hub, push the internal spacer to one side and drive out the right-hand bearing. Invert the wheel and remove the remaining bearing and the spacer.
3   Degrease and inspect the bearings, renewing them if obviously worn or pitted. Pack the bearings with grease, then fit them using a tubular drift or a large socket to drive them home squarely. Note that the plain, sealed sides should face outwards. Secure the bearings with their respective circlips, fitting the speedometer drive dog first, to the left-hand side. On the right-hand side, tap the new grease seal into place, taking care that it fits squarely into the hub bore.
4   Refit the discs, if these were removed, securing the Allen bolts to 2.3 kgf m (16.5 lbf ft). Note that the discs should be fitted with the chamfered hole side facing inwards. It is advisable to check disc runout as described in Section 10 of this Chapter.

7   Rear wheel bearings: removal and installation

1   Remove the rear wheel as described in Section 4 and 5. The bearings can then be dealt with in the same way as has been described in Section 6 for front wheel bearings. Note that the bearings used in the rear hub may not have integral grease seals, but should be fitted with the part numbered edge facing outwards. The right-hand bearing is retained by a circlip.
2   On the chain drive models only, the cush drive hub contains an additional bearing, circlip and grease seal. These can be dealt with in the same way as the wheel bearings.

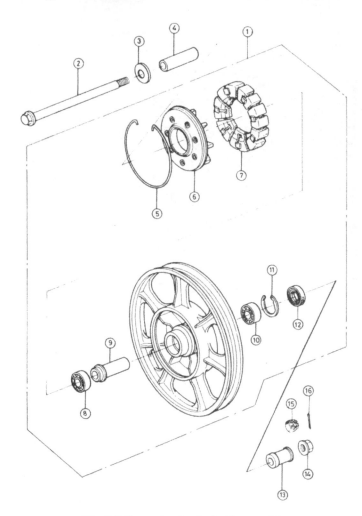

Fig. 5.3 Rear wheel – shaft drive models

1	Rear wheel	9	Spacer
2	Wheel spindle	10	Right-hand bearing
3	Washer	11	Circlip
4	Spacer	12	Grease seal
5	Wire retaining ring	13	Spacer
6	Cush drive hub	14	Nut – P model
7	Cush drive rubbers	15	Castellated nut – N model
8	Left-hand bearing	16	Split pin – N model

6.1a Remove wheel spindle to free spacer ...

6.1b ... and the speedometer drive gearbox

6.1c Remove circlip ...

6.1d ... to free the speedometer drive dog

6.1e Prise out and discard grease seal ...

6.1f ... and remove circlip

6.3a Bearings are fitted with sealed face outwards

6.3b Tap into place using large socket as drift

6.3c Do not omit to refit central spacer

7.1 Check condition of O-ring on hub boss

7.2a Remove cush drive bearing and seal for inspection

7.2b Grease bearing and install ...

7.2c ... then fit circlip to retain it

7.2d Fit grease seal followed by spacer

7.2e Remember to fit bearing spacer to hub ...

7.2f ... then offer up cush drive hub

8 Braking system: examination and pad renewal

The full procedure for general inspection of the braking system and front and rear pad renewal is described in Routine Maintenance under the 3 monthly/3000 mile service heading.

9 Brake discs: examination and renovation

1   Examine the brake discs for scoring, particularly the rear unit which is more vulnerable to accumulations of road dirt. Damaged discs will cause poor braking and will wear pads quickly, and should therefore be renewed. The disc thickness can be measured with a micrometer and should not be less than 4.5 mm (0.177 in) for the front discs or 6.0 mm (0.236 in) for the rear.
2   Check for warpage with the relevant wheel raised clear of the ground, using a dial gauge probe running near the edge of the disc. Warpage must not exceed 0.3 mm (0.118 in) when the disc is rotated. A warped disc will often cause judder during braking and will necessitate renewal.
3   The discs can be removed after the appropriate wheel has been removed from the machine. Each disc is retained by seven Allen bolts. When refitting the disc, ensure that it and the hub are clean and that the chamfered hole side of the disc faces inwards. Tighten the retaining bolts to 2.3 kgf m (16.5 lbf ft).

10 Brake hydraulic system overhaul: general information

1   Imprecise or spongy braking is indicative of the need to overhaul the hydraulic system. Though normally trouble-free, the various seals will eventually wear and begin to leak, and braking action will be noticeably impaired. Overhaul should be undertaken **before** poor braking develops to a point where total failure is likely.
2   Before attempting to dismantle any brake part clean the outside meticulusly. Drain the hydraulic fluid by connecting a length of plastic tubing to the bleed nipple on each caliper. Slacken the nipple by ½ to 1 turn, and slowly 'pump' the lever or pedal until all of the fluid has been expelled into a jar or drain bowl. Do not reuse old fluid.
3   Note: Hydraulic fluid will discolour and damage paintwork and many plastics. Take care not to allow it to contact either, and wipe up any spillages immediately.

11 Brake master cylinder: removal, overhaul and installation

Front
1   Drain the hydraulic system as described in Section 11. Remove the right hand mirror, then free the front brake light switch by depressing its locking pin using a small electrical screwdriver via the hole in the underside of the switch housing.
2   Pull back the banjo union dust cover and remove the union bolt

and washers. Wipe up any residual hydraulic fluid. Slacken the two master cylinder clamp bolts, remove the clamp half and lift the master cylinder away.

3 Remove the two screws which retain the reservoir cover, remove the cover and diaphragm and empty out any remaining fluid. Remove the brake lever locknut and pivot bolt and remove the lever. Remove the white plastic liner by pushing in the locking tabs which retain it. Pull out the piston assembly, separating the spring, dust seal and piston stop from the piston.

4 Examine the piston surface and master cylinder bore for signs of wear or corrosion. Renew both components if damaged in any way; new seals will not compensate for scoring and will wear out quickly. Check the primary and secondary seals for damage or swelling, renewing them unless in perfect condition. The cups are sold as a kit together with the piston and spring. Renew the dust seal at the same time to preclude road dirt entering the caliper body. Ensure that the supply port and the smaller relief port between the cylinder and reservoir are clear, especially where swollen or damaged cups have been noted.

5 All components can be measured for wear if required, and the readings checked against the figures given in Specifications.

Rear

6 Proceed as described above, noting the following exceptions. Drain the system and disconnect the banjo union at the caliper. Disconnect the hose from the reservoir and wipe up any spilled fluid. Slacken, but do not remove, the two master cylinder securing bolts. Remove the brake pedal. Prop up the right hand silencer and release the rear footrest mounting bolt, nut and flat washer.

7 Release the footrest plate by removing its two mounting bolts. Lift the assembly clear and disconnect the brake light switch spring. Unhook the brake pedal return spring. Remove the split pin, washer and clevis pin which retain the brake arm to the push rod. Remove the two master cylinder bolts and lift the cylinder away from the footrest plate.

8 Pull off the dust seal and remove it together with the push rod. Displace the retaining clip and continue dismantling and overhaul as described above.

Front and rear

9 Reassemble the master cylinder by reversing the dismantling sequence, ensuring that all components are kept spotlessly clean. Lubricate the piston, cups and cylinder bore with new hydraulic fluid during installation. Use new sealing washers on the banjo unions and tighten the union bolts to 3.0 kgf m (22 lbf ft). Note that when installing the rear brake pedal the line marked on the pedal should align with the dot on the end of the shaft. Refill and bleed the hydraulic system and check brake operation.

10 Check the rear brake pedal height before riding the machine. This should be measured from the top of the footrest rubber to the pedal foot and must be within the range shown below. If adjustment is required proceed as described above to gain access to the pushrod, located on the reverse side of the footrest plate. Slacken the lower of the two pushrod nuts (locknut) and turn the upper nut to alter the pedal height. Tighten the locknut on completion of adjustment.

KZ750 E and Z750 L1, L2	8 – 12 mm
Z750 E and KZ/Z750 R	13 – 17 mm
KZ/Z750 H	4 – 8 mm
KZ750 N and Z750 P	4 – 10 mm

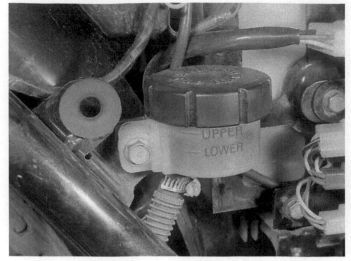

11.6 Rear reservoir is attached by single bolt to frame

11.7a Rear master cylinder can be removed complete with footrest plate

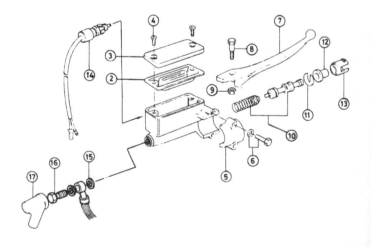

Fig. 5.4 Front brake master cylinder – KZ/Z 750 H, E and L

1	Master cylinder body	10	Piston and spring
2	Diaphragm	11	Piston stop
3	Cover	12	Dust seal
4	Screw – 2 off	13	Plastic liner
5	Clamp	14	Brake light switch
6	Bolt and washer – 2 off	15	Sealing washer – 2 off
7	Brake lever	16	Banjo union bolt
8	Pivot bolt	17	Dust cover
9	Locknut		

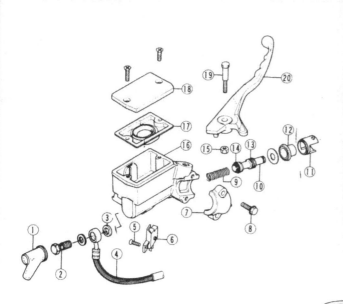

11.7b Footrest plate mounting bolts (arrowed)

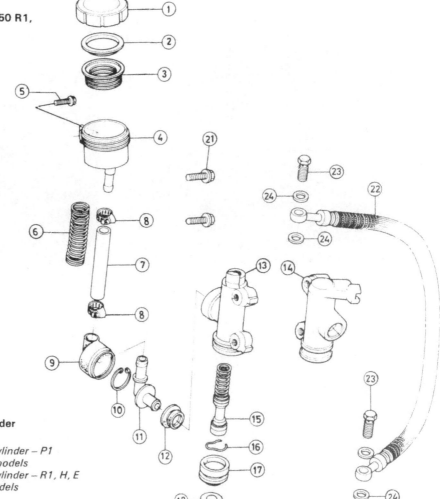

Fig. 5.5 Front brake master cylinder – KZ/Z 750 R1,
KZ 750 N1 and Z750 P1

1 Dust cover
2 Banjo union bolt
3 Sealing washer – 2 off
4 Hydraulic hose
5 Screw
6 Brake light switch
7 Clamp
8 Bolt – 2 off
9 Spring
10 Piston
11 Plastic liner
12 Dust seal
13 O-ring
14 Primary cup
15 Locknut
16 Master cylinder body
17 Diaphragm
18 Cover
19 Pivot bolt
20 Brake lever

Fig. 5.6 Rear brake master cylinder

1 Reservoir cap
2 Diaphragm plate
3 Diaphragm
4 Reservoir
5 Bolt
6 Spring
7 Hydraulic hose
8 Hose clamp – 2 off
9 Dust cover – P1 and
   N1 models
10 Circlip – P1 and
   N1 models
11 Union
12 Seal
13 Master cylinder – P1
   and N1 models
14 Master cylinder – R1, H, E
   and L models
15 Piston
16 Retaining clip
17 Dust seal
18 Pushrod
19 Clevis pin
20 Split pin
21 Bolt – 2 off
22 Hydraulic hose
23 Banjo union bolt – 2 off
24 Sealing washer – 4 off

12 Brake caliper: removal, overhaul and installation

1   Drain the hydraulic system (Section 11) and disconnect the hose(s) by releasing the union bolt(s). Remove the caliper from the caliper bracket (Sections 8 or 9). Pull off the caliper holder shafts and dust seals noting that one is of smaller diameter and has a friction sleeve fitted.

2   From the inside of the caliper body, remove the anti-rattle spring and the caliper piston dust seal. Wrap the caliper in rag to catch any residual hydraulic fluid, then displace the piston by directing compressed air into the hydraulic union thread. Take care to avoid trapping fingers as the piston emerges. Remove the piston and place it to one side, then remove the piston seal from its groove in the caliper bore using a small screwdriver. Take care to avoid scratching the bore surface.

3   Examine the piston surface and caliper bore for signs of wear or scoring, normally caused by the ingress of road dirt or corrosion. If the surfaces are undamaged, they can be checked for wear by direct measurement.

4   Check that the caliper holder shafts are a good sliding fit in their respective bores in the caliper bracket. If sloppy, renew the bracket, holder shafts and the single friction sleeve, together with the dust seals.

5   Before assembling the caliper, clean all parts using alchohol or hydraulic fluid only. Fit a new piston seal, and renew any other rubber component that is worn, damaged or swollen. Lubricate each part with hydraulic fluid during assembly. Check that the piston dust seal locates properly. Lubricate the two holder shafts wth PBC (Poly Butyl Cuprysil) grease. Complete assembly noting the specified torque figures. Refill and bleed the hydraulic system.

13 Brake hoses and pipes: renewal

1   Drain the hydraulic system (Section 11). Slacken the union bolts, noting the exact run of the faulty hose. Clean the unions, then refit the new hose using new sealing washers. Note in particular the notches in the three way union below the steering head. These should locate the banjo unions when fitted properly. Tighten the union bolts to 3.0 kgf m (22 lbf ft). Refill and bleed the system and check for leakage before using the machine.

12.1a Remove caliper from mounting bracket and lift out brake pads

12.1b Release caliper mounting bracket from fork

12.1c Pull out caliper holder shafts and dust seals

12.5 Do not omit caliper shims during assembly

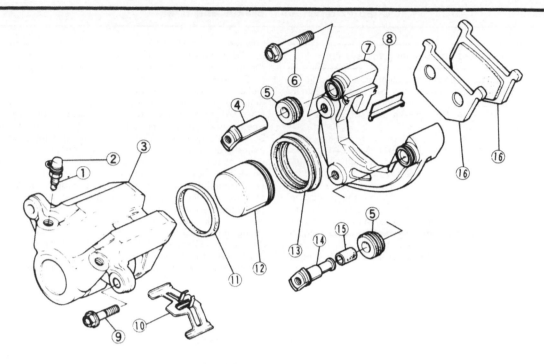

Fig. 5.7 Front brake caliper

1	Bleed nipple	5	Dust seal – 2 off	9	Bolt – 2 off	13	Piston dust seal
2	Bleed nipple cap	6	Bolt – 2 off	10	Anti-rattle shim	14	Caliper holder shaft
3	Caliper	7	Mounting bracket	11	Piston seal	15	Friction sleeve
4	Caliper holder shaft	8	Shim – 2 off	12	Piston	16	Brake pads

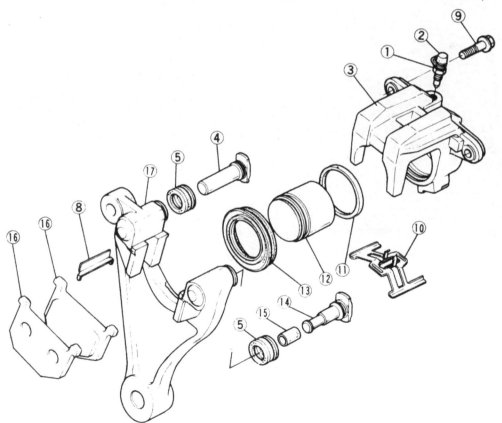

Fig. 5.8 Rear brake caliper

1	Bleed nipple	5	Dust seal – 2 off	11	Piston seal	15	Friction sleeve
2	Bleed nipple cap	8	Shim – 2 off	12	Piston	16	Brake pads
3	Caliper	9	Bolt – 2 off	13	Piston dust seal	17	Mounting bracket
4	Caliper holder shaft	10	Anti-rattle shim	14	Caliper holder shaft		

14 Changing the hydraulic fluid and bleeding the system

1   Connect a length of plastic tubing to the bleed valve, placing the free end in a glass jar to catch the expelled fluid. Slacken the valve by $\frac{1}{2} - 1$ turn, then operate the lever or pedal repeatedly until all hydraulic fluid has been expelled. On the front brake system, repeat this operation on the remaining caliper.

2   Remove the reservoir cap and diaphragm and fill to the maximum level with new hydraulic fluid conforming to DOT 3 or SAE J1703. Open the bleed valve, then operate the lever or pedal fully. Holding it in this position, close the bleed valve and release the lever or pedal. Repeat this sequence until fluid is expelled from the bleed valve. Remember to keep the reservoir topped up throughout this procedure.

3   To remove residual air from the system, first check that the reservoir is full. It is important never to let the level fall too low during bleeding. If this should happen, start the bleeding sequence again. Operate the lever or pedal a few times to ensure that any air inside the master cylinder is expelled. This will be indicated by bubbles rising to the fluid surface.

4   Place the cap over the reservoir and connect the bleed tube to the valve, placing the free end in a jar containing a small amount of fluid. Operate the pedal or lever several times and then hold it down. Open the valve and close it again quickly, then release the lever or pedal. Repeat the procedure until no more air bubbles emerge from the valve, remembering to monitor the fluid level in the reservoir. In the case of the front brake, carry out the bleeding sequence on the remaining caliper. Check that the brake operates correctly, then top up the reservoir and refit the cover or cap.

14.1 Connect bleed hose to valve as shown

15 Tyres: removal and replacement

1   It is strongly recommended that should a repair to a tubeless tyre be necessary, the wheel is removed from the machine and taken to a tyre fitting specialist who is willing to do the job or taken to an official Kawasaki dealer. This is because the force required to break the seal between the wheel rim and tyre bead is considerable and considered to be beyond the capabilities of an individual working with normal tyre removing tools. Any abortive attempt to break the rim to bead seal may also cause damage to the wheel rim, resulting in an expensive wheel replacement. If, however, a suitable bead releasing tool is available, and experience has already been gained in its use, tyre

removal and refitting can be accomplished as follows.

2   To remove the tyre from either wheel, first detach the wheel from the machine by following the procedure in this Chapter, Sections 3, 4 or 5 depending on whether the front or the rear wheel is involved. Deflate the tyre by removing the valve insert and when it is fully deflated, push the bead of the tyre away from the wheel rim on both sides so that the bead enters the centre well of the rim. As noted, this operation will almost certainly require the use of a bead releasing tool.

3   Insert a tyre lever close to the valve and lever the edge of the tyre over the outside of the wheel rim. Very little force should be necessary; if resistance is encountered it is probably due to the fact that the tyre beads have not entered the well of the wheel rim all the way round the tyre. Should the initial problem persist, lubrication of the tyre bead and the inside edge and lip of the rim will facilitate removal. Use a recommended lubricant, a dilute solution of washing-up liquid or french chalk. Lubrication is usually recommended as an aid to tyre fitting but its use is equally desirable during removal. The risk of lever damage to wheel rims can be minimised by the use of proprietary plastic rim protectors placed over the rim flange at the point where the tyre levers are inserted. Suitable rim protectors may be fabricated very easily from short lengths (4 – 6 inches) of thick-walled nylon petrol pipe which have been split down one side using a sharp knife. The use of rim protectors should be adopted whenever levers are used and, therefore, when the risk of damage is likely.

4   Once the tyre has been edged over the wheel rim, it is easy to work around the wheel rim so that the tyre is completely free on one side.

5   Working from the other side of the wheel, ease the other edge of the tyre over the outside of the wheel rim which is furthest away. Continue to work around the rim until the tyre is freed completely from the rim.

6   Refer to the following Section for details relating to puncture repair and the renewal of tyres. See also the remarks relating to the tyre valves in Section 17.

7   Refitting of the tyre is virtually a reversal of the removal procedure. If the tyre has a balance mark (usually a spot of coloured paint), as on the tyres fitted as original equipment, this must be positioned alongside the valve. Similarly, any arrow indicating direction of rotation must face the right way.

8   Starting at the point furthest from the valve, push the tyre bead over the edge of the wheel rim until it is located in the central well. Continue to work around the tyre in this fashion until the whole of one side of the tyre is on the rim. It may be necessary to use a tyre lever during the final stages. Here again, the use of a lubricant will aid fitting. It is recommended strongly that when refitting the tyre only a recommended lubricant is used because such lubricants also have sealing properties. Do not be over generous in the application of lubricant or tyre creep may occur.

9   Fitting the upper bead is similar to fitting the lower bead. Start by pushing the bead over the rim and into the well at a point diametrically opposite the tyre valve. Continue working round the tyre, each side of the starting point, ensuring that the bead opposite the working area is always in the well. Apply lubricant as necessary. Avoid using tyre levers unless absolutely essential, to help reduce damage to the soft wheel rim. The use of the levers should be required only when the final portion of bead is to be pushed over the rim.

10 Lubricate the tyre beads again prior to inflating the tyre, and check that the wheel rim is evenly positioned in relation to the tyre beads. Inflation of the tyre may well prove impossible without the use of a high pressure air hose. The tyre will retain air completely only when the beads are firmly against the rim edges at all points and it may be found when using a foot pump that air escapes at the same rate as it is pumped in. This problem may also be encountered when using an air hose, on new tyres which have been compressed in storage and by virtue of their profile hold the beads away from the rim edges. To overcome this difficulty, a tourniquet may be placed around the circumference of the tyre over the central area of the tread. The compression of the tread in this area will cause the beads to be pushed outwards in the desired direction. The type of tourniquet most widely used consists of a length of hose closed at both ends with a suitable clasp fitted to enable both ends to be connected. An ordinary tyre valve is fitted at one end of the tube so that after the hose has been secured around the tyre it may be inflated, giving a constricting effect. Another possible method of seating beads to obtain initial inflation is to press the tyre into the angle between a wall and the floor. With the airline attached to the valve additional pressure is then applied to the

TYRE CHANGING SEQUENCE - TUBELESS TYRES

Deflate tyre. After releasing beads, push tyre bead into well of rim at point opposite valve. Insert lever next to valve and work bead over edge of rim.

Use two levers to work bead over edge of rim. Note use of rim protectors.

When first bead is clear, remove tyre as shown.

Before installing, ensure that tyre is suitable for wheel. Take note of any sidewall markings such as direction of rotation arrows.

Work first bead over the rim flange.

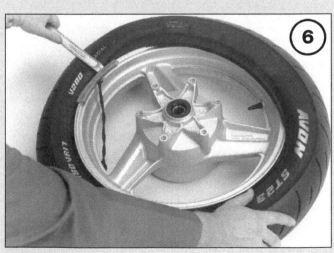

Use a tyre lever to work the second bead over rim flange.

tyre by the hand and shin, as shown in the accompanying illustration. The application of pressure at four points around the tyre's circumference whilst simultaneously applying the airhose will often effect an initial seal between the tyre beads and wheel rim, thus allowing inflation to occur.

11 Having successfully accomplished inflation, increase the pressure to 40 psi and check that the tyre is evenly disposed on the wheel rim. This may be judged by checking that the thin positioning line found on each tyre wall is equidistant from the rim around the total circumference of the tyre. If this is not the case, deflate the tyre, apply additional lubrication and reinflate. Minor adjustments to the tyre position may be made by bouncing the wheel on the ground.

12 Always run the tyre at the recommended pressures and never under- or over-inflate. The correct pressures for solo use are given in the Specification Section of this Chapter. If a pillion passenger is carried, increase the rear tyre pressure only as recommended.

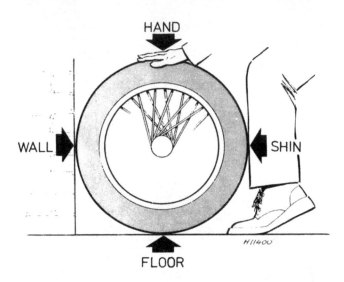

Fig. 5.9 Method of seating the beads on tubeless tyres

16 Puncture repair and tyre removal

1 The primary advantage of the tubeless tyre is its ability to accept penetration by sharp objects such as nails etc without loss of air. Even if loss of air is experienced, because there is no inner tube to rupture, in normal conditions a sudden blow-out is avoided.

2 If a puncture of the tyre occurs, the tyre should be removed for inspection for damage before any attempt is made at remedial action. The temporary repair of a punctured tyre by inserting a plug from the outside should not be attempted. Although this type of temporary repair is used widely on cars, the manufacturers strongly recommend that no such repair is carried out on a motorcycle tyre. Not only does the tyre have a thinner carcass, which does not give sufficient support to the plug, but the consequences of a sudden deflation are often sufficiently serious that the risk of such an occurrence should be avoided at all costs.

3 The tyre should be inspected both inside and out for damage to the carcass. Unfortunately the inner lining of the tyre — which takes the place of the inner tube — may easily obscure any damage and some experience is required in making a correct assessment of the tyre condition.

4 There are two main types of tyre repair which are considered safe for adoption in repairing tubeless motorcycle tyres. The first type of repair consists of inserting a mushroom-headed plug into the hole from the **inside** of the tyre. The hole is prepared for insertion of the plug by reaming and the application of an adhesive. The second repair is carried out by buffing the inner lining in the damaged area and

applying a cold or vulcanised patch. Because both inspection and repair, if they are to be carried out safely, require experience in this type of work, it is recommended that the tyre be placed in the hands of a repairer with the necessary skills, rather than repaired in the home workshop.

5 In the event of an emergency, the only recommended 'get-you-home' repair is to fit a standard inner tube of the correct size. If this course of action is adopted, care should be taken to ensure that the cause of the puncture has been removed before the inner tube is fitted.

6 In the event of the unavailability of tubeless tyres, ordinary tubed tyres fitted with inner tubes of the correct size may be fitted. Refer to the manufacturer or a tyre fitting specialist to ensure that only a tyre and tube of equivalent type and suitability is fitted, and also to advise on the fitting of a valve nut to the rim hole.

17 Tyre valves: description and renewal

1 It will be appreciated from the preceding Sections that the adoption of tubeless tyres has made it necessary to modify the valve arrangement, as there is no longer an inner tube which can carry the valve core. The problem has been overcome by fitting a separate tyre valve which passes through a close-fitting hole in the rim, and which is secured by a nut and locknut. The valve is fitted from the rim well, and it follows that the valve can be removed and replaced only when the tyre has been removed from the rim. Leakage of air from around the valve body is likely to occur only if the sealing seat fails or if the nut and locknut become loose.

2 The valve core is of the same type as that used with tubed tyres, and screws into the valve body. The core can be removed with a small slotted tool which is normally incorporated in plunger type pressure gauges. Some valve dust caps incorporate a projection for removing valve cores. Although tubeless tyres seldom give trouble, it is possible for a leak to develop if a small particle of grit lodges on the sealing face. Occasionally, an elusive slow puncture can be traced to a leaking valve core, and this should be checked before a genuine puncture is suspected.

3 The valve dust caps are a significant part of the tyre valve assembly. Not only do they prevent the ingress of road dirt into the valve, but also act as a secondary seal which will reduce the risk of sudden deflation if a valve core should fail.

18 Wheel balancing

1 The front wheel should be statically balanced, complete with tyre. An out of balance wheel can produce dangerous wobbling at high speed.

2 Some tyres have a balance mark on the sidewall. This must be positioned adjacent to the valve. Even so, the wheel still requires balancing.

3 With the front wheel clear of the ground, spin the wheel several times. Each time, it will probably come to rest in the same position. Balance weights should be attached diametrically opposite the heavy spot, until the wheel will not come to rest in any set position, when spun.

4 Balance weights are available from Kawasaki dealers in 10, 20 and 30 gram values. Note that in the case of 'H' models equipped with Bridgestone tyres, special weights **must** be used. These have a shorter clip section to prevent air leakage, and are identified by a circle around the weight marking. On no account use the earlier type on these machines.

5 To fit the balance weights, clip the hooked end over the bead and tap the weight home. Note that it may prove necessary to deflate the tyre slightly to allow the clip section to fit over the rim edge.

6 Although the rear wheel is more tolerant of out-of-balance forces than is the front wheel, ideally this too should be balanced if a new tyre is fitted. Because of the drag of the final drive components the wheel must be removed from the machine and placed on a suitable free-running spindle before balancing takes place. Balancing can then be carried out as for the front wheel.

Chapter 6 Electrical system

Refer to Chapter 7 for information relating to the 1983 on models

Contents

Specifications

Battery

	Chain drive models	Shaft drive models
Make	Furukawa	Yuasa
Type	FB 12A-A	SYB14L-AZ
Voltage	12 volts	12 volts
Capacity	12 Ah	14 Ah
Earth (ground)	Negative (–)	Negative (–)

Alternator

Rated output	17A ⊛ 10 000 rpm, 14 volts
Type	3-phase

Bulbs

Voltage	12 volts	12 volts
Wattages:	**US**	**UK**
Headlamp	60/55W	60/55W
Tail/brake	8/27W	5/21W
Parking (city) lamp	3.4W	4W
Turn signal	23W	21W
Turn signal/running lamp	23/8W	N/App
Instrument illumination	3.4W	3.4W
Warning lamps	3.4W	3.4W

Torque settings

Component	kgf m	lbf ft	lbf in
Alternator rotor bolt – 10 mm (see Chapter 1, Section 39)	7.00	51.0	–
Alternator rotor bolt – 12 mm (see Chapter 1, Section 39)	13.0	94.0	–
Alternator stator bolts	0.80	–	69.0
Neutral switch	1.50	11.0	–
Oil pressure switch	1.50	11.0	–
Starter motor and cover screws	0.55	–	48.0
Starter motor terminal nut	1.10	–	95.0
Turn signal mounting nuts – R1, P1	1.30	–	113.0

1 General description

The electrical system is based on a three-phase alternator mounted on the left-hand end of the crankshaft, the output from which is rectified and controlled by an electronic regulator/rectifier unit before being passed to the battery and main electrical circuit.

2 Electrical system: general information and preliminary checks

1   In the event of an electrical system fault, always check the physical condition of the wiring and connectors before attempting any of the test procedures described here and in subsequent Sections. Look for chafed, trapped or broken electrical leads and repair or renew these as necessary. Leads which have broken internally are not easily spotted, but may be checked using a multimeter or a simple battery and bulb circuit as a continuity tester. This arrangement is shown in the accompanying illustration. The various multi-pin connectors are generally trouble-free but may corrode if exposed to water. Clean them carefully, scraping off any surface deposits, and pack with silicone grease during assembly to avoid recurrent problems. The same technique can be applied to the handlebar switches.
2   A sound, fully charged battery is essential to the normal operation of the system. There is no point in attempting to locate a fault if the battery is partly discharged or worn out. Check battery condition and recharge or renew the battery before proceeding further.
3   Many of the test procedures described in this Chapter require that voltages or resistances be checked. This requires the use of some form of test equipment, and this in most cases can be a simple and inexpensive multimeter of the type sold by electronics or motor accessory shops.

If you doubt your ability to check safely the electrical system entrust the work to a Kawasaki dealer. In any event have your findings double checked before consigning expensive components to the scrap bin.

3 Battery: examination and maintenance

1   To check the battery thoroughly it is best to disconnect it and remove it from the machine. It is housed in a recess below the dualseat and is retained by a single metal strap. Batteries can be dangerous if mishandled. See 'Safety First' and note the precautions described for handling them. Wear overalls or old clothing in case of accidental acid spillage. Clean the outside of the battery carefully, and remove any deposits from the terminals, which should be coated with petroleum jelly prior to installation.
2   When new, the battery is filled with an electrolyte of dilute sulphuric acid having a specific gravity of 1.280 at 20°C (68°F). Subsequent evaporation, which occurs in normal use, can be compensated for by topping up with distilled or demineralised water only. Never use tap water as a substitute and do not add fresh electrolyte unless spillage has occurred.
3   The state of charge of a battery can be checked using a hydrometer.
4   The normal charge rate for a battery is between 10% and 30% of its rated capacity, thus for a 12 volt 12 ampere-hour unit charging should take place at 1.2 - 3.6 amp. Exceeding the higher figure could cause the battery to overheat, buckling the plates and rendering it useless. Few owners will have access to an expensive current controlled charger, so if a normal domestic charger is used check that after a possible initial peak, the charge rate falls to a safe level. If the battery becomes hot during charging **stop**. Further charging will cause damage. Note that cell caps should be loosened and vents unobstructed during charging to avoid a build-up of pressure and risk of explosion.
5   After charging, top up with distilled water as required, then check the specific gravity and battery voltage. Specific gravity should be above 1.250 and a sound, fully charged battery should produce 15 - 16 volts. If the recharged battery discharges rapidly if left disconnected it is likely that an internal short caused by physical damage or sulphation has occurred. A new battery will be required. A sound item will tend to loose its charge at about 1% per day.

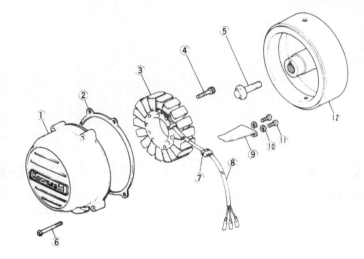

Fig. 6.1 Alternator

1	Casing	7	Grommet
2	Gasket	8	Wiring
3	Stator	9	Wiring clamp
4	Allen bolt – 3 off	10	Spring washer – 2 off
5	Bolt	11	Screw – 2 off
6	Screw	12	Rotor

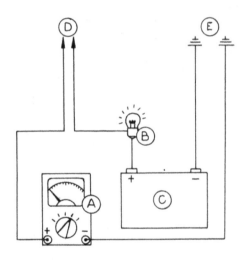

Fig. 6.2 Simple testing arrangement for checking the wiring

A	Multimeter	D	Positive probe
B	Bulb	E	Negative probe
C	Battery		

4 Checking the charging system

1   Before attempting to test the output of the charging system, check the wiring connections and battery condition (Sections 2 and 3). Open or remove the dualseat to gain access to the battery terminals. Set the multimeter on the 0-20 volts dc scale, and attach the negative (−) probe to the negative terminal and the positive (+) probe to the positive terminal. Start the engine, and note the meter reading at about 4000 rpm. If all is well a reading of about 14.5 volts should be indicated.
2   A reading significantly in excess of 14.5 volts indicates a possible defective regulator/rectifier unit or loose or broken wiring connections. Check these and repeat the test to find out if the problem has been resolved. At idle, battery voltage will be shown on the meter. If this does not increase as the engine speed rises the alternator or regulator/rectifier may be at fault or disconnected.

5 Checking the alternator

1 Trace and disconnect the three yellow alternator output leads. Set the meter to the 250 volts ac scale and connect the probes to any two of the output leads. Start the engine and measure the voltage at about 4000 rpm. Note the reading, then repeat the test until all combinations of leads have been checked (three tests in all). A reading of about 50 volts should be obtained in each case, in which case the fault must lie with the regulator/rectifier unit.

2 A reading significantly lower than that shown above indicates a fault in the alternator itself, and the alternator winding resistances should be measured to discover the nature of the fault. With the engine off, measure the resistance between each pair of leads, making three tests as described above. The multimeter should be set on the ohm x 1 scale. A sound winding will give a reading of 0.48 - 0.72 ohm. If infinite resistance is shown, the windings are open (broken), whilst a much lower reading or zero resistance indicates a short. In both instances the alternator stator must be renewed.

3 Set the meter on its highest resistance range, normally ohm x 1000 or kilo ohms, and check for insulation between each alternator lead and earth (ground). Anything less than infinity is indicative of a short between the stator core and its windings, again requiring renewal.

6 Checking the regulator/rectifier unit

1 Check that the ignition switch is off and remove the left-hand side panel. Release the electrical panel cover to gain access to the regulator/rectifier, a finned alloy unit mounted on the underside of the battery carrier. Trace and disconnect the red/white lead and the six-pin connector from the unit.

2 Set the multimeter on the ohm x 10 or ohm x 100 scale, and measure the resistance between the red/white lead and each of the three yellow leads. Note the reading, then reverse the meter probes and repeat. Once these six tests have been completed, repeat the sequence using the black lead in the connector in place of the red/white lead, another six tests. In each test, a very high resistance should be shown in one direction, with a very low reading if the meter probes are reversed. The actual resistance figures are not important, but a large difference between the two readings indicates that that particular diode is functioning normally. If with any pair of leads a similar reading is shown in both directions, a diode has failed and the unit must be renewed.

3 To check the regulator unit, three 12 volt car or motorcycle batteries and a 12 volt bulb rated at 3-6W will be required. Using the accompanying illustration for reference, connect the topmost battery with the test lamp as shown that is, with the positive (+) battery terminal to one of the yellow leads and the negative (–) terminal to the black lead via the test bulb. The bulb should remain off at this stage.

4 Connect the remaining two batteries in series as shown to produce a 24 volt source, connecting the positive (+) terminal to the brown lead. Read the notes below before connecting the negative (–) lead. **Important note** Do not use a meter or bulb of a different wattage in place of the test lamp specified. It acts both as an indicator and as a current limiter. On no account apply more than 24 volts to the unit and do not apply even this voltage for more than a few seconds. The unit may be destroyed if these precautions are not observed.

5 Touch the negative (–) lead from the 24 volt source **briefly** against the black terminal of the connector. If the regulator stage is functioning normally, the test lamp should light. Repeat the test with the 12 volt positive (+) lead connected to each of the remaining yellow leads in turn. It should be noted that whilst the above tests will usually reveal a regulator fault, the sequence is not infallible. If the tests indicate a sound unit, no other charging system faults can be found but the problem persists, it will be necessary to check the unit by substitution.

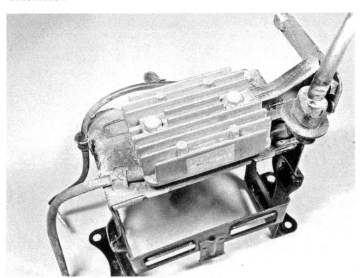

6.1 Regulator/rectifier is mounted on underside of battery tray

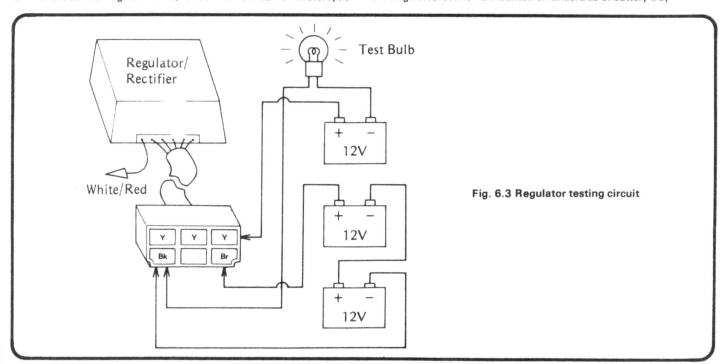

Fig. 6.3 Regulator testing circuit

7 Fuses: location and renewal

1   The electrical system is protected by between two and five fuses, depending on the model and market designation. Refer to the wiring diagrams at the end of the Chapter for details of specific arrangements. The fuses are housed in a plastic holder housed behind the left-hand side panel and electrical panel cover.

2   The function of a fuse is to introduce an intentional weak link in a circuit so that it will 'blow' before the expensive electrical components are damaged. Fuses do age and may occasionally fail through this or vibration, but the system must be checked for possible faults if the replacement fails soon afterwards. Avoid using fuses of the wrong rating unless circumstances make this unavoidable. Remember that using the wrong fuse or wrapping the blown fuse in metal foil will **not** protect the electrical system. Fit the correct fuse as soon as possible, and replace any spare fuse used at the same time.

8 Starter circuit tests

1   The starter system is of robust construction and will rarely malfunction. In the event of a fault, always check that the battery is in good condition noting that a failing battery may operate the general electrical system adequately whilst being unable to produce the heavy starting current. Look also for broken, chafed or corroded wiring before proceeding further.

Starter solenoid (relay)

2   Remove the left-hand side panel and electrical panel cover to gain access to the relay, then detach the heavy starter motor lead from the relay terminal. Set the multimeter to the ohm x 1 scale and connect one lead to the starter motor lead terminal and the other to earth (ground). Switch on the ignition, pull in the clutch lever and press the starter button. An audible click from the relay should be accompanied by a zero ohm reading if the unit is sound.
   Note: On machines equipped with a side-stand interlock switch, make sure that the stand is retracted during the check.

3   If the relay clicks normally but the meter still indicates that the contacts are open, the unit must be considered defective and renewed. If there is no sign of activity from the relay check that the starter switch circuit is operating correctly.

Starter switch circuit

4   Disconnect the black lead and the yellow/red lead from the relay. Set the multimeter on the 0-20 volts dc scale, and connect the negative (–) probe to the yellow/red lead and the positive (+) probe to the black lead. Switch on the ignition, pull in the clutch lever and press the starter button. If battery voltage is shown, but the relay would not work when tested as described above, it should be considered defective and renewed. It is a sealed unit and cannot be repaired. If battery voltage is not shown, check the wiring, connections and interlock switches to locate the fault.

Starter button test

5   Remove the fuel tank (Chapter 2) and disconnect the 4-pin connector and the separate black lead from the right-hand switch cluster. Set the meter on the ohm x 1 scale and connect the probes to the yellow and the black leads. Press the starter button and check that zero resistance is indicated. If not, the starter button contacts are faulty and should be cleaned by spraying aerosol contact cleaner into the switch housing. If this fails, renew the switch cluster.

Clutch interlock switch test

6   Remove the fuel tank (Chapter 2) and disconnect the two black leads from the clutch interlock switch. Set the multimeter to the ohm x 1 and test for continuity when the clutch lever is pulled in. If a reading of zero ohms is not obtained, the switch is defective and should be renewed. The switch can be freed by depressing its locating pin with a small screwdriver and pulling it out of the clutch lever housing. In an emergency, bypass the switch by joining the two black leads together. Renew the switch as soon as possible.

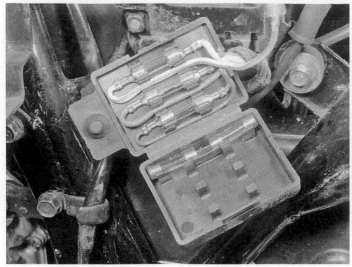

7.1 Fuse holder is housed behind left-hand side panel

8.2 Starter solenoid (relay) location

9 Starter motor: removal, overhaul and installation

1   In the event of a starter fault, check first that the battery is fully charged and that the solenoid is operating correctly (Section 8). Remove the engine sprocket cover or front gear case cover as appropriate. Remove the two starter motor mounting bolts and pull the motor clear of the crankcase. Once it is partly clear, turn the motor body to gain access to the terminal. Slide back the rubber boot and remove the nut to free the starter cable.

2   Place the motor on a clean workbench and dismantle it, following the photographic sequence which accompanies this Section. Lay out each part in sequence as a guide during reassembly. Clean the motor components using a non-greasy high flash-point solvent. It will be noted that the motor shown in the photographs is of the later, four brush type. Earlier models made use of a two brush motor of similar construction, and this may be dismantled as follows.

3   Remove the two long retaining screws and lock washers and remove the starter motor end covers, leaving the planetary gears in place in the right-hand cover. Remove the thin gasket, end plate, thick gasket, thrust washers and armature from the right-hand end of the motor. Moving to the left-hand end, release the screw which retains the field coil lead to the brush plate and remove the plate and brushes. **Do not** attempt to remove the field coil windings from the motor body.

Two brush motors

4 Pull back and displace the brush springs to allow the carbon brushes to be measured. Renew them if they are 6 mm (0.236 in) or less in length. The brush springs should exert a pressure of 560 - 680 grams. In practice, the brush springs can be considered serviceable if they press firmly on the brushes. Clean the commutator surface with fine abrasive paper to restore a smooth, polished surface. Clean out the grooves between the commutator segments, using a hacksaw blade ground to the correct width. Each segment should be straight sided, with an undercut of 0.5 - 0.8 mm (0.020 - 0.032 in). If the depth of undercut is less than 0.2 mm (0.008 in) the armature should preferably be renewed. It is possible to re-cut the grooves using the modified hacksaw blade mentioned above, but this requires care and patience. Do not cut into the segment material, or leave the groove anything other than square sided. See the accompanying illustration for details.

5 Set the multimeter on the ohm x 1 scale, then check the resistance between each commutator segment and its neighbour. A very high or infinite resistance indicates an open circuit and the armature must be renewed. Next, set the multimeter on its highest resistance scale, normally ohm x 1000 (kilo ohms) and check the resistance between the armature core (shaft) and each of the commutator segments. There should be no conductivity shown in this test, any reading indicating a partial or complete short circuit, again necessitating renewal.

6 Place the motor body on the workbench with the brushes towards you and the starter motor lead terminal to the right-hand side. With the meter on the ohm x 1 range, check for continuity between the positive (+) brush (located on the left-hand side, opposite the terminal) and the terminal. If a reading close to zero ohms is not obtained, the field coil windings are open (broken) and the assembly should be renewed. Now set the meter on its highest range and check for resistance between the positive (+) brush and the motor body. Anything other than infinite resistance indicates a short circuit, again requiring renewal.

Four brush motors

7 These should be dealt with as described above, except for the following points. Brush spring tension has been increased to 740 - 860 grams, but again in practice the springs can be considered acceptable if they bear firmly upon the brushes. Measure the resistance between each positive (+) brush (attached to the motor body) and the body, with the meter set on the ohm x 1000 scale. No reading should be shown. Next, set the meter to ohm x 1 and measure the resistance between the two positive (+) brushes. Unless the reading is at or close to zero ohms, renew the brushes and leads to correct the open (broken) circuit.

8 Moving to the brush plate and the negative (–) brushes, set the meter on the ohm x 1 scale and check for resistance between the two negative brushes. If a high or infinite resistance is shown, the brush plate assembly should be renewed. Set the meter on the ohm x 1000 scale and measure the resistance between each brush holder and the brush plate. There should be no conductivity between the two, any reading indicating the need for renewal.

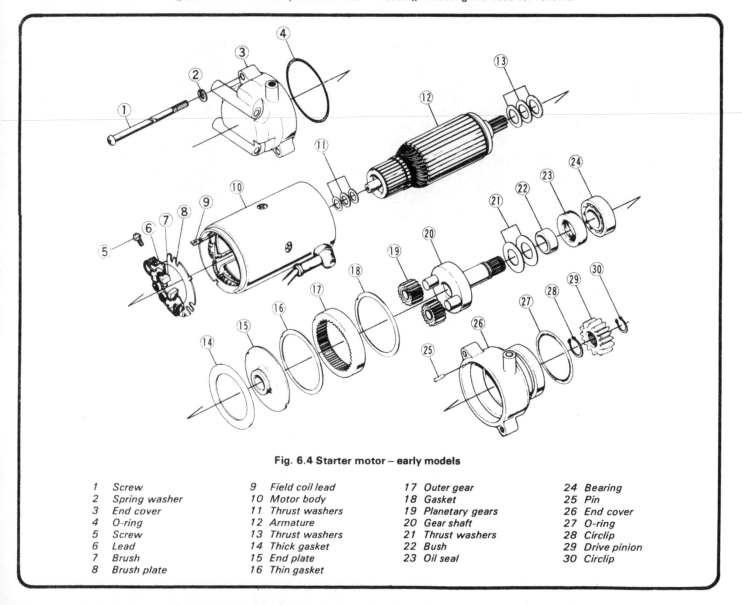

Fig. 6.4 Starter motor – early models

1	Screw	9	Field coil lead	17	Outer gear	24	Bearing
2	Spring washer	10	Motor body	18	Gasket	25	Pin
3	End cover	11	Thrust washers	19	Planetary gears	26	End cover
4	O-ring	12	Armature	20	Gear shaft	27	O-ring
5	Screw	13	Thrust washers	21	Thrust washers	28	Circlip
6	Lead	14	Thick gasket	22	Bush	29	Drive pinion
7	Brush	15	End plate	23	Oil seal	30	Circlip
8	Brush plate	16	Thin gasket				

9.2a Remove screws and lift clear the reduction drive

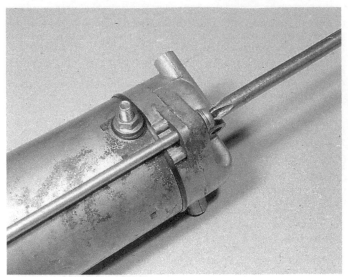

9.2b Gear teeth should be greased during overhaul

9.2c Remove the brush cover ...

9.2d ... noting shim and washer arrangement

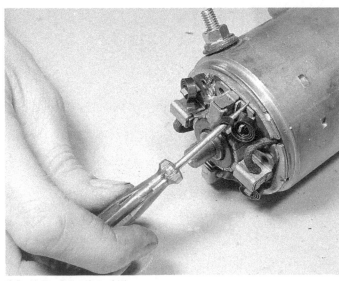

9.2e Unhook brush springs ...

9.2f ... and disengage as plate assembly is removed

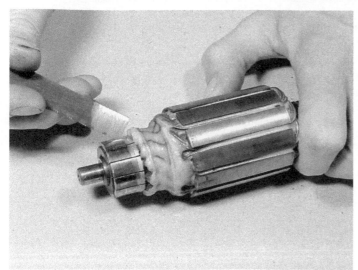

9.4a Undercutting segments using ground-down hacksaw blade

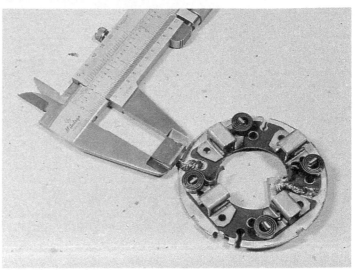

9.4b Check brush length and renew if at or near limit

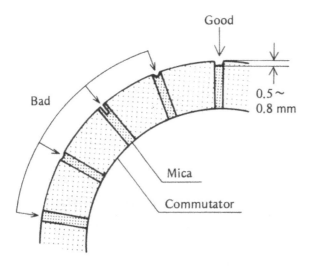

Fig. 6.5 Correct re-cutting of commutator grooves

10 Neutral switch: location and testing

1 Remove the engine sprocket cover or front gear case cover as appropriate. Disconnect the neutral switch lead and unscrew the switch from the crankcase. Set the multimeter to the ohm x 1 scale and measure the resistance between the neutral switch terminal and the spring-loaded pin. If the switch is in good condition there should be little or no resistance. Set the meter to the 20 volts dc scale and connect the positive (+) probe to the switch lead and the negative (−) probe to earth (ground). Switch on the ignition and note the meter reading. If battery voltage is not shown, a fault lies in the neutral switch wiring or the indicator bulb has failed.

11 Ignition switch: location and testing

1 The combined ignition switch and steering lock unit is mounted on the steering head, to which it is secured by two bolts. To test the switch remove the headlamp unit and separate the 6-pin connector from the switch. Using a multimeter set on the ohm x 1000 range as a continuity tester, check the various switch terminal connections in each of the three positions (the 'Lock' position is irrelevant in this context), as shown in the relevant wiring diagram. If the switch is not working correctly it will be necessary to obtain and fit a new unit. No repair is possible. The switch is retained by two Allen bolts to the underside of the top yoke, and can be removed after the headlamp has been detached to gain access.

10.1a Neutral switch location (arrowed)

10.1b Contact should protrude as shown on inside of casing

12 Headlamp and reserve lighting system: testing

1   The US H1, H2 and H3, the E2 and E3, and the R1 models together with the shaft drive N1 model employ a reserve lighting system which automatically switches in the remaining bulb filament in the event of headlamp failure. For details of the headlamp and lighting system of all other models, see Section 14.

2   To check the operation of the reserve lighting system, first release the headlamp unit by removing the retaining screws. Disconnect the headlamp connector and make up three insulated test leads which should be connected between the headlamp and the wiring connector as shown in the accompanying illustration. Set the dipswitch to the low beam position and turn on the ignition switch to operate the lights. Disconnect the test lead from the red/yellow low beam lead and check that main beam comes on, though more dimly than normal, and that the headlamp failure warning lamp is lit. Reconnect the test lead and switch to main beam. Disconnect the test lead from the red/black main beam lead and check that low beam comes on together with the failure warning lamp.

3   To test the dipswitch, remove the fuel tank and disconnect the red/black, red/yellow, blue/yellow and blue leads from the dipswitch. Use the multimeter as a continuity tester to check that the appropriate switch terminals are connected in the two switch positions, as shown in the relevant wiring diagram. If the switch proves faulty, dismantle and clean the contacts using fine abrasive paper and aerosol contact cleaner. If this fails, renew the switch unit. For details on testing the reserve lighting unit, see Section 13.

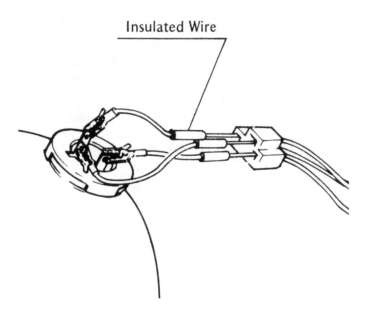

Insulated Wire

Fig. 6.6 Reserve lighting operation test

13 Reserve lighting unit: testing

1   Unlock and open the seat and remove the right-hand side panel. Separate the 6-pin connector from the reserve lighting unit. Set the meter on the 20 volts dc range, connecting the negative (–) probe to earth (ground). Use the positive (+) probe to test the wiring to the unit as described below.

System fails to select remaining filament after one has failed
2   Check the voltage on the blue/orange lead. If about 12 volts is shown, both bulb filaments have failed or the black/yellow lead is broken or disconnected. If less than 12 volts is shown, check the voltage on the blue lead. If about 12 volts is shown the reserve lighting unit is defective. If 0 volts is shown the ignition switch or wiring is faulty.

Both bulb filaments are selected
3   Check the voltage on the blue/orange lead. If 12 volts, the reserve lighting unit is defective. If 0 volts, the dipswitch or wiring is at fault.

Main beam is not dimmed when selected after low beam failure
4   Check the voltage on the red/black lead. If 12 volts, the reserve lighting unit has failed. If 0 volts, the wiring harness is broken or disconnected.

Failure warning lamp inoperative
5   Check the voltage on the light green/red lead. If 12 volts, the bulb has failed or is disconnected. If zero volts, the reserve lighting unit has failed.

14 Headlamp and lighting system: testing

1   All models except those covered by Section 12 are equipped with a conventional manual lighting system. In the event of a failure, check the operation of the lighting switch and dipswitch as follows. Remove the fuel tank and disconnect the brown/white lead, brown lead and blue lead from the lighting switch. Using a multimeter set on the resistance scale (ohms) as a continuity tester, check that the appropriate switch terminals are connected at the various switch positions, referring to the relevant wiring diagram.

2   The dipswitch can be checked in the same way, this time tracing the red/black, blue and red/yellow leads and checking the switch contacts. A fault in the headlamp flasher, or pass, switch can be traced by checking the red/black and brown leads and the switch for continuity.

15 Headlamp bulb/sealed beam unit: renewal and adjustment

1   The headlamp unit comprises a lens and reflector unit, either round or rectangular, secured in the headlamp shell by two screws. To free the unit from the shell, remove the two short screws which pass through the lower edge of the rim, noting that the alignment screw(s) should not be disturbed. Lift and disengage the unit and pull off the wiring connector, and on European models, the front parking lamp or 'city lamp' lead.

2   Pull off the rubber boot from the back of the bulb and release the bulb retainer. Depending on the type of bulb used a ring-type retainer may be fitted, this being a bayonet fitting, or a wire spring retainer may be used. Free the retainer and remove the bulb.
   Note: A quartz halogen bulb is fitted. Never touch the quartz glass envelope with the hands; it will be damaged by oil or skin acids. When fitting a new bulb, check that it is positioned correctly and secure the retainer. Refit the rubber boot, ensuring that the moisture drain channel faces downward and that it seats fully against the back of the reflector.

3   Refit the headlamp in the shell and check that the horizontal alignment screw in the front edge of the rim is set so that the beam shines straight ahead. If vertical adjustment is necessary, slacken the headlamp mounting nuts inside the shell and the vertical adjustment nut below the unit. Angle the headlamp to comply with local legislation, noting that the rider should be seated normally, then secure the nuts.

4   European models are equipped with a low wattage parking (city) lamp incorporated in the headlamp unit. This can be renewed when the headlamp unit has been detached as described above. Pull out the bulbholder from the rubber grommet which retains it to gain access to the bayonet fitting bulb.

16 Tail/brake lamp: bulb renewal

1   Release the lens retaining screws and remove the lens. The bulb or bulbs are removed by pushing inwards slightly and then twisting anti-clockwise to free the bayonet pins. Always fit a replacement bulb of the correct wattage. When refitting the lens, check that the seal is in place and take care not to overtighten the retaining screws.

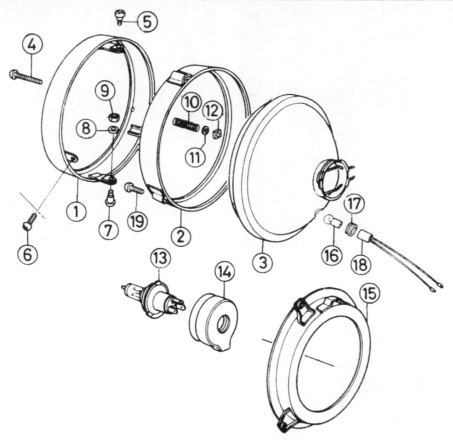

Fig. 6.7 Headlamp – KZ/Z750 L, H, N and E

1	Outer rim	6	Screw – 2 off	11	Washer	16	Parking lamp bulb
2	Inner rim	7	Screw	12	Nut		– UK models
3	Reflector	8	Washer – 2 off	13	Headlamp bulb	17	Grommet
4	Adjusting screw	9	Nut – 2 off	14	Bulb retainer	18	Bulbholder
5	Screw	10	Spring	15	Backing ring		

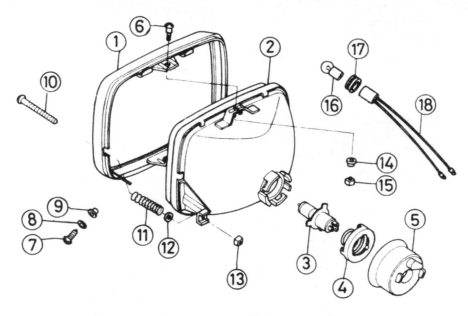

Fig. 6.8 Headlamp – KZ/Z750 R1 and P1

1	Rim	6	Screw – 2 off	11	Spring	16	Parking lamp bulb –
2	Reflector	7	Screw – 2 off	12	Washer		UK models
3	Headlamp bulb	8	Spring washer – 2 off	13	Nut	17	Grommet
4	Bulb retainer	9	Collar – 2 off	14	Collar – 2 off	18	Bulbholder
5	Rubber boot	10	Adjusting screw	15	Nut – 2 off		

15.3a Disconnect headlamp wiring connector

15.3b Pull off rubber boot to gain access to bulb

15.3c Release retainer to free bulb. **Do not** touch quartz envelope

15.4 Horizontal adjustment screw

15.5 European models have parking (city) lamp

16.1 Note that some models employ twin tail/brake bulbs

17 Tail/brake lamp circuit: general description and testing

1 The H1 and H2 models were fitted with a brake light failure warning circuit which uses the low fuel level warning lamp to indicate failure of the brake lamp filament or a broken connection in the brake lamp circuit. All other models employ twin bulbs in the tail lamp to obviate the need for the warning circuit. In the event of failure of the brake lamp in either system, first check that the bulb has not blown (see Section 16) then proceed as follows.

2 To check that the front brake switch is functioning normally, remove the headlamp unit from the shell and disconnect the brown/blue leads from the switch. Connect the probes from a multimeter set on the ohm x 1 scale to the switch leads and check that no resistance is shown when the lever is operated. If the switch is faulty it must be renewed, the sealed construction precluding repair or cleaning. To free the switch use a small screwdriver to push in the locating tab via the hole in the underside of the lever assembly.

3 The rear brake switch is located inboard of the right-hand footrest plate and is operated via a spring by the brake pedal. The operation of the switch can be checked as described above, noting that it is possible to adjust the position of the switch, and thus the point at which it comes oh, by moving it up or down in relation to the mounting bracket.

4 The brake light failure warning circuit (H1 and H2 only) is controlled by a switch unit housed behind the left-hand side panel. If the system is working normally, the warning lamp should come on when the brake is applied and go off when it is released. If the brake lamp bulb is blown or disconnected, the warning lamp should still come on when the brake is applied, but will flash when released.

5 If a fault occurs, it is preferable to check the switch unit by substituting a new item. If this cannot be arranged, check the system wiring and switch unit with a multimeter, referring to the accompanying illustration for details of meter settings, connections and appropriate readings. Remember to turn the ignition switch off before taking resistance readings.

18 Turn signal and hazard warning circuit: testing

1 In the event of a fault in the turn signal system, check that a blown bulb is not the cause of the problem before proceeding further. The lenses are retained by two screws, and the bulb is of the bayonet fitting type. Note that the US/Canadian market models employ twin filament bulbs in the front lamps, which double as running lights. It is important to ensure that the replacement bulb is of the correct wattage (see Specifications) or the unbalanced load on the system will cause erratic operation.

2 As a general guide, turn signal problems affecting the whole system are usually attributable to the relay, wiring or switch, whilst if the fault is confined to one side, the relay and its supply can be considered sound. In this case examine the bulbs, lamp wiring and switch for faults. If the system fails totally, check first that the battery is in good condition and fully charged, and that all wiring and connections are sound.

3 Remove the left-hand side panel and open the electrical panel cover. Pull off the brown lead and orange/green lead from the relay, and check the resistance between the two relay terminals. This should be close to zero ohms if the relay is sound. If a higher resistance is found, renew the relay.

4 If the relay is sound, set the multimeter to the 20 volt dc range and connect the positive (+) probe to the brown lead and the negative (−) probe to the orange/green lead. Switch on the ignition and note the meter reading when the turn signal switch is moved to the 'L' and 'R' positions. If battery voltage is not shown in both positions check the switch contacts and wiring. If no voltage is shown in either position, check the fuse, ignition switch and wiring.

5 If both turn signal lamps come on when selected, but do not flash, or flash very slowly, check that the battery is charged and that all wiring connections are secure. Next, check that the bulbs are of the correct wattage. If the above checks fail to reveal the cause of the fault, renew the relay.

6 If only one of the two lamps comes on, and fails to flash or flashes weakly, check that the inoperative bulb is not blown and is of the correct wattage. Other possible causes are broken or disconnected wiring or a poor earth connection. If neither lamp comes on, check that the switch and switch wiring is sound.

7 The lamps should flash at 60-120 flashes per minute. If this speeds up excessively, the regulator/rectifier may be faulty, the relay may be faulty or the bulb wattages incorrect.

Hazard warning system

8 A hazard warning system is fitted to the E1 and E2 models, the H1 and H2 models, the US/Canadian E3, H3 and R1 models and the N1. In the event of a suspected fault, start by testing the turn signal system as described above.

9 The hazard warning switch can be checked by tracing the wiring from the left-hand switch cluster to the 6-pin or 9-pin connector

Meter Range	Connections	Brake	Reading
20V DC	Meter (+) Lead → Blue Lead / Meter (−) Lead → Chassis Ground	Apply	Battery Voltage
		Release	0 V
	Meter (+) Lead → Green/White Lead / Meter (−) Lead → Chassis Ground	———	Battery Voltage
x 1 Ω	Black/Yellow Lead ↔ Chassis Ground	———	0 Ω

Meter Range	Connections	Brake	Reading
20V DC	Meter (+) Lead → Yellow Lead / Meter (−) Lead → Chassis Ground	Apply	Battery Voltage
		Release	0 V
	Meter (+) Lead → Green/White Lead / Meter (−) Lead → Chassis Ground	Apply	0 V
		Release	Battery Voltage

Fig. 6.9 Brake light failure warning system test table – KZ/Z750 H1 and H2

beneath the fuel tank. Separate the connector and check the switch connections using a multimeter as a continuity tester. See the wiring diagram for details of switch position.

10 The hazard warning relay is located on the electrical panel, just forward of the turn signal relay. To check the relay, disconnect the leads and measure the resistance across the two terminals. If this is significantly lesser or greater than 60 ohms, renew the relay.

11 If the switch and relay prove to be serviceable, check for battery voltage between the grey lead from the main harness and the grey lead from the handlebar switch with the ignition switch at the 'Park' and 'On' positions and the hazard switch on. Repeat the test with the meter connected to the green lead from the switch unit as described above. If battery voltage is not shown in both positions, check the fuse, hazard warning switch and wiring.

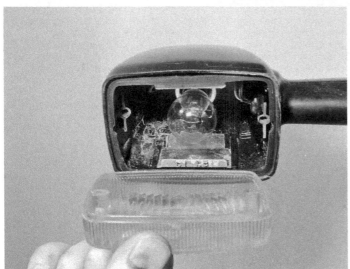

18.1 Turn signal bulb renewal

18.3 Turn signal relay is located in electrical panel

19 Automatic turn signal cancelling system: testing

Distance sensor

1   Remove the headlamp unit and trace the distance sensor wiring from the speedometer, separating it at the four-pin connector. Release the lower end of the speedometer drive cable by unscrewing the knurled retaining ring. Connect multimeter probes to the red lead and

light green sensor lead and select one of the ohms scales to check continuity. Slowly rotate the speedometer inner cable and note how many times the sensor switches on and off during each revolution. If the sensor is working correctly, there should be four on/off pulses per revolution. Failing this, it will be necessary to obtain a new speedometer or to remember to switch the turn signals off manually. No repair is possible.

Turn signal and selector switches

2   Remove the fuel tank and trace the wiring from the left-hand switch cluster back to the 9-pin connectors. Refer to the switch diagrams shown in the main wiring diagrams at the end of this Chapter and check the continuity of the turn signal switch contacts, and also those of the selector switch (H2 model only). If an open or short circuit is discovered, check the switch wiring for damage and dismantle and clean the switch contacts. If the problem persists, renew the left-hand switch cluster. Note that individual parts are not available for the switch clusters, making repair impracticable.

3   To check the operation of the solenoid which resets the turn signal switch, connect a test lead to the battery positive (+) terminal. Set the turn signal switch to the left or right, then touch momentarily the test lead on the white/green lead to the solenoid. If the solenoid fails to reset the switch from both the left and right positions, renew the switch assembly. Note that battery voltage should not be applied to the solenoid for more than a second or so, or the solenoid windings may be burned out.

System wiring checks

4   Remove the right-hand side panel and locate the 6-pin connector which will be found below the battery tray and the regulator/rectifier unit. Using the accompanying table, check the voltage on the various leads with the switches set as described. Note that this test sequence requires the system to be intact, with all connections made normally. If a discrepancy is noted, make a careful check of all connectors and leads. If this fails to resolve the fault, renew the control unit. This is located to the front of the battery tray.

20 Low fuel level warning system: testing

H1 model

1   In the event of a fault check that the battery is fully charged, and that the brake light failure circuit is functioning normally. This latter check is necessary because the same warning lamp is used for both circuits.

2   If the fuel level is low but the warning lamp fails to come on, trace and disconnect the sender leads at the 2-pin connector below the fuel tank. Check the voltage on the sender leads on the harness side of the connector. With the multimeter set to the 20 volts dc range, connect the positive (+) probe to the green/white lead and the negative (−) probe to the black/yellow lead. If, with the ignition switched on, battery voltage is indicated, the fault lies with the sensor. If no reading is shown, check the wiring for breaks or loose connections.

3   If the warning lamp stays on irrespective of the amount of fuel in the tank, and the brake light failure circuit is sound, the sender unit can be assumed to be faulty. Intermittent flashing of the warning lamp is usually due to fuel surge in the tank during braking or acceleration. Other possible causes are damaged wiring earthing against the frame or a regulator/rectifier fault.

4   If the sender unit is to be renewed, remove and drain the fuel tank (see Chapter 2), then release the two screws which retain it to the underside of the tank. Use a new O-ring when fitting the new unit and check for leaks before refitting the tank.

E2, E3, H2, H3, L1 and L2 models

5   These models employ an identical fuel level warning system to that described above, but with the addition of a 'self-checker' unit which switches the warning lamp on for a few seconds when the ignition is turned on. This arrangement serves to indicate whether or not the warning lamp bulb is sound. After about 3 minutes, if the fuel level is low, the bulb will come on again to indicate this. Check the system as described above in the event of a fault, noting that if the self-checker unit is unplugged and the system works normally it should be renewed. The unit is housed behind the right-hand side panel and is connected to the system via a three-pin connector.

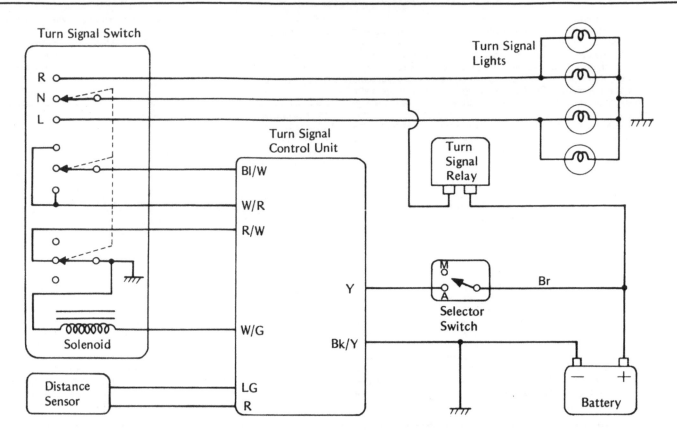

Fig. 6.10 Automatic turn signal cancelling system circuit

Meter Range	Connections*		Ignition Switch	Selector Switch Position	Turn Signal Switch Position	Reading
25V DC	Meter (+) →	Yellow, Blue/White	ON	A	Any (R, L, Neu.)	Battery voltage
			OFF	M	Any	0 V
	Meter (+) → White/Red		ON	A	R or L	Battery voltage
			OFF	M	Neutral	0 V

*Connect the meter negative (−) lead to ground.

Fig. 6.11 Automatic turn signal cancelling system wiring check table

21 Low fuel and oil level warning system: testing – N1

1   The N1 model employs an integrated fuel and oil level warning system. When the ignition is first switched on, a 'self-checker' unit automatically lights the two warning bulbs for about 3 seconds, and then switches them off. After a delay of up to three minutes, the self-checker unit switches in the two level sensors, which will light the appropriate warning bulb if the fuel or oil level is low.

2   In the event of a fault, check that the battery is fully charged and that the fuel tank level is well above the half-way mark. The engine oil level should also be checked and topped up if necessary. If either bulb fails to come on, check that it has not blown before moving on to the tests described below.

System wiring tests

3   Remove the right-hand side panel and locate the self-checker unit. It is mounted near the bottom of the battery holder. Disconnect the self-checker leads at the connector. Trace and disconnect the fuel level sender leads at the connector below the rear of the fuel tank. Make up a test lead from a length of insulated wire with both ends bared for about ½ inch. Switch on the ignition and use the test lead to connect the terminals described below, noting the response from the warning lamps. The object of these tests is to simulate the operation of the sensors by bridging the appropriate terminals with the test lead.

4   Starting with the 6-pin connector from the self-checker unit, bridge the male terminals of the green/white lead and black/yellow lead. If all is well, the warning lamp should come on. Now repeat the test on the same leads at the fuel level sensor. If the lamp still comes on, the fault lies with the fuel level sensor unit.

5   To check the oil level sensor circuit, connect the test lead between the male terminals of the yellow/blue lead and blue/red lead at the 6-pin connector. If the circuit is sound but the oil level switch is at fault, the warning lamp will come on. If in the above tests the warning lamp(s) fail to operate, check the wiring, connectors, bulbs and sensors.

6   The power supply to the self-checker unit can be checked using a multimeter on the 20 volts dc scale. Connect the positive (+) probe to the brown lead and the negative (−) probe to the black/yellow lead on the male terminal side of the 6-pin connector. Switch on the ignition and check that battery voltage is shown. If this is not the case, check the leads for damage.

Fuel level sensor tests

7   Remove and drain the fuel tank (Chapter 2) and release the two screws which retain the sensor to the underside of the tank. Fill a jar with fuel so that the cylindrical thermistor can be submerged in fuel for tests purposes, taking normal precautions to avoid fire risks. Connect the sensor to the main harness via its two-pin connector and place the thermistor in the jar of fuel. Switch on the ignition and observe the warning lamp, which should come on for about 3 seconds and then go off. Remove the jar of fuel to expose the sensor to air. After an interval of between 20 seconds and 3 minutes, the warning lamp should come on. If the sensor fails to operate normally and the rest of the system appears to function correctly, renew the sensor.

Oil level switch test

8   Drain the engine oil and remove the switch from the underside of the sump by releasing the two retaining bolts. Using a multimeter on the ohm x 1 scale, connect one probe to the switch lead and the other to the switch body. When held upright, a reading of infinite resistance should be shown, whilst when the switch is inverted, less than 0.5 ohm should be indicated. If the switch does not conform to the above, or works erratically, it must be renewed.

22 Computer warning system: general description – P1 and R1 models

1   The P1 and R1 models differ from the rest of the range in that a sophisticated microprocessor-controlled monitor system is incorporated in the instrument panel. A series of red LCD (liquid crystal display) segments warn the rider if the side stand is down, the oil level is low or the battery electrolyte level is low. A nine-segment display monitors fuel level, the bottom segment flashing to indicate that the fuel level is low. In addition to the above, a separate LED (light emitting diode) flashes to attract the rider's attention when any of the warning segments comes on.

2   The microprocessor monitors the system via sensors, and switches on the appropriate segments as necessary. It includes an automatic checking sequence which runs through the various system functions whenever the ignition is switched on.

23 Computer warning system: initial checks

1   If a fault develops in the system, always check the more obvious possible causes before dismantling anything. Start by switching on the ignition and watching the display panel during the automatic self-checking procedure. If nothing at all happens it can be assumed that the power supply to the system has failed. Refer to Section 24 for further details. If the self-checking sequence is completed normally, the fault can be assumed to lie in one of the sensors or the associated wiring. Refer to Section 25 for further details. If one or more of the LCD segments fail to come on, or work erratically the display panel and microprocessor may be at fault. Refer to Section 26 for further details.

24 Computer warning system: power supply tests

1   Turn the ignition switch off and remove the headlamp unit to gain access to the red 6-pin connector. Separate the connector, set the multimeter to the 25 volts (or 20 volts) scale, and connect the meter probes to the supply side of the connector as shown in the accompanying illustration. If the power supply conforms to the readings shown, but the display still fails to operate in the self-checking mode, refer to Section 26. If the power supply is not working, trace the brown lead and black/yellow lead back until the fault is located. If the fault appears

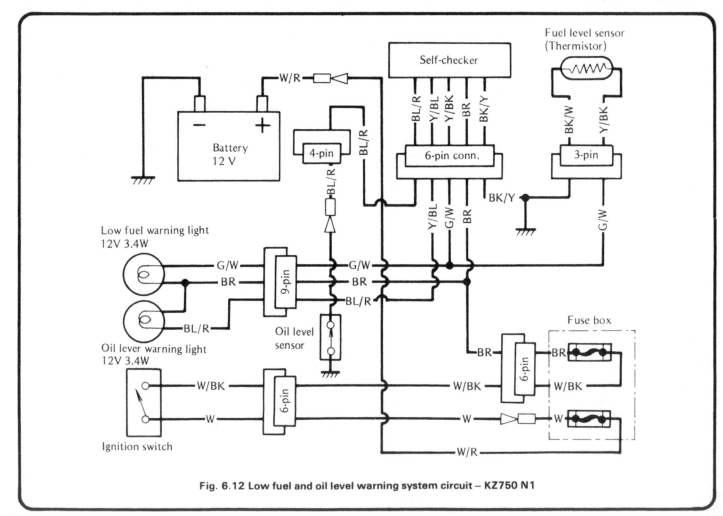

Fig. 6.12 Low fuel and oil level warning system circuit – KZ750 N1

to be confined to one sensor, leave the headlamp unit out and refer to Section 25.

25 Computer warning system: wiring and sensor tests

1 With the headlamp unit removed and the red 6-pin connector separated as described in Section 24, check that the kill switch is set to the 'Run' position. The following tests should be made on the female terminal (main harness) half of the connector. Note that the ignition switch must be **on** for the side stand switch test only. If the readings obtained conflict with those shown in the accompanying table, trace back through the wiring, checking for damaged wires or loose connectors.

Side stand switch

2 Set the multimeter to the ohm x 1 scale and connect one probe to each of the switch terminals. On UK and European models a reading of zero ohms should be shown when the stand is deployed with the weight of the motorcycle bearing on it, or in the fully retracted position. If the stand is down but with no weight on it, infinite resistance should be shown. In the case of US and Canadian versions, zero ohms should be shown with the stand retracted and infinite resistance when it is deployed. If the readings obtained do not agree with those described, check that the switch is mounted properly and that the operating spring has not been damaged. If the switch proves faulty it must be renewed, though in an emergency the two switch leads can be joined together as a 'get-you-home' measure.

Oil level sensor

3 The oil level sensor is a float-type switch and can be checked as described in Section 21, paragraph 8.

Battery electrolyte sensor

4 The battery is equipped with a sensor which consists of a small electrode fitted in place of one of the cell caps. If the electrolyte falls below a specific level the electrode is exposed to the air and the lack of current triggers the warning panel. Check that the electrolyte is above the minimum mark and that the sensor is fitted next to the arrow mark on the battery casing. Connect the positive (+) probe of the multimeter to the sensor lead and set the meter to the 10 volts dc range. Touch the negative (-) probe to earth (ground) and check that at least 6 volts is shown. If a low reading is obtained, remove the sensor and wash it with copious quantities of water. When dry, clean the electrode section with a wire brush, then refit it and perform the voltage check again. Note that if that particular cell of the battery fails, it may not be possible to persuade the sensor to operate correctly. Check the battery as described in Section 3, and renew it if necessary.

Fuel level sender

5 The sender unit comprises a variable resistance (rheostat) operated by a float on a pivoting arm. As the fuel level and float fall inside the tank, the microprocessor interprets the increased resistance and translates this to the LCD unit by switching off successive display segments.
6 To check the sender unit, remove and drain the tank, then release the sender unit by releasing the securing screws. Take care not to damage the sealing gasket or bend the float arm. Check that the float assembly moves smoothly and evenly with no signs of sticking. Set a multimeter on the appropriate resistance range and check the operation of the variable resistor. At the full (highest) position a reading of 1 – 5 ohm should be indicated. This should increase smoothly to a value of 103 – 117 ohm in the empty (lowest) position. If the sender does not conform to this, or is erratic in its operation, it should be renewed.

Meter Range	Connections	Meter Reading (Criteria)
25 V DC	○Meter (+) → Brown wire ○Meter (−) → Black/yellow wire	○0 V when ignition switch is off. ○Battery voltage when ignition switch is on.

Fig. 6.13 Computer warning system power supply testing table

Wire	Meter Range	Connections	Meter Reading (Criteria)
Side stand warner	25V DC	○Meter (+) → Green/white wire ○Meter (−) → Black/yellow wire	○Battery voltage when side stand is up. ○0 V when side stand is down.
Oil level warner	x 10 Ω	○One meter lead → Blue/red wire ○Other meter lead → Black/yellow wire	○Less than 0.5 Ω when engine oil level is higher than "lower level line" next to the oil level gauge. ○∞ Ω when engine oil level is much lower than the "lower level line".
Battery electrolyte level warner	10V DC	○Meter (+) → Pink wire ○Meter (−) → Black/yellow wire	○More than 6 V when electrolyte level is higher than "lower level line". ○0 V when electrolyte level is lower than "lower level line".
Fuel gauge and low fuel warner	x 10 Ω	○One meter lead → White/yellow wire ○Other meter lead → Black/yellow wire	○1 – 117 Ω

Fig. 6.14 Computer warning system wiring and sensor testing table

25.2 European-type side-stand switch

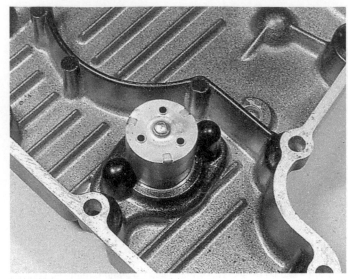
25.3 Oil level sensor is housed inside sump

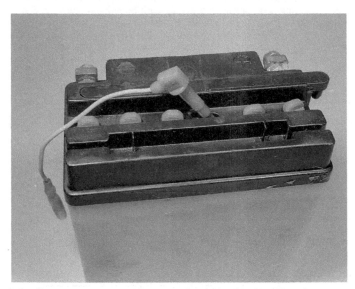
25.4 Note position of level sensor – it must be fitted to this cell

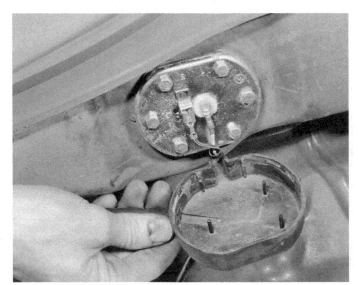
25.6a Remove tank, open cover and release bolts

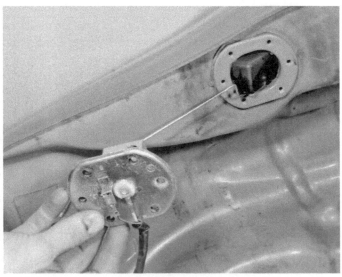
25.6b Sender unit can now be removed for testing

26 Computer warning system: microprocessor and display tests

1   Prepare a 12 volt power source (the machine's battery is ideal for this and may be left in situ if long test leads are used) and six test leads. Remove the headlamp and connect the test leads to the display side of the red six-pin connector as described below.
2   **Power supply.** Connect a test lead from the battery positive (+) terminal to the brown lead. Connect a test lead from the battery negative (-) terminal to the black/yellow lead.
3   **Side stand switch.** Connect a test lead from the battery positive (+) terminal to the green/white lead.
4   **Oil level sensor.** Connect a test lead from the battery negative (-) terminal to the blue/red lead.
5   **Battery level sensor.** Connect a test lead from the battery positive (+) terminal to the pink lead.
6   **Fuel level sensor.** Connect a test lead from the battery negative (-) terminal to the white/yellow lead.
7   Briefly disconnect and reconnect one of the power supply leads. The system should then go into the self-checking mode, after which all warning segments should go off leaving the fuel gauge display showing a 'full tank'. Now disconnect each of the sensor test leads in turn and check that the appropriate warning segments and the red LED begin to flash. Note that in the case of the fuel level system a delay circuit is incorporated, and a pause of up to 12 seconds will be noted before the microprocessor acknowledges a change.
8   If any of the sensor circuits fails to provoke the appropriate display the unit must be considered faulty and renewed. This procedure is described in Section 25 of this Chapter.

27 Combined tachometer/voltmeter: testing – N1, P1 and R1 models

1   The three models covered by this Section are equipped with electronic tachometers driven from the ignition primary windings. In addition to this function, a secondary scale and circuit allows them to function as voltmeters when a change-over switch is pressed down. Before performing any other tests in the event of a suspected fault, check that the instrument panel is secure and that all damper rubbers are in place and in good condition. The battery must be fully charged and the ignition system working normally.
2   Remove the handlebar fairing (R1 only) then release the headlamp unit to gain access to the instrument panel connector block (9-pin). In the case of the N1 model, remove the cover from the base of the tachometer. On the other models remove the instrument panel bottom cover.
3   Set a multimeter to the 25 volts (or 20 volts) scale, and connect the positive (+) probe to the brown lead terminal and the negative (-) probe to the black lead terminal. The meter should show zero volts when the engine is stopped and 2 – 4 volts when it is running. If no reading is obtained, trace back and check the two supply leads.
4   Disconnect the green, red and yellow leads from the changeover switch. Using a multimeter as a continuity tester, check that the green and red leads are connected when the switch is released and that the yellow and red leads are connected when it is pressed. If the switch is faulty, renew it.
5   If the above tests show the wiring and switch to be functioning normally it must be assumed that the meter itself is faulty. If possible, check this by substituting a new meter.

28 Combined tachometer/voltmeter: testing – E and H models

1   The E and H models are fitted with a voltmeter mounted either in the lower section of the tachometer face or in the warning lamp panel. To test the meter, remove the headlamp unit and separate the 6-pin (E models) or 3-pin (H models) connector from the tachometer. Using a multimeter, check the voltmeter resistance by measuring between the brown lead and the black/yellow lead. If a reading outside the normal 60 – 80 ohm is obtained, the meter assembly must be renewed.

29 Instrument panel: removal and dismantling – R1 and P1 models

1   The above models employ a one-piece instrument console mounted on the top yoke. The accompanying photographic sequence describes its removal and overhaul.

30 Instrument panel and warning lamps: removal – all models except R1 and P1

1   Slacken the drive cable retaining ring (where appropriate) to free the cable from the underside of the instrument to be removed. Remove the two domed nuts and free the bottom cover. Pull the instrument head upwards and clear of the rubber mounts. Unplug the bulbholders and lift the instrument clear.
2   The warning lamp panel can be removed after its two securing screws have been released. The bulbs are all rated at 12 volts 3.4w and are a bayonet fitting.

26.1 Locate and separate red 6-pin connector (arrowed)

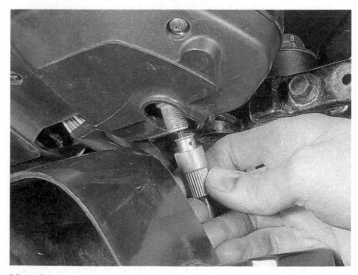

29.1a Disconnect speedometer cable and wiring at connectors in headlamp shell

29.1b Remove mounting bolts to free instrument panel

29.1c Release screws and remove bottom cover ...

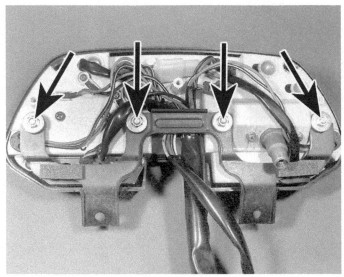

29.1d ... then remove mounting bracket

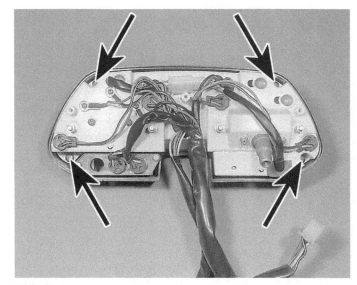

29.1e Remove screws to free panel unit from case

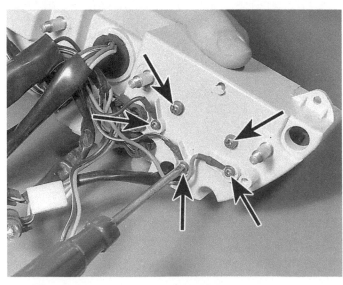

29.1f Release screws to free tacho/volt meter

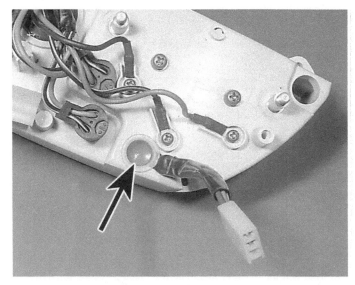

29.1g Prise out plastic plug ...

29.1h ... to allow wiring to be pulled clear

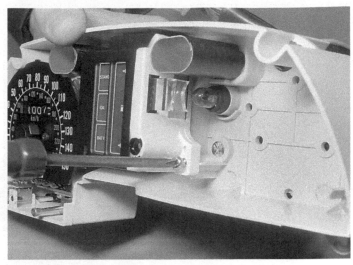

29.1i Microprocessor/display unit is screwed to case

29.1j Do not attempt to dismantle unit

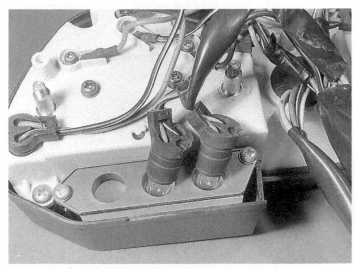

29.1k Warning lamps are push-fit in panel

31 Handlebar switches: removal and maintenance

1   Most of the electrical system is controlled by handlebar mounted switches. The right-hand switch incorporates the throttle twistgrip assembly and for security is located by a projection which engages in a hole in the handlebar. The switch unit halves can be separated by unscrewing the two securing screws, thus exposing the internal switch components. Note that the screws often differ in length and this should be noted to ensure correct reassembly.

2   Limited access to the switch contacts is possible, and cleaning is best confined to the application of a multi-purpose cleaner and lubricant such as WD40. The internal switch components are not available individually, so if cleaning fails to cure a faulty switch it will be necessary to renew the cluster.

32 Horn: location and adjustment

1   A single or twin horn arrangement is fitted according to the model. Each horn is mounted on a resilient steel bracket. If the horn fails to operate, or works feebly, it can be adjusted by slackening the locknut and turning the adjuster screw in or out by a small amount until the best sound is obtained. If this fails to restore it to normal working condition, a new unit or units must be fitted.

31.1 Separate switch halves to gain access for cleaning

The Z750 L4 model

The Z750 P4 (GT750) model

The ZX750 A1 (GPz) model

The ZX750 A3 (GPz) model

Chapter 7 The 1983 on models

Contents

Specifications

The following specifications relate to the 1983 on models, and are given where they differ from the specifications at the beginning of Chapters 1 to 6. Owners of ZX750 A and KZ750 F1 models should refer back to the R1 and N1 information respectively, unless shown below. All models are identified by their suffix letter (see Section 1 for further details) and the following specifications apply to all models unless otherwise stated.

Specifications relating to Chapter 1

Engine

Compression ratio	9.5 : 1
Maximum horsepower:	
H4	55.2 kW (75 PS) @ 9500 rpm
L3, L4	58.8 kW (80 PS) @ 9000 rpm
A1, A2, A3, A5 (UK), A1 (US)	63.3 kW (86 PS) @ 9500 rpm
A2, A3, (US)	62.5 kW (85 PS) @ 9500 rpm
N2, F1	75.0 HP @ 9500 rpm
P2, P3	78.0 HP @ 9500 rpm
P4, P5	Not available

Maximum torque:
  H4, N2, F1, P2, P3 .... 6.4 kgf m (46.3 lbf ft) @ 7500 rpm
  P4, P5 .... Not available
  L3, L4 .... 6.7 kgf m (48.5 lbf ft) @ 7500 rpm
  A1, A2, A3, A5 .... 6.9 kgf m (50.0 lbf ft) @ 7500 rpm

Crankshaft and big-end assembly – H4, L3, L4, A1, A2, A3, A5
Big-end side clearance service limit .... 0.50 mm (0.0197 in)
Main bearing to journal clearance service limit .... 0.08 mm (0.0032 in)
Main bearing journal diameter:
  Unmarked .... 35.984 – 35.992 mm (1.4167 – 1.4170 in)
  Marked 'O' .... 35.993 – 36.000 mm (1.4170 – 1.4173 in)
  Service limit .... 35.96 mm (1.4157 in)
Crankshaft endfloat service limit .... 0.35 mm (0.0138 in)

Piston
Ring groove width service limit:
  Top ring groove .... 1.12 mm (0.0440 in)
  2nd ring groove .... 1.31 mm (0.0516 in)
  Oil ring groove .... 2.60 mm (0.1024 in)

Piston rings
Ring/groove clearance service limit:
  Top .... 0.17 mm (0.0067 in)
  2nd .... 0.16 mm (0.0063 in)
Ring thickness service limit:
  Top .... 0.90 mm (0.0354 in)
  2nd .... 1.10 mm (0.0433 in)

Valves
Valve clearances – A2, A3, A5 .... 0.13 – 0.23 mm (0.0051 – 0.0091 in)
Valve stem diameter service limit – H4, L3, L4, A1, A2, A3, A5 ...... 6.94 mm (0.2732 in)
Valve/guide clearance service limit – H4, L3, L4, A1, A2, A3, A5 ... 0.33 mm (0.0118 in)

Valve timing
H4, L3, L4, N2, F1, P2, P3, P4, P5:
  Inlet opens at .... 30° BTDC
  Inlet closes at .... 60° ABDC
  Duration .... 270°
  Exhaust opens at .... 60° BBDC
  Exhaust closes at .... 30° ATDC
  Duration .... 270°
A1, A2, A3, A5 models:
  Inlet opens at .... 38° BTDC
  Inlet closes at .... 68° ABDC
  Duration .... 286°
  Exhaust opens at .... 68° BBDC
  Exhaust closes at .... 38° ATDC
  Duration .... 286°

Camshafts – H4, L3, L4, A1, A2, A3, A5
Cam lobe height service limits:
  Inlet and exhaust – H4, L3, L4 .... 36.15 mm (1.4232 in)
  Inlet – A1, A2, A3, A5 .... 36.65 mm (1.4429 in)
  Exhaust – A1, A2, A3, A5 .... 35.65 mm (1.3839 in)
Journal diameter service limit .... 21.91 mm (0.8625 in)
Bearing surface ID service limit .... 22.14 mm (0.8717 in)
Bearing/journal clearance service limit .... 0.23 mm (0.0091 in)

Clutch – H4, L3, L4, A1, A2, A3, A5
Friction plate thickness service limit .... 3.4 mm (0.1339 in)
Plain plate warpage service limit .... 0.3 mm (0.0118 in)
Clutch spring free length service limit .... 33.9 mm (1.3346 in)

Transmission
Primary reduction ratio .... 2.550 : 1 (27/23 x 63/29)
Final reduction ratio:
  H4 .... 2.461 : 1 (32/13)
  L3, L4 .... 2.538 : 1 (33/13)
  A1, A2, A3, A5 .... 2.533 : 1 (38/15)
  N2, F1, P2, P3, P4, P5 .... 2.522 : 1 (15/22 x 37/10)
Overall reduction ratio (in top gear):
  H4 .... 5.492 : 1
  L3, L4 .... 5.664 : 1
  A1, A2, A3, A5 .... 5.652 : 1
  N2, F1, P2, P3, P4, P5 .... 5.629 : 1

Supplementary torque settings – L4, A1, A2, A3, A5

Component	lbf ft	kgf m	lbf in
Alternator rotor bolt – 12 mm	94	13.0	–
Gearbox sprocket nut	72	10.0	–
Crankcase bolts – L4, A2, A3, A5:			
6 mm	–	1.2	104
8 mm	22	3.0	–
Cylinder head cover bolts – L4, A2, A3, A5	–	1.2	104
Gearbox sprocket nut – L4, A2, A3, A5	72	10.0	–
Oil pressure switch – L4, A2, A3, A5	11.0	1.5	–
Oil pressure relief valve – L4, A2, A3, A5	14.5	2.0	–

Specifications relating to Chapter 2

Fuel tank capacity – overall

	Litre	Imp gal	US gal
H4	12.4	2.73	3.28
L3, L4	21.7	4.77	5.73
A1, A2, A3, A5	19.0	4.18	5.02
N2, F1	14.8	3.26	3.91
P2	23.4	5.15	6.18
P3	24.3	5.35	6.42
P4, P5	Not available		

Note: reserve capacities – not available

Carburettors

	H4	L3, L4	A1, A2, A3, A5
Make	Keihin	Mikuni	Mikuni
Type	CV34	BS34	BS34
Main jet	n/app	110	110
Primary main jet	65	n/app	n/app
Secondary main jet	90	n/app	n/app
Needle jet	n/app	Y-9	Y-8
Jet needle (US)	N10A	4BE4	4BC6
Jet needle (UK)	n/app	4BE3	4BC7
Needle clip position, grooves from top (UK models)	n/app	3rd	3rd
Needle clip position, grooves from top (US models)	fixed	fixed	fixed
Pilot jet	35	37.5	37.5
Pilot screw setting, turns out (UK)	n/app	2.0	2.0
Pilot screw setting, turns out (US)	fixed	fixed	fixed
Fuel level	4 ± 1 mm	3 ± 1 mm	3 ± 1 mm
Float height	21 ± 2 mm	18.6 ± 2 mm	18.6 ± 2 mm

Carburettors

	N2	P2 to P5	F1
Make	Keihin	Mikuni	Keihin
Type	CV34	BS34	CV34
Main jet	n/app	110 R	100
Primary main jet	65	n/app	n/app
Secondary main jet	90	n/app	n/app
Needle jet	N426-01B36	Z-2	N426-01B36
Jet needle (US)	N10A	n/app	N10C
Jet needle (UK)	n/app	4BE3	n/app
Needle clip position, grooves from top (UK models)	n/app	3rd	n/app
Needle clip position, grooves from top (US models)	fixed	n/app	fixed
Pilot jet	35	37.5	35
Pilot air jet	110	300	n/app
Main air jet	n/app	n/app	60
Pilot screw setting, turns out (UK)	n/app	2.0	n/app
Pilot screw setting, turns out (US)	fixed	n/app	fixed
Fuel level	4 ± 1 mm	3 ± 1 mm	0.5 ± 1 mm
Float height	n/av	n/av	14.0 mm

Specifications relating to Chapter 3

Ignition system

Type	Electronic
Advance range:	
A2, A3, (California)	10° BTDC @ 1200 rpm, 40° BTDC @ 3600 rpm
L4	10° BTDC @ 1050 rpm, 35° BTDC @ 3800 rpm
All other models	10° BTDC @ 1050 rpm, 40° BTDC @ 3600 rpm

Pickup coil resistances

Pickup coil resistances ..... See text

Spark plugs

	NGK	ND
Type, normal use:		
L3, L4 (UK)	BR8ES	W24ESR-U

P2, P3, P4, P5 (UK)	BR8ES	W24ESR-U
H4, L3 (US)	B8ES	W24ES-U
A1, A2, A3, A5 (UK)	BR9ES	W27ESR-U
A1, A2, A3 (US)	B9ES	W27ES-U
Type, low-speed riding:		
L3, L4 (UK)	BR7ES	W22ESR-U
P2, P3, P4, P5 (UK)	Not available	
H4, L3 (US)	B7ES	W22ES-U
A1, A2, A3, A5 (UK)	BR8ES	W24ESR-U
A1, A2, A3 (US)	B8ES	W24ES-U

Specifications relating to Chapter 4

Front forks
Travel:
H4	180 mm (7.1 in)
L3, L4, A1, A2, A3, A5	150 mm (5.9 in)
N2, P2, P3, P4, P5	160 mm (6.3 in)
F1	Not available

Air pressure range:
H4	0.5 – 1.0 kg/cm² (7.1 – 14.0 psi)
L3, L4	0.6 – 0.9 kg/cm² (8.5 – 13.0 psi)
A1, A2, A3, A5	0.4 – 0.6 kg/cm² (5.7 – 8.5 psi)
F1, N2, P2, P3, P4, P5	0.5 – 0.7 kg/cm² (7.1 – 10.0 psi)

Fork oil capacity (per leg):
H4	312.0 ± 4 cc
L3, L4	297.0 ± 4 cc
A1, A2, A3, A5	248.5 ± 4 cc
N2	293.0 ± 2.5 cc
P2, P3, P4, P5	300.0 ± 2.5 cc
F1	297.0 ± 2.5 cc

Fork oil level:
H4	438 ± 2 mm (extended)
L3, L4	103 ± 2 mm (compressed)
A1, A2, A3, A5	185 ± 2 mm (compressed)
N2	457 ± 2 mm (extended, spring removed)
P2, P3, P4, P5	418.5 ± 2 mm (extended, spring removed)
F1	418.0 ± 2 mm (extended, spring removed)

Rear suspension
Type:
H4, L3, L4, F1	Swinging arm, twin oil-damped coil spring suspension units
N2, P2, P3, P4, P5	Swinging arm, twin interconnected air-assisted suspension units
A1, A2, A3, A5	Uni-Trak rising-rate, single air-assisted suspension unit with adjustable damping

Wheel travel:
H4	95 mm (3.7 in)
L3, L4	111 mm (4.4 in)
A1, A2, A3, A5	130 mm (5.9 in)
N2	98 mm (3.9 in)
P2, P3, P4, P5	100 mm (3.9 in)
F1	Not available

Rear suspension unit air pressure range:
A1, A2, A3, A5	0.5 – 3.0 kg/cm² (7.1 – 43 psi)
N2	1.5 – 3.0 kg/cm² (21 – 43 psi)
P2, P3, P4, P5	1.5 – 4.0 kg/cm² (21 – 57 psi)

Rear suspension unit air chamber volume:
N2	105 ± 2.5 cc
P2, P3, P4, P5	97.5 ± 2.5 cc

Rear suspension unit oil quantity:
N2	396 ± 2.5 cc
P2, P3, P4, P5	355 ± 2.5 cc
Rear suspension unit oil grade – N2, P2, P3, P4, P5	SAE 5W

Supplementary torque settings – KZ/Z750 L

Component	kgf m	lbf ft	lbf in
Front fork air valve	0.8	–	69
Front fork wheel spindle clamp nuts	1.4	10	–
Front fork upper pinch bolt	2.1	15	–
Steering stem clamp bolt nut	2.1	15	–
Handlebar clamp bolts	1.9	13.5	–

Supplementary torque settings

Component	kgf m	lbf ft	lbf in
Steering stem top bolt	4.3	31	–
Steering stem clamp bolt nut – A2, A3, A5	2.1	15	–

	kgf m	lbf ft	lbf in
Steering stem adjuster locknut – A2, A3, A5	0.5	–	43
Handlebar clamp bolts – A1	1.0	–	87
Handlebar clamp bolts – A2, A3, A5	1.8	13	–
Handlebar holder bolts	7.5	54	–
Front fork air valves	0.8	–	69
Front fork top plugs	2.3	16.5	–
Front fork clamp bolts:			
Upper	2.0	14.5	–
Lower	3.8	27	–
Anti-dive valve mounting bolts	0.7	–	61
Anti-dive plunger mounting bolts	0.45	–	39
Front wheel spindle nut – A2, A3, A5	6.0	43	–
Front wheel spindle clamp bolt	2.0	14.5	–
Rear suspension unit air valve	0.8	–	69
Rear suspension unit air hose union	1.2	–	104
Rear suspension unit mountings – A1:			
Upper	3.8	27	–
Lower	7.0	51	–
Rear suspension unit mountings – A2, A3, A5	3.0	22	–
Uni-Trak rocker arm pivot shaft nut	7.0	51	–
Tie-rod nuts:			
Upper	3.8	27	–
Lower	7.0	51	–
Rear wheel spindle	12	87	–
Rear brake torque arm nuts	3.0	22	–

Specifications relating to Chapter 5

Tyres

	Front	Rear
Type	Tubeless	Tubeless
Sizes:		
H4	3.25H19 4PR	130/90-16 67H
L3 (US)	100/90-19 57H	120/90-18 65H
L3, L4 (UK)	100/90V19	120/90V18
A1, A2, A3 (US)	110/90-18 61H	130/80-18 66H
A1, A2, A3, A5 (UK)	110/90V18	130/80V18
P2, P3, P4, P5	100/90-19 57H	120/90-18 65H
N2, F1	100/90-19 57H	130/90-16 67H

Tyre pressures

	H4	L3, L4
Front	25 psi	28 psi
Rear, up to 97.5 kg/215 lb loading	21 psi	32 psi
Rear, 97.5 – 165 kg/215–364 lb loading	25 psi	36 psi

	A1, A2, A3 (US)
Front	28 psi
Rear, up to 97.5 kg/215 lb loading	32 psi
Rear, 97.5 – 180 kg/215 – 397 lb loading	36 psi

	A1, A2, A3, A5 (UK)
Front, up to 150 kg/331 lb loading	28 psi
Front, above 210 kph/130 mph	32 psi
Rear, up to 97.5 kg/215 lb loading	32 psi
Rear, 97.5 – 180 kg/215 – 397 lb loading	36 psi
Rear, above 97.5 kph/130 mph	41 psi

	N2, F1	P2, P3, P4, P5
Front	25 psi	28 psi
Rear, up to 97.5 kg/215 lb loading	25 psi	32 psi
Rear, 97.5 – 180 kg/215 – 397 lb loading	32 psi	36 psi

Rear brake pedal height

A1, A2, A3, A5	50.5 – 54.5 mm (1.98 – 2.14 in)
L3, L4	14 – 18 mm (0.55 – 0.70 in)

Supplementary torque settings – A1, A2, A3, A5

Component	kgf m	lbf ft	lbf in
Front wheel spindle nut – A2, A3, A5	6.0	43	–
Front wheel spindle clamp bolt	2.0	14.5	–
Rear wheel spindle nut	12.0	87	–
Rear brake torque arm nuts	3.0	22	–
Caliper bleed valves	0.8	–	69
Brake hose union bolts	3.0	22	–
Brake hose clamp screws – A2, A3, A5	0.1	–	9.0
Brake lever pivot bolt – A2, A3, A5	0.3	–	26
Brake lever pivot locknut – A2, A3, A5	0.6	–	52
Caliper holder shaft bolts – A2, A3, A5	1.8	13	–
Caliper mounting bolts (front and rear)	3.3	24	–

Supplementary torque settings – A1, A2, A3, A5 continued

Disc mounting Allen bolts – A2, A3, A5	2.3	16	–
Front master cylinder clamp bolts – A2, A3, A5	0.9	–	78
Rear caliper torque link nut	3.0	22	–
Chain adjuster clamp bolts	3.3	24	–

Specifications relating to Chapter 6

Battery

Type:	
H4, L3, L4	FB12A-A
A1, A2, A3, A5, P2, P3, P4, P5	SYB14L-A2
N2, F1	YB14L-A2
Voltage – all models	12V
Capacity:	
H4, L3, L4	12 Ah
A1, A2, A3, A5, P2, P3, P4, P5, N2, F1	14 Ah
Electrolyte level sensor resistance – A1, A2, A3, A5	600 – 750 ohms

Alternator

Rated output – L3, L4, A1, A2, A3, A5	17A @ 8000 rpm, 14V
Unregulated output – A1, A2, A3, A5	about 50V @ 4000 rpm

1 Introduction

Since this manual was published, the KZ/Z750 models have continued to appear in modified form. In addition, there have been a few major revisions, and these are described in greater detail in the main text of this Chapter. If no mention is found in this Chapter refer back to Chapters 1 to 6.

A supplementary model summary is shown below, and relates to the additional models covered by this update Chapter. This should be used in conjunction with the main introductory text (see p6).

1983 Z750 L3 (UK), KZ750 L3 (US)
Frame no. (UK) KZ750 R – 014501 on
Frame no. (US) JKAKZDF1*DA014501 to 017200
Engine no. (UK and US) KZ750EE093001 – 128100

The L3 model is a straightforward continuation of the basic L2 UK version described in the main text. The main recognition features are the adoption of the 1982 model GPz tank, seat and side panels, more prominent decals and a rectangular headlamp. Mechanical changes include the fitting of an oil cooler, Mikuni carburettors, new engine mountings, an electronic tachometer and tapered roller steering head bearings.

1983 KZ750 H4 – LTD (US only)
Frame no. JKAKZDHI*DA039501 to DA041001
Engine no. KZ750EE093001 to 095727

The 1983 version of the LTD model, the H4 has little in the way of features to distinguish it from the previous model. Main cosmetic alterations are to the engine covers.

1983 ZX750 A1 – GPz
Frame no. (UK) ZX750 A – 000001 on
Frame no. (US) JKAZXDA1*DA000001 to 018000
Engine no. (UK and US) KZ750EE110001 on

Rather confusingly, the A1 was the first of the series of 'proper' GPz machines, and is significantly different to the previous year's R1 model, in spite of the latter's GPz tag. In retrospect, the R1 is probably best regarded as a half-way move towards the GPz range proper. The A1 features a major styling revision, with a humped fuel tank which flows into the side panels to form a unified whole. A small fairing and three-spoke cast wheels are fitted. More important modifications include the fitting of anti-dive front forks and Kawasaki's rising-rate, single-shock rear suspension arrangement, Uni-Trak.

1983 KZ750 N2 – Spectre (US only)
Frame no. JKAKZDN1*DA007401 to DA009965
Engine no. KZ750NE008901 to 016363

The 1983 edition of the Spectre model can be distinguished by its two-tone paint scheme and other detail cosmetic changes from the previous model.

1983 Z750 P2 – GT750 (UK only)
Frame no. KZ750P-003201 on
Engine no. KZ750NE008901 to 011400, KZ750NE011701 on

The 1983 version of the GT750 shaft-drive sports tourer. Largely unchanged from the P1 version.

1983 KZ750 F1 – LTD Shaft (US only)
Frame no. JKAKZDF1*DA000001 to 003155
Engine no. KZ750NE011701 to 016229

The shaft-drive LTD model appeared for one year only before being replaced in 1984 by the ZN700 model. Similar to the Spectre model, but with oil-damped rear suspension units instead of the air assisted units.

1984 to 1987 Z750 L4 (UK only)
Frame no. KZ750R-017201 on
Engine no. KZ750EE128101 on

The 1984 version of the previous year's L3, and almost identical in appearance other than the revised graphics. The major mechanical change is the fitting of a revised ignition system.

1984 ZX750 A2 – GPz
Frame no. (UK) ZX750A-018001 on
Frame no. (US – Made in Japan) JKAZXDA1*EA018001 on
Frame no. (US – Made in US) JKAZXDA1*EB500001 to EB504000
Engine no. (UK and US) KZ750EE128101 to 145000

The GPz continued with little change from the previous year's A1 model. Main recognition points are the revised fairing, paint scheme and graphics. US models are fitted with an emission control system.

1984 Z750 P3 – GT750 (UK only)
Frame no. KZ750P-005301 to 006400
Engine no. KZ750NE016601 to 018400

The 1984 version of the GT750 shaft-drive sports tourer. Mechanical modifications include the fitting of a revised ignition system.

1985 to 1987 ZX750 A3 – GPz (UK), 1985 ZX750 A3 – GPz (US)

Frame no. (UK) ZX750A-025501 on
Frame no. (US – Made in Japan) JKAZXDA1*FA025501 on
Frame no. (US – Made in US) JKAZXDA1*FB504101 to 505600
Engine no. (UK and US) KZ750EE145001 on
  This version of the GPz model can be distinguished by its full sports fairing with belly pan, and revised paint scheme.

1985 to 1987 Z750 P4 – GT750 (UK only)

Frame no. KZ750P-006401 on
Engine no. KZ750NE018401 on
  The 1985 version of the GT750 shaft-drive sports tourer. Remained on sale in the UK until superseded by the P5 during 1987.

1987 to 1989 Z750 P5 – GT750 (UK only)

  The 1987-91 version of the GT750 shaft-drive sports tourer. Cosmetic revisions only. The initial frame and engine numbers are not available for this model.

1988 ZX750 A5 – GPz (UK only)

Frame no. ZX750A-028714 on
Engine no. KZ750EE149715 on
  The 1988 model has no significant changes from the 1985 to 1987 ZX750 A3.

Note: *The asterisk (*) in the above frame numbers of US machines indicates a digit within the number which varies according to the particular machine.*

---

2 Cylinder block: modifications

  There have been a number of changes to the cylinder blocks of the later models, and in general these have little effect on working procedures. The small damper blocks fitted to the earlier machines were omitted on later models, and the modified exhaust port stud arrangement described in Section 26 of Chapter 1 was incorporated on all subsequent machines. The detail changes relating to the prevention of oil leakage discussed in Sections 19, 42 and 43 of Chapter 1 apply equally to the models covered in this update Chapter.

---

3 Oil cooler: modifications – ZX750 A and KZ/Z750 L models

1   The later GPz machines covered in this Chapter retain the oil cooler arrangement first used on the earlier GPz model, the KZ/Z750 R1. Note that the oil cooler radiator is now retained by two rubber-bushed mounting bolts instead of four.
2   The KZ/Z750 L3 and L4 models are fitted with an oil cooler similar to that fitted to the R1 models. Unlike the ZX models, the oil cooler is retained by four rubber-bushed mounting bolts.
3   Refer to Chapter 1, Section 5 for oil cooler removal and Section 48 for refitting.

---

4 Clutch release mechanism: ZX750 A models

1   The ZX750 models feature a revised clutch release mechanism operated by an external arm to the rear of the right-hand engine cover. The clutch itself is largely unchanged, but the pushrod arrangement is of the rack and pinion type. The gearbox input shaft no longer carries the clutch pushrod assembly, and its sealing O-ring is therefore omitted.

Adjustment

2   Before commencing adjustment, and especially after the clutch outer cover has been disturbed, check that the clutch release arm is set at the correct operating angle. When correctly positioned the arm should lie at approximately 30° below horizontal. If necessary, remove the cover so that the toothed end of the shaft can be repositioned in relation to the rack to obtain the correct angle.
3   Clutch cable free play adjustment is carried out using the in-line adjuster at the lower end and the lever adjuster at the upper end. Start by screwing the upper adjuster fully home in the lever stock. Set the free play between the lever stock and blade to 2 – 3 mm (0.08 – 0.12 in) using the lower adjuster, and tighten the locknuts firmly. Subsequent fine adjustment can be made at the lever adjuster. Check that the dust boots are intact and in position, and check clutch operation before riding the machine.

Overhaul

4   The main clutch components can be dealt with as described in Chapter 1, Sections 10 and 30. Note that the release mechanism can be removed as follows. The release rack and its bearing can be displaced and removed once the pressure plate has been detached. The release arm and shaft are carried in needle roller bearings in the cover, the assembly being retained by a circlip. Once the circlip has been released, the arm and shaft can be withdrawn for examination.
5   Check the condition of the shaft, rack and bearings, renewing worn or damaged parts. Grease the bearings prior to assembly, then install the various parts by reversing the removal sequence.

3.1 Oil cooler radiator is secured by two bolts and single rubber-bushed tab (ZX750 A models)

4.2 The revised clutch operating mechanism. Note angle of release arm (see text)

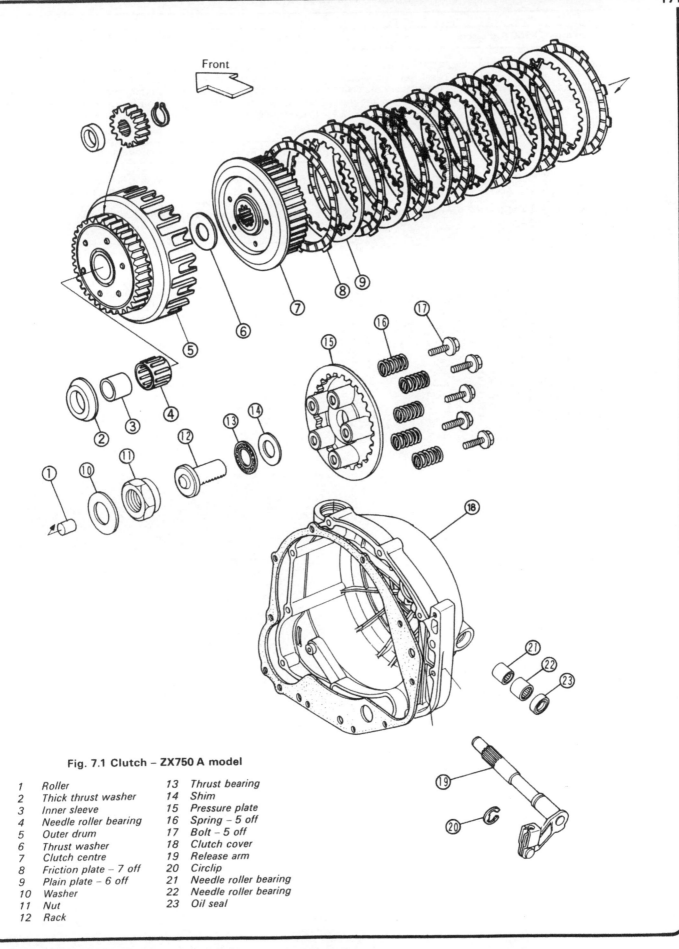

Fig. 7.1 Clutch – ZX750 A model

1	Roller	13	Thrust bearing
2	Thick thrust washer	14	Shim
3	Inner sleeve	15	Pressure plate
4	Needle roller bearing	16	Spring – 5 off
5	Outer drum	17	Bolt – 5 off
6	Thrust washer	18	Clutch cover
7	Clutch centre	19	Release arm
8	Friction plate – 7 off	20	Circlip
9	Plain plate – 6 off	21	Needle roller bearing
10	Washer	22	Needle roller bearing
11	Nut	23	Oil seal
12	Rack		

5 Gearchange linkage: ZX750 A models

Like the previous KZ/Z750 R1 model, the ZX750 A machines are equipped with a remote, rearset gearchange pedal which operates through an external linkage. On the R1, a reversed pedal arrangement was fitted, the pedal being attached to a pivot post screwed into the outer cover. On the ZX750 A models, this system is simplified, a conventional forward-facing pedal is connected by an adjustable operating rod to a short link. The link attaches directly to the splined end of the gearchange shaft.

5.1 The simplified rearset gear linkage as fitted to the ZX750 A models

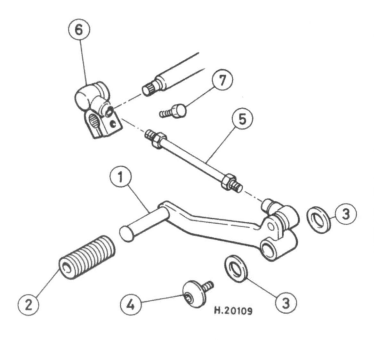

Fig. 7.2 Gearchange lever linkage – ZX750 A model

1	Gearchange lever	5	Adjustable rod
2	Rubber	6	Link
3	Washer – 2 off	7	Pinch bolt
4	Special screw		

6 Exhaust system: mountings – ZX750 A models

Apart from the revised exhaust port stud diameters and lengths (see Section 2) the exhaust system is basically similar to that used on earlier models. Note that the silencer mounting arrangement is simplified; the separate mounting plates featured on the R1 were deleted in favour of brackets integral with the silencers.

7 Exhaust system: KZ/Z750 L models

The later L models are fitted with a new exhaust system which is basically similar to that fitted to the R1 model. The outer cylinder pipes and the balance pipes form a single assembly, the inner pipes and the silencer being attached to it via stubs and clamps.

8 Engine mountings: KZ/Z750 L models

The KZ/Z750 L3 and L4 models are fitted with rubber-bushed engine mountings to reduce engine vibration transmitted to the frame and rider. For further details, refer to the information given in Section 47 of Chapter 1 for the 750 R1 models.

9 Fuel tank: mounting

1   Note that in the case of the ZX750 A models, the method of securing the rear of the tank is revised; the single retaining bolt is replaced by a mounting bracket which secures a projecting tab at the rear of the tank. The bracket is retained by two nuts.
2   The fuel tank mounting arrangement is slightly revised on the KZ/Z750 L3 and L4 models, but the single bolt fixing is retained.

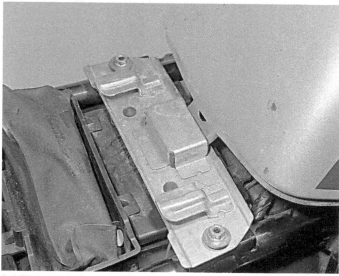

9.1 Tank is held by metal bracket on ZX machines

10 Evaporative emission control system: 1984 on California models

Note: *Fuel, especially in vapour form, is extremely flammable and may explode if in contact with a heat source. When working on any part of the fuel or emission control system, work only in a well-ventilated area, well away from any source of heat, flame or sparks. Do not smoke.*

1   Later models sold in California are equipped with a full evaporative emission control system to comply with local laws limiting the emission of unburnt hydrocarbons, in the form of fuel vapour, into the atmosphere.

2   Fuel vapour expelled from the fuel tank due to expansion is passed through a hose to a separator. When the engine is running, any liquid fuel is routed back to the tank from the separator, while fuel vapour is passed back through the air cleaner case to be fed through the engine and burnt.

3   When the machine is parked, vapour is stored in a canister to be drawn through the engine when it is next run. The system is shown in the accompanying line drawing, and a hose routing diagram is attached to all machines equipped with the system.

Hose inspection

4   The hoses should be checked periodically (every 3000 miles/5000 km), and any that are damaged in any way should be renewed. The hoses are colour-coded to aid identification. When fitting hoses, make sure that they are securely connected.

Separator test

5   Examine the separator unit after removing it from the machine and cleaning it thoroughly. If it is cracked or otherwise damaged, renew it.

6   With the separator installed normally, disconnect the blue breather hose from the separator stub and inject about 20 cc of fuel into the separator.

7   Disconnect the fuel return hose (red) at the tank end, placing the open end into a container positioned level with the top of the tank.

8   Start the engine and allow it to idle. If the separator is working correctly, the fuel will be expelled into the container. If this does not occur, fit a new separator.

Canister inspection

9   The canister should not normally require attention during the life of the machine, unless it gets damaged or becomes contaminated by liquid fuel, solvents or water. Disconnect the hoses and free the rubber strap to allow the canister to be lifted away for inspection. If the canister is cracked or split, or if it has become contaminated, fit a new one.

10 **Caution:** when filling the fuel tank never allow the fuel level to rise into the filler neck, otherwise liquid fuel may enter the emission control system, due to heat expansion.

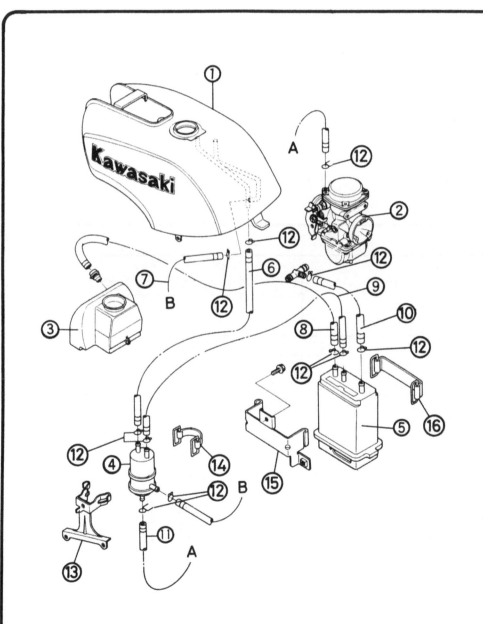

1	Fuel tank
2	Carburettor
3	Air cleaner case
4	Separator
5	Canister
6	Breather hose (blue)
7	Fuel return hose (red)
8	Purge hose (green)
9	Breather hose (blue)
10	Breather hose (yellow)
11	Vacuum hose (white)
12	Pipe clips
13	Separator holder
14	Retaining strap
15	Canister holder
16	Retaining strap

Fig. 7.3 Evaporative emission control system components – 1984 on US models

11 Carburettor: modifications – Z750 P, KZ/Z750 L and KZ750 F1 models

Z750 P models

1   When installing the carburettors on the later Z750 P models, note that the vent hoses should be routed, as indicated in the accompanying line drawing. Make sure that the hose ends are secured in the clamps provided.

KZ/Z750 L models

2   The L3 and L4 models are equipped wth Mikuni BS34 carburettors instead of the Keihin instruments fitted to the L2. Refer to Chapter 2 for further details, noting the revised specifications at the beginning of this Chapter.

KZ750 F1 model

3   This model is fitted with the Keihin type carburettors shown in Fig. 2.2. Note however, that items 38 and 39, the primary main jet and bleed pipe are not fitted. Refer to the specifications at the beginning of this Chapter.

12 Ignition system: general description – ZX750 A, Z750 P3, P4, P5 and Z750 L4 models

1   These models feature a revised electronic ignition system in which the mechanical automatic timing unit (ATU) used on the earlier machines was discontinued in favour of an electronic advance circuit, housed within the ignitor unit. The new arrangement removes the only mechanical aspect of the ignition system, and with it the need for periodic maintenance.

2   The ignition circuit is shown in the accompanying circuit diagram and this relates to all models employing the revised system. Note that in the case of the L4, P3, P4 and P5 models, the junction box (item 10) is not applicable. See wiring diagram for details.

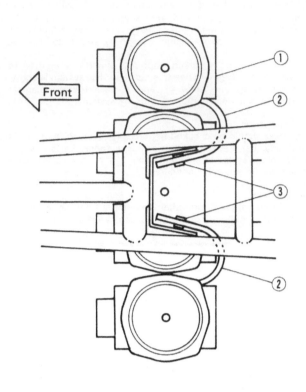

Fig. 7.4 Carburettor vent hose routing – Z750 P model

| A | Vent hoses | B | Clamps |

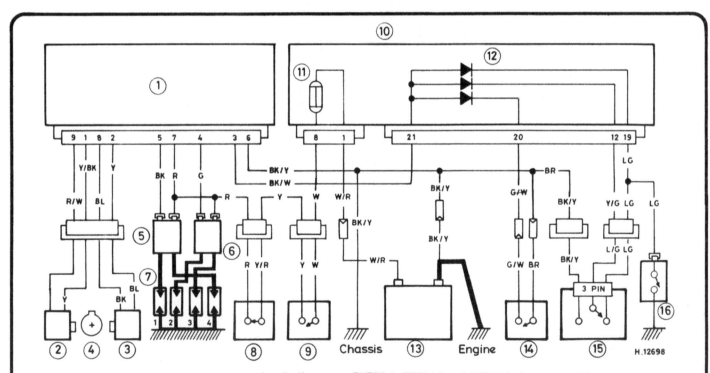

Fig. 7.5 Ignition system circuit diagram – ZX750 A, Z750 L4 and Z750 P3, P4, P5 models

1	Ignitor	6	Ignition HT coil – cylinders 2 and 3	10	Junction box – except Z750 L4, P3, P4 and P5	13	Battery
2	Pickup – cylinders 1 and 4	7	Spark plugs	11	Fuse	14	Side stand switch
3	Pickup – cylinders 2 and 3	8	Engine stop switch	12	Diodes	15	Starter lockout switch
4	Rotor	9	Ignition switch			16	Neutral switch
5	Ignition HT coil – cylinders 1 and 4						

13 Ignition system: fault diagnosis – ZX750 A, Z750 P3, P4, P5 and Z750 L4 models

1 Before commencing work, note that the ignitor unit will not withstand short circuits or reversed battery polarity, and any such occurrence will almost certainly damage the unit. To prevent such damage, never disconnect the battery or any other electrical connection while the ignition is switched on. When removing or installing the battery, be careful to observe polarity. Always disconnect the black negative (–) lead first, and reconnect it last, to avoid accidental short circuits.

2 In the event of a suspected ignition fault, check through the system in a methodical fashion as described in the following Sections. Checks should be carried out in the following sequence:

   a) *Check physically all connections and wiring in the ignition system. Remake or renew any damaged or corroded connections and renew frayed or broken leads.*
   b) *Check the ignition timing*
   c) *Check power supply to the ignitor unit*
   d) *Check the ignition coil*
   e) *Check the pickup coils*
   f) *Check the ignitor unit*
   g) *Check the ignition and engine stop switches*

14 Ignition system: timing check – ZX750A, Z750 P3, P4, P5 and Z750 L4 models

1 The ignition timing is checked using a stroboscopic timing lamp as described in Section 3 of Chapter 3. Note the advance range and engine speed details given in the specifications at the beginning of this Chapter.

2 The 'F' mark should align at idle speed, whilst at the specified engine speed the '1.4' line should coincide with the fixed index mark. Failure of the ignition to advance is unlikely unless the ignitor unit is damaged. Note that for obvious reasons, Section 4 of Chapter 3 can be disregarded for these models.

15 Ignition system: power supply to ignitor unit check – ZX750 A, Z750 P3, P4, P5 and Z750 L4 models

1 With the ignition switched off, disconnect the 10-pin connector from the ignitor unit. Using a multimeter set to the 25 volts dc range, connect the positive probe to the red lead terminal in the connector

and the negative (–) probe to the black/yellow terminal. Switch on the ignition and check that the engine stop switch is set to the 'Run' position.

2 If normal battery voltage is *not* shown, check the following areas:

   a) *Wiring and connectors*
   b) *Ignition switch and wiring*
   c) *Engine stop switch and wiring*
   d) *Junction box components – described later in this Chapter (except Z750 L4, P3, P4 and P5 models)*
   e) *Main fuse*
   f) *Diodes (except L4, P3, P4 and P5 models)*

16 Ignition system: ignition coil and pickup coils check – ZX750 A, Z750 P3, P4, P5 and Z750 L4 models

1 Check the ignition coils as described in Chapter 3, Section 9.
2 Disconnect the 4-pin block connector at the ignitor unit. Using a multimeter set to the ohm x 100 scale check the resistance of each pickup coil. Connect the multimeter probes between the first pair of wires shown below and note the reading. Repeat the test between the second pair of wires. If the readings obtained differ greatly from the values given the complete pickup unit is faulty and must be renewed.

Model	Resistance	Wire colours
ZX750 A	380 – 560 ohm	R/W to Y, Bk to Bl
Z750 L	360 – 540 ohm	Bk to Bl, Y to R
Z750 P	380 – 560 ohm	Bk to Bl, Y to R

17 Ignition system: ignitor unit test – ZX750 A, Z750 P3, P4, P5 and Z750 L4 models

1 This test requires the use of a multimeter. Note that Kawasaki recommend the use of their own meter (Kawasaki Hand Tester 57001 – 983) and caution that other meters may give slightly different readings from those specified. *Note also that any tester which subjects the unit to high test currents will damage the unit.*

2 Switch off the ignition, then disconnect and remove the ignitor. Zero the meter, then connect the meter probes to each pair of terminals in turn, according to the accompanying table. Compare the readings obtained with those shown. If significantly different, it is likely that the unit will require renewal. If possible, check this by substitution.

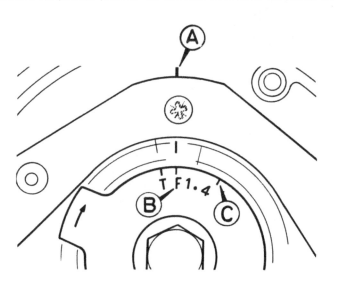

Fig. 7.6 Ignition timing marks – ZX750 A, Z750 P3, P4, P5 and Z750 L4 models

A   *Crankcase index mark*   C   *Rotor advance mark (1.4)*
B   *Rotor F mark*

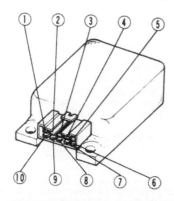

	Tester (+) Lead Connection									
Terminal Number	1	2	3	4	5	6	7	8	9	10
1		D	D	D	D	D	D	D	D	∞
2	D		D	D	D	D	D	D	D	∞
3	C	C		B	B	B	B	B	B	∞
4	∞	∞	∞		∞	∞	∞	∞	∞	∞
5	∞	∞	∞	∞		∞	∞	∞	∞	∞
6	C	C	B	A	A		A	O	O	∞
7	C	C	B	A	A	A		A	A	∞
8	C	C	B	A	A	O	A		O	∞
9	C	C	B	A	A	O	A	O		∞
10	∞	∞	∞	∞	∞	∞	∞	∞	∞	

Tester (−) Lead Connection

Fig. 7.7 Ignitor unit test table – ZX750 A, Z750 P3, P4, P5 and Z750 L4 models

Results

O	No resistance	C	25 – 75 K ohm
A	0.3 – 4.2 K ohm	D	125 – 375 K ohm
B	6.6 – 21.4 K ohm	∞	Infinity

Note – results expected when using Kawasaki hand tester

(57001-983)

18 Ignition system: switch connections check – ZX750 A, Z750 P3, P4, P5 and Z750 L4 models

1 An intermittent fault in either the ignition switch or the engine stop switch can cause ignition faults, and the operation of the switches and their associated leads and connections should be checked.
2 Bear in mind that on ZX750 A and Z750 P models the side stand switch, the starter lockout (clutch) switch and the neutral switch are all interconnected with the ignition system (see Fig. 7.5). If any switch is working intermittently or if the connecting leads are loose or damaged, ignition failure or intermittent misfiring may result.
3 To eliminate the above switches, temporarily bypass them by connecting an earth lead between the battery negative terminal and the black/white terminal on the ignitor (terminal 3). If this resolves the problem, check each of the above switches and its wiring to locate the fault.

19 Front forks: modification – KZ750 H4 model

The forks fitted to the H4 model are broadly similar to those fitted on previous H models. Note that the fork top bolts are modified slightly, and the air valves are positioned vertically in them.

20 Front fork: modifications – KZ/Z750 L models

1 The forks fitted to the L3 and L4 models are broadly similar to those fitted to the L2 model. Note that the fork top bolts are modified slightly, and the air valves are fitted vertically in them.
2 Note also that two rebound springs are fitted instead of one, and that many of the torque wrench settings have been revised, see Specifications.

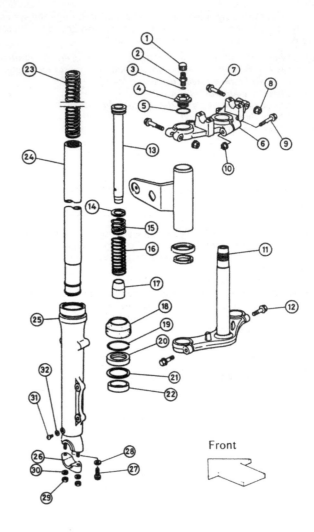

Front

Fig. 7.8 Front forks – KZ/Z750 L3 and L4 models

1	Cap	17	Damper rod seat
2	Air valve	18	Dust cover
3	O-ring	19	Circlip
4	Top bolt	20	Oil seal
5	O-ring	21	Washer
6	Top yoke	22	Bottom bush
7	Pinch bolt – 2 off	23	Spring
8	Nut – 2 off	24	Stanchion
9	Pinch bolt	25	Lower leg
10	Nut	26	Spindle clamp
11	Steering stem/bottom yoke	27	Allen bolt
12	Pinch bolt – 2 off	28	Sealing washer
13	Damper rod	29	Nut – 2 off
14	Damper rod piston ring	30	Spring washer – 2 off
15	Short rebound spring	31	Drain screw
16	Long rebound spring	32	Sealing washer

21 Steering head bearings: KZ/Z750 L models

These models are fitted with tapered roller steering head races in place of the cup and cone type fitted to the L2 model. For further details refer to Sections 7 and 10 of Chapter 4.

22 Anti-dive front forks: ZX750 A models

1   The ZX750 A models are equipped with revised forks incorporating an anti-dive arrangement interconnected with the front brake hydraulic system. A union in the front brake pipe connects the hydraulic system to a junction block bolted to the rear of the fork lower legs.
2   Hydraulic pressure is fed to the anti-dive valve assembly. This allows the fork damping effect to be increased while the front brake is applied. The degree of anti-dive effect is adjustable by way of a knurled control knob at the bottom of the anti-dive unit.

Inspection
3   The anti-dive unit and the related hydraulic pipes and unions should be checked for signs of leakage at regular intervals. If the pipes appear cracked or if there are indications of fluid leakage, renew them at once. Renew the metal brake pipe if it becomes corroded. If the anti-dive unit develops a fork oil or hydraulic fluid leak, it must be renewed. The anti-dive unit cannot be repaired if damaged or leaking.

Brake plunger test
4   Remove the two Allen bolts and disconnect the brake plunger assembly from the anti-dive unit, leaving the hydraulic pipe connected. Operate the front brake lever lightly, and note whether the plunger is pushed out from the plunger housing. If the unit is working normally, the plunger should come out 2 mm (0.08 in), and should push back into the housing under finger pressure once the brake lever is released.

Anti-dive valve test
5   The anti-dive valve unit can be tested after the fork legs have been removed from the yokes. Dealing with one leg at a time, remove the two Allen bolts and disconnect the plunger unit from the anti-dive valve, leaving the brake pipe connected. Remove the fork top bolt and withdraw the fork spring. Remove the front wheel, brake caliper, front mudguard and the brake hose union from the lower leg. Remove the fork leg, complete with anti-dive valve, and tape over the air equalizer hole in the stanchion to prevent loss of damping oil during the test.
6   Holding the fork leg upright push the stanchion up and down. The movement should be smooth and progressive, and should become much harder when the anti-dive valve rod is depressed with a finger. If the valve does not operate normally, or if it is jammed, it must be renewed.
7   Note that Kawasaki regard the anti-dive plunger assembly and the metal brake pipe as service items, and recommend that they should be renewed at two-yearly and four-yearly intervals respectively.

Fork overhaul
8   Although the forks are complicated somewhat by the use of the anti-dive arrangement, the basic overhaul procedure remains broadly similar to that described in Section 11 of Chapter 4. Note that it will be necessary to disconnect the brake hose union block and the anti-dive units from each fork leg prior to its removal. Note also that the internal arrangement of the fork legs differs in detail from earlier types, particularly in the case of the damper components. For details, compare the accompanying line drawing to those shown in Chapter 4. After reassembly is complete, check that both anti-dive unit adjusters are set to the same position, and that fork and anti-dive action are normal.

22.1a General view of the anti-dive components

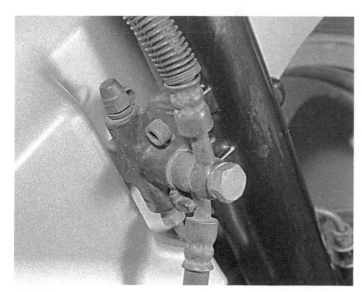

22.1b Hydraulic union block is attached to rear of fork lower leg

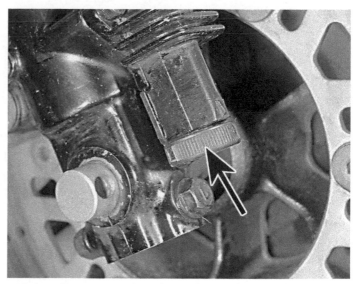

22.2 Anti-dive effect is adjusted using knob at base of valve unit

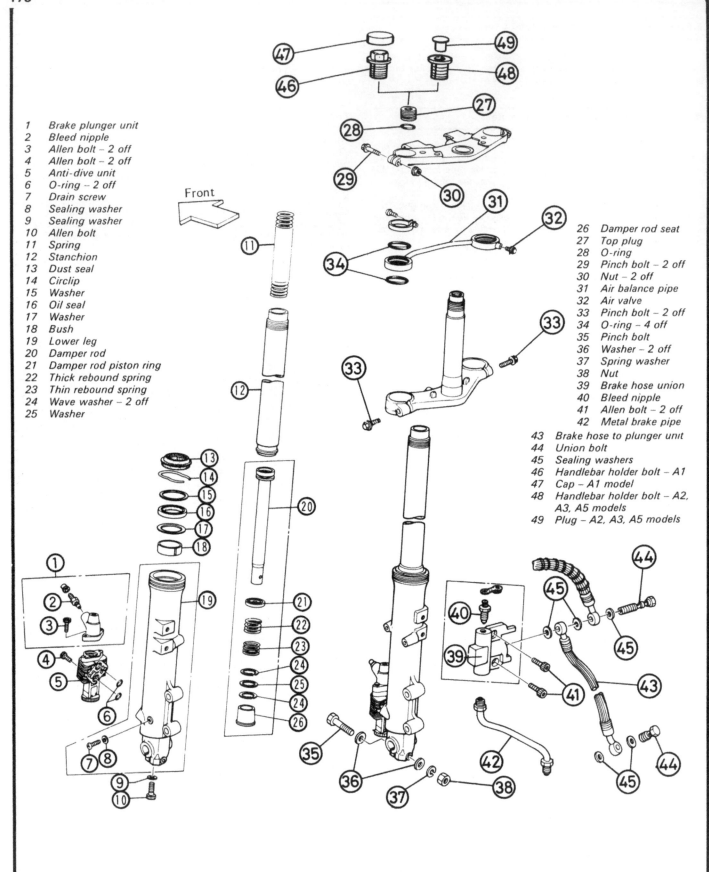

1   Brake plunger unit
2   Bleed nipple
3   Allen bolt – 2 off
4   Allen bolt – 2 off
5   Anti-dive unit
6   O-ring – 2 off
7   Drain screw
8   Sealing washer
9   Sealing washer
10 Allen bolt
11 Spring
12 Stanchion
13 Dust seal
14 Circlip
15 Washer
16 Oil seal
17 Washer
18 Bush
19 Lower leg
20 Damper rod
21 Damper rod piston ring
22 Thick rebound spring
23 Thin rebound spring
24 Wave washer – 2 off
25 Washer

26 Damper rod seat
27 Top plug
28 O-ring
29 Pinch bolt – 2 off
30 Nut – 2 off
31 Air balance pipe
32 Air valve
33 Pinch bolt – 2 off
34 O-ring – 4 off
35 Pinch bolt
36 Washer – 2 off
37 Spring washer
38 Nut
39 Brake hose union
40 Bleed nipple
41 Allen bolt – 2 off
42 Metal brake pipe
43 Brake hose to plunger unit
44 Union bolt
45 Sealing washers
46 Handlebar holder bolt – A1
47 Cap – A1 model
48 Handlebar holder bolt – A2,
    A3, A5 models
49 Plug – A2, A3, A5 models

Front

Fig. 7.9 Front forks – ZX750 A model

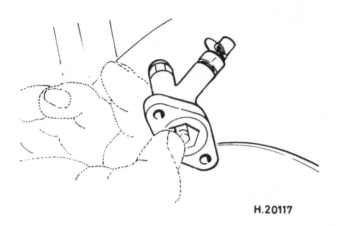

H.20117

Fig. 7.10 Anti-dive unit plunger test – ZX750 A model

23 Uni-Trak rear suspension: ZX750 A models

Note: *Prior to removing the suspension unit, linkage or swinging arm ensure that the machine is positioned on the centre stand, with the front securely tied down to prevent it toppling.*

1   The ZX750 models differ significantly from the remaining models in the range in their use of Kawasaki's Uni-Trak rear suspension system. In this arrangement a single air-assisted suspension unit is connected to the swinging arm via a bellcrank linkage, the relative movement of the linkage providing a progressive, rising-rate suspension effect. Variable air pressure and a four-position damper adjuster give a wide range of variation to the suspension characteristics.

Adjustment
2   Adjustment of the rear suspension unit is carried out from behind the right-hand side panel, where the remote air valve and damper adjustment plunger are located. Adjustment should be carried out with the machine on its centre stand.
3   To check the air pressure use the Kawasaki air pressure gauge (part number 52005-1003) or a similar unit designed specifically for use on suspension systems. Do not use a tyre pressure gauge because the inevitable air loss which occurs when using gauges of this type will give rise to inaccurate readings.
4   Note that air or nitrogen gas only should be used in the suspension unit, and that adjustment should be carried out with the unit at room temperature. Note that the reading shown after riding the machine will be higher than normal due to the increased temperature of the unit. On no account exceed the maximum safe pressure of 5.0 kg/cm² (71 psi) or the suspension unit seal may be damaged.
5   The recommended pressure for one rider, assuming no luggage and good road surface, is 0.5 kg/cm² (7.1 psi). This may be varied to any pressure up to the maximum working pressure of 3.0 kg/cm² (43 psi) to accommodate a passenger and/or luggage, or to cope with more difficult road surfaces.
6   Damping adjustment is controlled from a plunger mechanism located near the air valve, and this can be pulled out to increase the damping effect or pushed in to reduce it. As a general guide, use a soft setting (position 1 or 2) for normal solo use on good road surfaces. Increase the setting (3 or 4) to cope with poor road surfaces, high speed use or when touring with a passenger and/or luggage.

Suspension unit – removal and refitting
7   The suspension unit can be removed without disturbing the rest of the rear suspension components. Start by removing both side panels. Slacken the hose bracket nut until the air valve assembly can be disengaged from the frame lug. Slacken the locknut on the damper adjuster rod and unscrew the damper adjuster from the top of the suspension unit.
8   Free the tie rod pivot bolts from the swinging arm and pivot the tie

rods clear of the suspension unit lower mounting bolt. Slacken and remove the mounting bolt to free the lower end of the unit. The rocker arm and tie rods can now be manoeuvred clear of the unit.
9   Remove from the left-hand side the suspension unit upper mounting bolt securing nut. Withdraw the bolt from the right-hand side whilst supporting the suspension unit with one hand. Lower the unit downwards and remove it from between the swinging arm.
10 Before refitting the unit, check, clean and grease the various pivots, sleeves and bearings as described below. Renew the dust seals as a precautionary measure to avoid premature wear of the bearings and sleeves.
11 Refit the unit by reversing the removal operation, leaving the pivot bolts finger-tight until all are in position. Tighten all fasteners to the prescribed torque setting. Refit the air valve and the damper adjuster, and set the damper rate and air pressure to the prescribed settings.

Suspension linkage and swinging arm – removal and refitting
12 Remove the suspension unit as described above. Remove the rear wheel and place it to one side. Free the rocker arm pivot from the frame and remove the rocker arm and the tie rods. Disconnect the tie rods from the rocker arm to await cleaning and examination.
13 The swinging arm can now be removed from the frame in a similar manner to that described in Chapter 4, Section 15.
14 Before refitting the linkage and the swinging arm unit check, clean and grease the various pivots, sleeves and bearings as described below. Renew all dust seals as a precautionary measure to avoid premature wear of the bearings and sleeves.

Swinging arm and suspension linkage – examination and renovation
15 Once the suspension linkage components have been removed from the machine, carefully clean off all accumulated road dirt before dismantling the linkage assembly. Given the close association of the linkage and the swinging arm it is advisable to check and overhaul both assemblies together.
16 Note that Kawasaki recommend that the linkage components are checked and lubricated every 10 000 km (6000 miles). In view of the profusion of needle roller bearings within the assembly, this advice should not be ignored, and the advised interval should be reduced as appropriate where the machine is used under especially adverse conditions.
17 Note that regular checking and lubrication will prolong considerably the life of the bearings. Before commencing the overhaul, note that all seals should be renewed as a matter of course, even if the bearings and sleeves do not require attention.
18 Starting with the swinging arm, remove the dust seal, circlip and the journal ball bearing from the right-hand side of the pivot bore. Displace and remove the pivot sleeve. Wash all parts, including the needle roller bearings which will still be in the swinging arm bore at this stage, in solvent to remove all traces of old grease.
19 In the case of the suspension linkage components, remove the dust seals and sleeves and wash all parts thoroughly in solvent. Note that the bearings should not be removed at this stage.
20 After all parts have been cleaned carefully, check for signs of wear or damage. Look for pitting, scoring or corrosion of the needle roller bearing sleeves. If these are damaged they should be renewed, together with the bearings which carry them. Temporarily refit the sleeves in their bearings and check for play. If movement can be felt, renew the bearings and sleeves as a set. Check the tie rod spherical bearings for wear or damage. If these feel loose or have seized, renew them.
21 Use a simple drawbolt arrangement as shown in Fig. 7.13 to remove damaged bearings. Note that it will be necessary to make up the drawbolt to a size suitable for either the spherical bearings or the needle roller bearings. On no account attempt to remove a good bearing; it will almost certainly be damaged during removal.
22 Use the drawbolt system to fit the new bearings in their bores; do not attempt to drive them in with a drift or they will be damaged. Pack all bearings with molybdenum disulphide grease. Use the same grease on the bearing sleeves, then fit them into their bearings. Fit new seals to all bearings. In the case of the rocker arm needle roller bearing seals, note that there are two lips to each seal. Fit the smaller diameter lip towards the inside of the recess.

23.6 Rear damping adjustment is controlled from pull-out knob

23.7a Slacken nut (arrowed) to free air valve from bracket

23.7b Release locknut and unscrew the damper adjuster

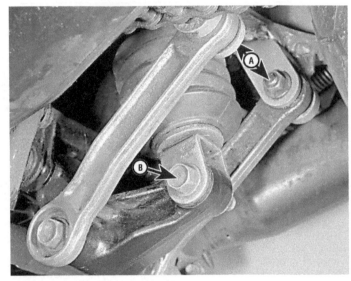

23.8 A: Tie-rod pivot bolts   B: Suspension unit lower mounting bolt

23.9 Upper mounting bolt can be reached from left-hand side

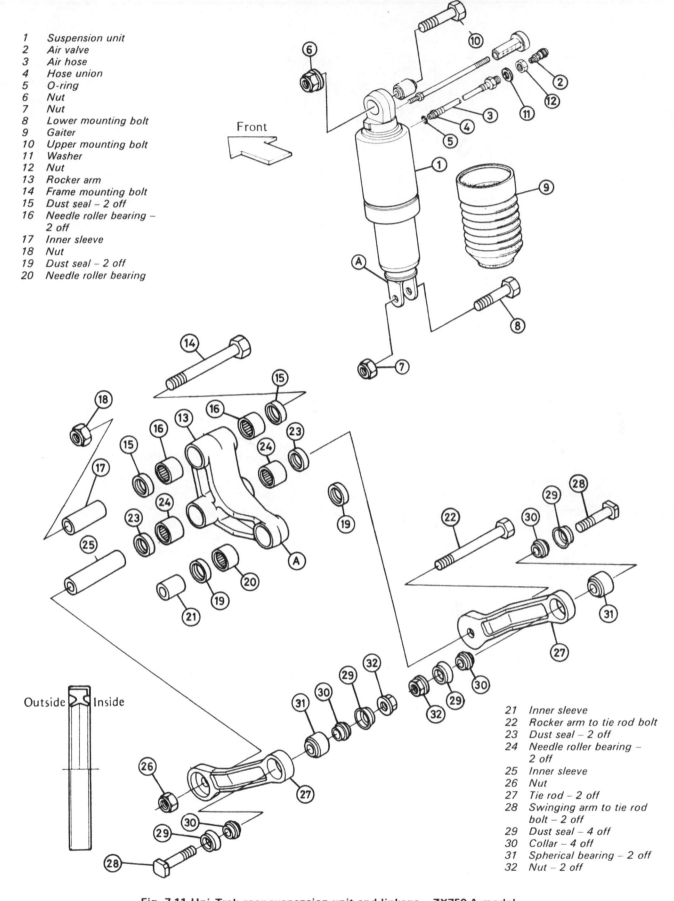

1   Suspension unit
2   Air valve
3   Air hose
4   Hose union
5   O-ring
6   Nut
7   Nut
8   Lower mounting bolt
9   Gaiter
10 Upper mounting bolt
11 Washer
12 Nut
13 Rocker arm
14 Frame mounting bolt
15 Dust seal – 2 off
16 Needle roller bearing –
    2 off
17 Inner sleeve
18 Nut
19 Dust seal – 2 off
20 Needle roller bearing

Front

Outside Inside

21 Inner sleeve
22 Rocker arm to tie rod bolt
23 Dust seal – 2 off
24 Needle roller bearing –
    2 off
25 Inner sleeve
26 Nut
27 Tie rod – 2 off
28 Swinging arm to tie rod
    bolt – 2 off
29 Dust seal – 4 off
30 Collar – 4 off
31 Spherical bearing – 2 off
32 Nut – 2 off

Fig. 7.11 Uni-Trak rear suspension unit and linkage – ZX750 A model

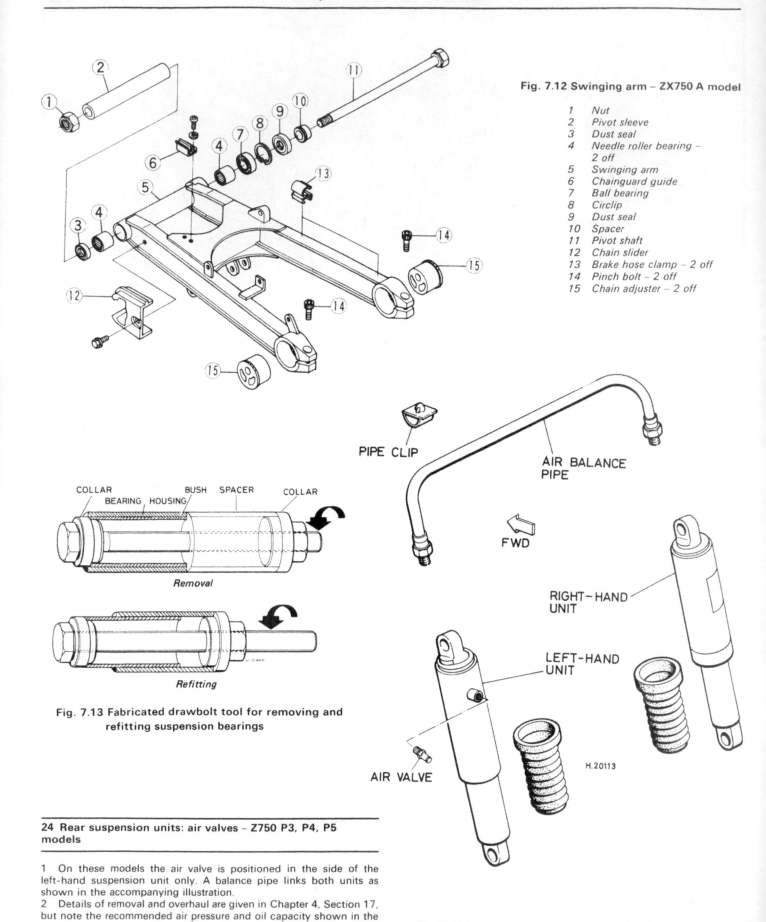

Fig. 7.12 Swinging arm – ZX750 A model

1   Nut
2   Pivot sleeve
3   Dust seal
4   Needle roller bearing –
    2 off
5   Swinging arm
6   Chainguard guide
7   Ball bearing
8   Circlip
9   Dust seal
10 Spacer
11 Pivot shaft
12 Chain slider
13 Brake hose clamp – 2 off
14 Pinch bolt – 2 off
15 Chain adjuster – 2 off

PIPE CLIP

AIR BALANCE
PIPE

FWD

RIGHT-HAND
UNIT

LEFT-HAND
UNIT

AIR VALVE

H.20113

COLLAR
BEARING HOUSING   BUSH SPACER   COLLAR

Removal

Refitting

Fig. 7.13 Fabricated drawbolt tool for removing and
refitting suspension bearings

**24 Rear suspension units: air valves – Z750 P3, P4, P5
models**

1   On these models the air valve is positioned in the side of the
left-hand suspension unit only. A balance pipe links both units as
shown in the accompanying illustration.
2   Details of removal and overhaul are given in Chapter 4, Section 17,
but note the recommended air pressure and oil capacity shown in the
Specifications section of this chapter.

Fig. 7.14 Rear suspension units – Z750 P3, P4, P5 models

25 Final drive chain adjusters: ZX750 A models

1   In place of the drawbolt type chain adjusters, fitted to all other chain drive models, are eccentric wheel spindle holders which are retained by pinch bolts in the fork end.
2   Check for chain free play with the machine on its centre stand. Turn the rear wheel to find the tightest point of the chain, and check the up and down free play. If this exceeds 45 mm (1.8 in) or is less than 35 mm (1.4 in), the chain free play should be adjusted.
3   Slacken the rear wheel spindle nut and the two adjuster pinch bolts, then use a screwdriver blade in the spindle end to rotate the spindle and eccentric blocks to obtain the required free play. Check that the wheel remains in alignment. To facilitate this, alignment marks are engraved on the flanged ends of the blocks, and these should be aligned with the index mark on the swinging arm. Ensure that the same mark is aligned on each side of the machine.
4   When adjustment is complete, tighten the pinch bolts to 3.3 kgf m (24.0 lbf ft) and the wheel spindle nut to 12.0 kgf m (87.0 lbf ft). Use a new split pin to secure the spindle nut.

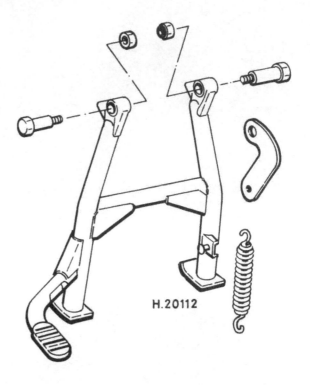

Fig. 7.15 Centre stand – ZX750 A and KZ/Z750 L3, L4 models

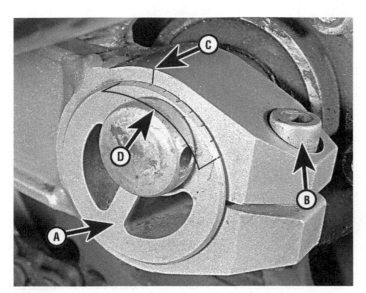

25.3 The rear wheel eccentric block adjusters (A), pinch bolts (B), swinging arm index mark (C) and adjuster wheel alignment mark range (D)

2   The fairing mounting arrangement can best be understood by referring to the accompanying line drawings. Note that although the external appearance of the A1 and A2 type fairings are similar, the mounting system was revised on the A2. The A3 and A5 versions were revised completely, the new fairing featuring lower sections and a belly pan to form a full sports fairing. Needless to say, yet another mounting arrangement was required.
3   When refitting the fairing ensure that the various washers and collars are fitted in the correct order, see the accompanying illustrations. Beware of overtightening, which will cause the fairing material to fracture.

26 Centre stand: examination – ZX750 A and KZ/Z750 L models

The centre stand fitted to these models differs from previous versions in its mounting arrangement. The pivot tube is replaced by two pivot bolts and nuts. These should be checked for security each time the stand pivots are lubricated. Apart from this, the remarks in Section 18 of Chapter 4 can be applied.

27 Fairing assembly: ZX750 A models

1   The ZX750 A models are all equipped with a fairing, the detail design of which altered according to the model year. In all cases this is a rather more elaborate affair than the small handlebar fairing fitted to the KZ/Z750 R1 model, being a frame-mounted, rather than a fork-mounted design.

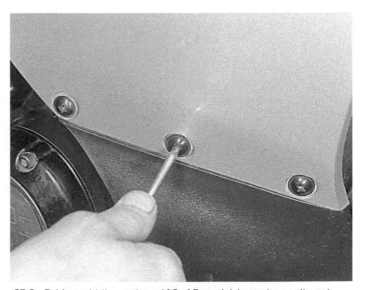

27.2a Fairing middle sections (A3, A5 models) can be unclipped after releasing the three screws along lower edge

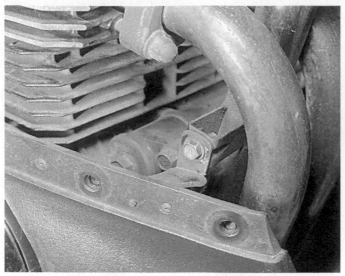

27.2b Fairing lower section, or belly pan, is retained by bolt at front edge ...

27.2c ... and screws at rear corners

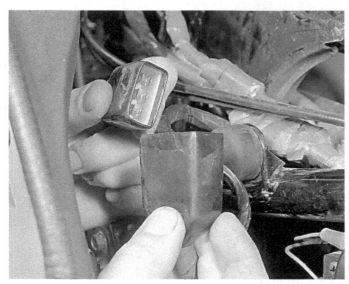

27.2d Separate fairing wiring connector (left-hand side)

27.2e Remove fairing subframe bolts and horn unit

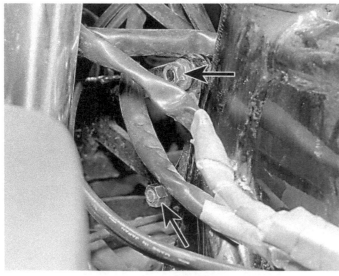

27.2f Release subframe to steering head nuts ...

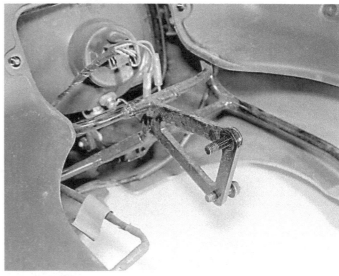

27.2g ... and lift fairing away. Note upper mounting is a stud, lower one a bolt

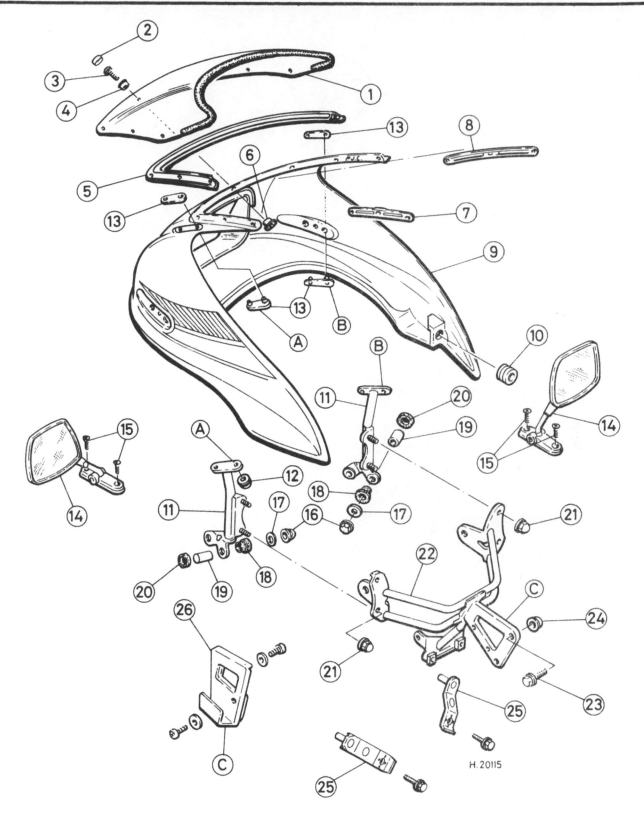

Fig. 7.16 Fairing – ZX750 A1 model

1	Screen	8	Right-hand plate	15	Screw – 4 off	22	Fairing mounting bracket
2	Cap – 7 off	9	Fairing	16	Nut – 4 off	23	Bolt – 2 off
3	Screw – 7 off	10	Grommet – 2 off	17	Washer – 4 off	24	Nut – 2 off
4	Collar – 7 off	11	Mirror mounting – 2 off	18	Damping rubber – 4 off	25	Rear mounting bracket –
5	Trim	12	Nut – 4 off	19	Spacer – 4 off		2 off
6	Nut – 7 off	13	Damping rubber – 4 off	20	Damping rubber – 4 off	26	Cable guide
7	Left-hand plate	14	Mirror – 2 off	21	Nut – 4 off		

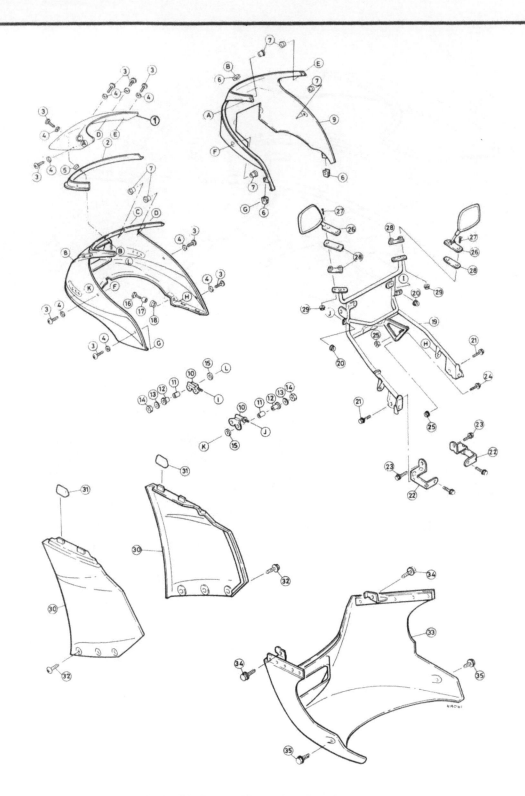

Fig. 7.17 Fairing – ZX750 A2, A3 and A5 models

1	Screen	11	Spacer – 4 off	21	Bolt – 4 off	30	Fairing middle sections*
2	Trim	12	Damping rubber – 4 off	22	Rear mounting bracket –	31	Damping rubber – 4 off*
3	Screw – 11 off	13	Washer – 4 off		2 off	32	Screw – 6 off*
4	Nylon washer – 11 off	14	Nut – 4 off	23	Bolt – 2 off	33	Fairing lower section (belly
5	Collar	15	Damping rubber – 4 off	24	Bolt		pan)*
6	Nut – 3 off	16	Bolt – 2 off	25	Nut – 2 off	34	Screw – 2 off*
7	Nut – 8 off	17	Spacer – 2 off	26	Mirror – 2 off	35	Bolt – 2 off*
8	Fairing upper section	18	Damping rubber – 2 off	27	Screw – 4 off		
9	Fairing inner	19	Fairing mounting bracket	28	Damping rubber – 4 off	*ZX750 A3 and A5 only
10	Turn signal mounting	20	Nut – 2 off	29	Nut – 4 off		
	bracket – 2 off						

28 Wheels: modifications – ZX750 A and Z750 P models

1 The ZX750 A models are fitted with cast aluminium alloy wheels with a revised spoke design as part of the general cosmetic update. The general arrangement is similar to the earlier models, but note the following:

Fitting and removing balance weights – ZX750 A

2 Balance weights are normally fitted whenever a new tyre is fitted onto the rim to counteract any small imbalance in the tyre, and thus to eliminate the high speed vibration that this would cause. Kawasaki produce balance weights designed specifically for the ZX750 models, available in 10 gram, 20 gram and 30 gram weights.
3 When fitting the weights, lubricate the rim, the weight and the tyre bead with soapy water to ease fitting. *Do not use oil or any other similar substance as a lubricant, or the tyre will be damaged.*
4 The accompanying sequence of line drawings illustrate the procedure for fitting and removing weights. Check the security of any balance weights when making regular checks of wheel condition. If a weight seems loose or badly fitted, remove and discard it and fit a new one of the same weight.
5 When tyres are fitted by an independent supplier, make sure that the wheel is balanced correctly, and that the weights fitted are suitable for this type of wheel. It is preferable to use original Kawasaki weights to avoid problems later. Do not forget that a weight flung off at speed becomes a dangerous projectile.

Rear wheel removal and refitting – ZX750 A

6 The rear wheel removal and refitting procedure is broadly similar to that described in Chapter 5, Section 4, but is influenced by the redesigned swinging arm assembly. During wheel removal, note that the spindle is fitted from the left-hand side.
7 Note also that the eccentric adjuster blocks which provide drive chain adjustment carry the wheel spindle in the ends of the swinging arm. Make sure that the blocks are pushed fully home in the swinging arm ends during installation, and check that chain free play adjustment and wheel alignment are correct before tightening the pinch bolts which secure the blocks.

Bearing removal and refitting – ZX750 A

8 The procedure for removing, checking and refitting the wheel bearings is as described in Chapter 5. In the case of the rear wheel, note that the spindle now passes through the hub from the left-hand side and is secured by a washer, castellated nut and split pin on the right. The spacer and bearing arrangement remain as shown in Fig. 5.2.

Rear wheel spindle – Z750 P2, P3, P4, P5

9 The rear wheel spindle on later Z750 P models is secured by a castellated nut and split pin instead of the plain flanged nut shown in Fig. 5.3. The rear wheel assembly remains otherwise unchanged.

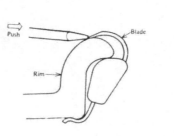
Removal with tyre removed

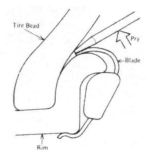
Removal with tyre installed

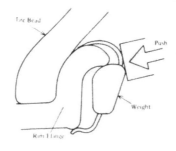
Installation

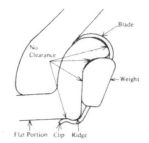
Correct fitted position

Fig. 7.18 Wheel balance weight fitting details – ZX750 A model

29 Braking system: modifications – ZX750 A models

1 The braking system in general is similar to that used on the earlier models, the major exception being the inclusion of the anti-dive system described earlier in this Chapter. The adoption of the anti-dive components has occasioned the fitting of extra hydraulic pipes, and thus a close check should be kept on these during regular inspections of the hydraulic system.
2 The anti-dive components have complicated the brake bleeding procedure somewhat. When bleeding the system, start by bleeding air from the caliper and the anti-dive valve bleed valves, then from the junction block bleed valve.
3 The anti-dive plunger unit and the metal pipe which connects it to the hydraulic system are service items. Renew either part if there are signs of leakage, corrosion or other damage. Renew the plunger unit every 2 years and the metal pipe and hoses every four years, regardless of appearance.
4 Other changes to the system include minor revisions to the master cylinders and the fitting of new calipers. These alterations have no material effect on overhaul procedures. For details, refer to Chapter 5, noting the line drawings of the new calipers which accompany this text.
5 A final point worth noting in the case of the earlier ZX750 models in particular, is that the rear caliper hose should be checked regularly for signs of chafing against the clips which secure it to the swinging arm. If there are signs of damage, renew the hose at once. Check the position of the clips, and reposition them, where necessary, as follows:
6 The rear clip should be positioned 15 mm (0.59 in) forward of the welded end of the swinging arm. The front clip should be positioned so that the distance between the rear edge of the rear clip and the rear edge of the front clip is 150 mm (5.90 in). On some early models, the front clip was positioned further forward. If necessary, reposition the clip as described.

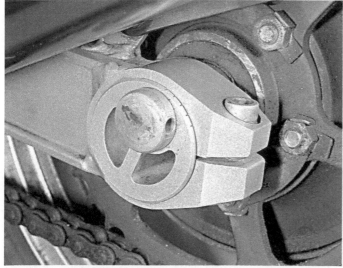
28.7 Check that adjuster block flange is tight against the swinging arm

29.2 Use bleed valve on anti-dive valve and caliper first, then union bleed valve

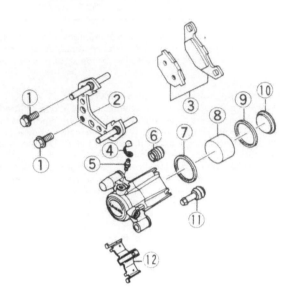

Fig. 7.19 Front brake caliper – ZX750 A model

1	Bolt – 2 off	7	Fluid seal
2	Mounting bracket	8	Piston
3	Brake pads	9	Dust seal
4	Bleed nipple cap	10	Piston cap
5	Bleed nipple	11	Dust cover
6	Dust cover	12	Anti-rattle spring

31 Electrical system: modifications – general

1   There have been numerous detail changes to the electrical systems fitted to the models covered in this manual. Most of the alterations are in line with styling changes, or to suit local market requirements, and are of little overall significance. Many of these are covered by the wiring diagrams which will be found at the end of this Chapter.

2   There have also been a few more major revisions, and these are discussed in detail in the Sections which follow. Items such as revised lamp unit designs are not covered here given the self-evident nature of the changes.

32 Junction box assembly: general description and testing – ZX750 A models

General description and component location

1   The junction box is housed behind the left-hand side panel and provides a centralised location for most of the electrical components normally found in scattered locations around the frame.

2   At the top left-hand corner of the panel is an accessory terminal rated at 10A maximum (there is a second accessory feed in the form of spare white/blue (+) and yellow/black (−) leads inside the headlamp shell, also rated at 10A maximum).

3   The junction box cover is retained by two screws. Once the cover has been detached, access can be gained to the bank of fuses ranged across the top of the panel. These are of the cartridge type and **must** be replaced with fuses of the correct type and rating. The amperage of each fuse is clearly marked on its outer edge. To remove and refit the fuses, a special puller is located to the right of the fuse bank. Spare fuses are housed below the puller tool (1 x 10A and 1 x 30A).

4   The junction box panel also carries the turn signal relay, plus a group of relays and diodes relating to the various electrical circuits. These are shown in the accompanying photographs.

Junction box panel removal and refitting

5   Before removing the panel or its relays, disconnect the battery to avoid any risk of accidental short circuits. Disconnect the negative (−) lead first, then the positive (+) lead.

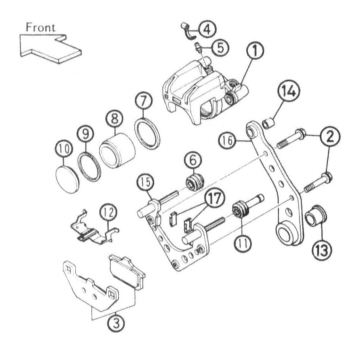

Fig. 7.20 Rear brake caliper – ZX750 A model

1	Caliper	10	Piston cap
2	Bolt – 2 off	11	Dust cover
3	Brake pads	12	Anti-rattle spring
4	Bleed nipple cap	13	Collar
5	Bleed nipple	14	Collar
6	Dust cover	15	Mounting bracket
7	Fluid seal	16	Torque stay
8	Piston	17	Pad stopper spring – 2 off
9	Dust cover		

30 Braking system: modifications – KZ/Z750 L models

The master cylinder was revised for the later L models, the main change being a redesigned piston profile and a change to a plunger-type brake light switch. These changes have no significant effect on overhaul or repair procedures, and the rest of the braking system remains similar to the earlier type. For details, refer to Chapter 5.

6   Remove the junction box cover (two screws). Release the two junction box panel holding bolts and lift the assembly away from the frame. Grasp the wiring connector, squeezing together the ends of the locking tangs to allow the wiring to be unplugged.
7   The relays can be removed after squeezing together the ends of the locking tangs and pulling the unit straight out of its socket. When refitting a relay, check that it is correctly aligned with its socket, then push it home squarely until a click indicates that it is locked in place.
8   The panel can be refitted by reversing the removal sequence. Note that the wiring connectors should be pushed fully home until they lock in place.

Testing the relay diodes

9   The headlamp relay (US models only) and the starter relay each have a set of three diodes interconnected with them, the diodes being used to prevent backfeeding in the relay circuits. If a fault is suspected, carefully remove the diode assembly in question from the panel and check it as follows.
10   Each of the three diode elements should conduct electricity in one direction only. Check this by setting a multimeter to the resistance scale and zeroing the meter needle. Connect the meter probes to each element in turn, and note the reading shown, then reverse the probes and repeat the check. For each of the three diodes, the meter should indicate zero, or very low resistance in one direction, and infinite, or very high resistance in the other. If any one of the three diodes reads high or low in both directions, fit a new diode assembly.

Testing the main, starter and headlamp relays

11   The three relays can be checked by removing them from the junction box panel and connecting the machine's battery and a multimeter set on the resistance (ohms x 1) scale as shown in the accompanying diagram. The meter should indicate zero resistance when the battery supply is connected, and infinite resistance when it is disconnected. If the relay does not operate normally, fit a new unit. Note that the headlamp relay and its diode block are not used on UK market models.

Testing the turn signal relay

12   Remove the turn signal relay from the junction box panel and connect the battery and one or more turn signal bulbs as shown in the accompanying diagram. Count the number of cycles (flashes) per minute and compare the result with the table shown. If the relay fails to operate at the specified rate, or works erratically, fit a new relay. If the relay works normally, but the fault persists, check the associated wiring, switch and bulbholders for poor connections.

Testing the junction box circuit board

13   Remove the junction box panel and disconnect all fuses, diodes, relays and connectors. Check all terminals for damage or corrosion and correct as necessary. Using a multimeter set on the resistance (ohms x 1) scale, check for continuity between the various terminals. Note that any pair of similarly numbered terminals should conduct (continuity), whilst any two dissimilar terminals should be isolated (no continuity). If the results obtained indicate an open or short circuit, fit a new junction box unit.

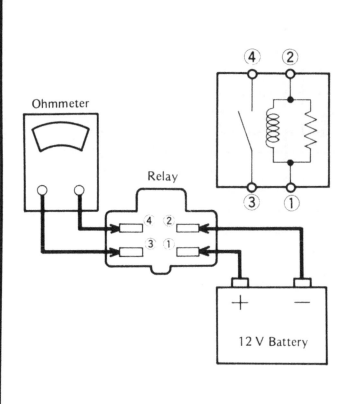

① and ② : Relay Coil Terminals
③ and ④ : Relay Switch Terminals

Fig. 7.21 Junction box main, starter and headlamp relay test – ZX750 A model

Turn Signal Relay

Turn Signal Lights

12 V Battery

Load			Flashing Times (c/m*)
The Number of Turn Signal Lights	Wattage (W)		
1	21 – 23		More than 150
2	42 – 46		
3	63 – 69		75 – 95
4	84 – 92		

∗ : Cycle(s) per minute

Fig. 7.22 Junction box turn signal relay test – ZX750 A model

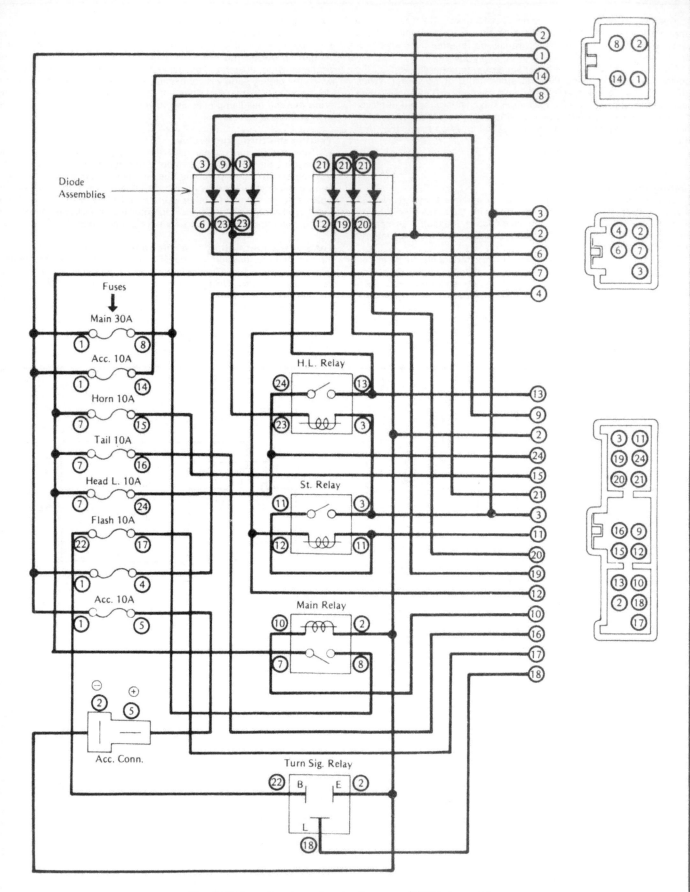

Fig. 7.23 Junction box circuit diagram – ZX750 A model

33 Regulator/rectifier unit: testing – ZX750 A models

The regulator test procedure is as described in Chapter 6, Section 6, paragraph 3 onwards. Note, however, that a revised test circuit is applicable to the ZX750 A models, and this is shown in the accompanying line drawing. No alternative rectifier test is available for these models. If a rectifier fault is suspected, the unit must be tested by substitution.

34 Starter circuit: testing – ZX750 A models

1   Check the starter circuit and starter components as described in Chapter 6, Section 8, noting the following points and referring to the accompanying circuit diagram.

2   To check the supply to the starter relay, remove the left-hand side panel and separate the two-pin connector from the relay. Note that it is essential that the battery is in good condition and fully charged prior to this test. Position the machine on the centre stand and connect a multimeter set to the 25 volt dc range between the yellow/red wire terminal (+ positive probe) and the black/yellow wire terminal (− negative probe). Turn the ignition switch to the 'On' position, check

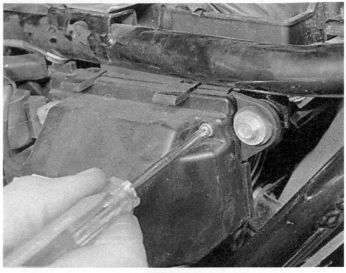

33.3a Junction box cover is retained by two screws

33.3b Use special puller to withdraw fuses

33.4a Turn signal relay plugs into panel holder

33.4b Other relays fit in a similar fashion

33.4c Note diode block(s) on either side of relays

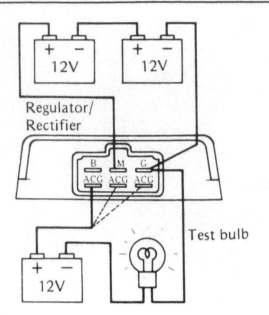

that the engine stop switch is in the 'Run' position, select neutral, and press the starter button. If the supply is correct, battery voltage should be indicated. If no reading is shown, check the wiring and connections, then check the following:

a)   Junction box
b)   Main fuse
c)   Starter circuit relay
d)   Ignition switch
e)   Engine stop (kill) switch
f)   Starter button
g)   Starter lockout switch
h)   Neutral switch
i)   Associated wiring and connectors

3   For details of the wire colours and related switch connections, refer to the wiring and circuit diagrams.

Fig. 7.24 Regulator test circuit – ZX750 A model

B	Battery terminal	ACG	Alternator terminals
M	Monitor terminal		(Yellow wires)
G	Earth terminal		
	(Black/yellow wire)		

Junction Box

Engine Stop/
Starter Switch

Starter
Circuit
Relay

Fuse 30A

Y/R   R   BK   BK

4 P or 6 P

12   11   3

18 P

8   1

4 P

Ignition
Switch

Y   R   BK   BK

Y/G   BK   Y/R

W   W/R

Y   Y

6 P

W   W

BK/Y

Y/G

BK/Y

Y/R   BK/Y

W/R

LG

9 P

2 P

BK/Y

2 P

BK/Y   Y/G   LG

Starter
Motor

Starter
Lockout
Switch

Neutral
Switch

Starter
Relay

Battery

BK	Black	R	Red
G	Green	W	White
LG	Light green	Y	Yellow

Fig. 7.25 Starter circuit diagram – ZX750 A model

35 Headlamp circuit: testing – ZX750 A US models

1 The US ZX750 A models feature a relay in the headlamp circuit which controls the switching of the lamp. When the ignition switch is turned on, the headlamp remains off until the engine is started. The circuit then remains on until the ignition switch is turned off. If the engine is stalled during riding, the headlamp is turned off while the starter motor is in operation.

2 In the event of a fault in the system, it should be checked as follows, making reference to the accompanying circuit diagram. Note that it is assumed that the charging and starting systems are working normally. Set a multimeter to the 25 volt dc range. Locate the 6-pin connector to the reserve lighting unit, and without separating it, connect the positive (+) meter probe to the blue lead terminal and the negative (−) probe to the black/yellow lead terminal.

3 Switch on the ignition and check that battery voltage is indicated on the meter. If the starter switch is operated, the meter should indicate

zero volts.

4 If the above test gives the expected results, the fault can be assumed to be in one of the following areas:

a) Wiring or connections
b) Bulb(s)
c) Dip (dimmer) switch
d) Reserve lighting device (see Chapter 6, Section 13)

5 If the above test failed to give the expected results, check the following areas:

a) Junction box
b) Fuses (main and headlamp)
c) Relays (main and headlamp)
d) Diode assembly
e) Ignition switch
f) Wiring and connections (see accompanying switch connection diagram)

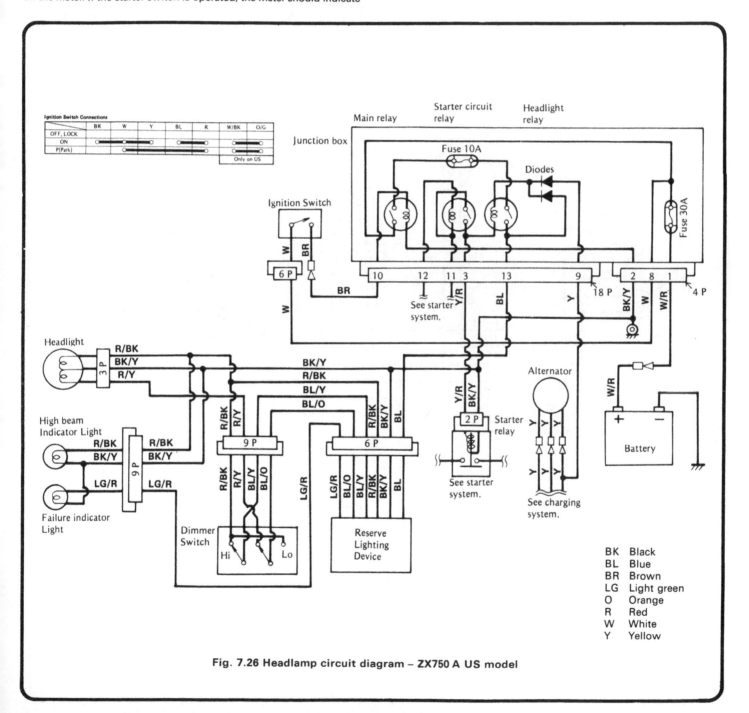

Fig. 7.26 Headlamp circuit diagram – ZX750 A US model

36 LCD fuel gauge and warning unit: testing – ZX750 A models

1   The fuel gauge and warning system fitted to the ZX750 A models is similar to that used on the earlier R1 GPz model, but differs in some respects. Refer to Sections 22 onwards in Chapter 6, and to the accompanying circuit diagram, noting the following points.
2   The 'ENG SW' function (indicating the position of the engine stop switch) is deleted, the two-segment red LCD remaining as a warning that the side stand is down, in which case it flashes. A separate warning lamp housed in the instrument console also flashes when one of the LCD segments flashes.

Power supply test

3   Refer to Section 24 of Chapter 6, noting the following. The test is performed at the female side of the gauge/warning unit block connector, with the connector halves separated. Set the meter to the 25 volts dc range and connect the meter positive (+) probe to the brown/white lead, and the negative (−) probe to the black/yellow lead. With the ignition switch on, battery voltage should be indicated, with zero volts being shown when it is switched off. If the test fails to produce these readings, check the wiring, connections and fuse (see circuit diagram).

Gauge and warning unit tests

4   Refer to Chapter 6, Section 26, noting the following revisions to the test procedure:
5   The test is performed with the gauge and warning unit removed from the fuel tank panel. You will also require seven test leads; two for connection to the battery supply, four to simulate the four sensor leads, and one to operate the warning lamp. Take great care not to cause short circuits when connecting up the test leads.
6   Connect one lead between the green/yellow terminals of the male and female halves of the fuel gauge/warning unit connector to supply the warning lamp.
7   Connect the four sensor simulating leads as follows:

Circuit	Connections
Side stand switch	Green/white lead terminal to battery (−)
Oil level sensor	Blue/red lead terminal to battery (−)
Battery electrolyte sensor	Pink lead terminal to battery (+)
Fuel gauge/warning unit	White/yellow lead terminal to battery (−)

8   At this stage of the test, the nine fuel gauge segments should appear on the LCD panel. The warning lamp, and the stand, oil and battery warning LCDs should remain unlit (see the accompanying illustration).

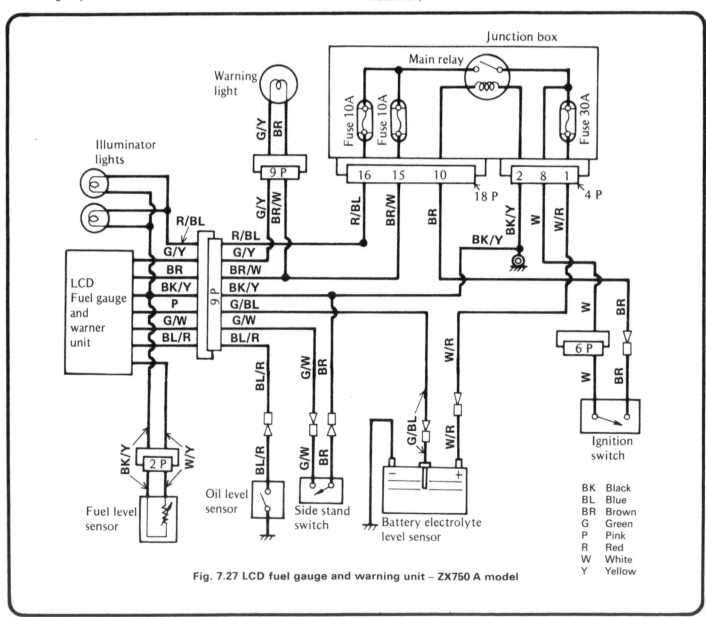

Fig. 7.27 LCD fuel gauge and warning unit – ZX750 A model

9 Now connect the unit to the battery, being very careful to avoid short circuits. Connect the battery positive (+) terminal to the brown terminal in the connector, and the battery negative (−) terminal to the connector black/yellow terminal.

10 With the unit supplied from the battery, the self-checking sequence is initiated. Wait for the checking procedure to finish, then check that display returns to the condition shown before the battery was connected. If the self-checking process fails to start, does not complete or if there is a display fault, fit a new fuel gauge/warning unit.

11 Disconnect each of the sensor simulating leads in turn. In each case, the warning lamp and the relevant LCD should flash (see accompanying illustration). Note that in the case of the fuel gauge display, the first segment (marked E) will flash, but note that there is a timing circuit to delay and thus stabilise the fuel gauge display. This means that each segment requires 3 − 12 seconds to appear or disappear, and the first segment will require a further 3 − 7 seconds before it starts to flash. If any one of the warning or fuel gauge functions fails to work normally, the unit must be renewed.

Wiring, connector and sensor test

12 Refer to Chapter 6, Section 25, noting the following test table which should be applied instead of Fig. 6.14.

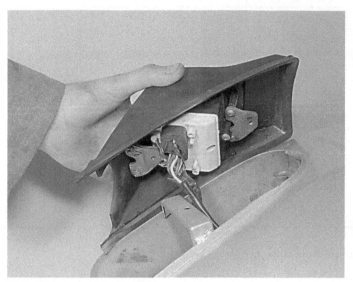

36.5a Remove tank fuel/warning panel ...

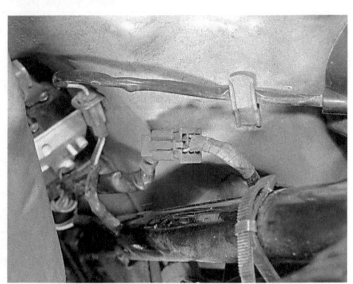

36.5b ... after freeing wiring connectors below tank (ZX750A models)

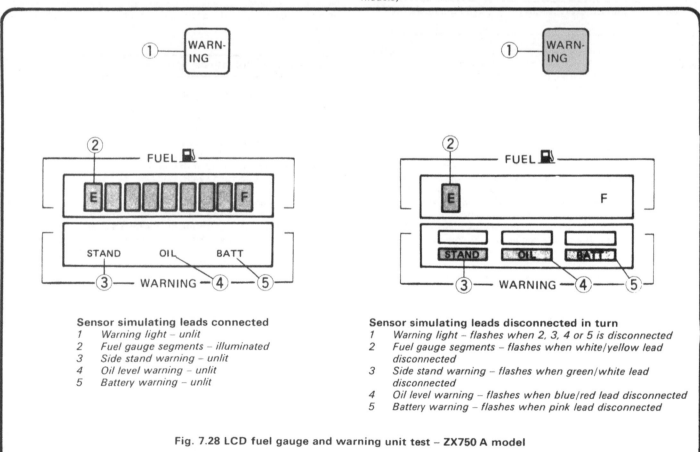

Sensor simulating leads connected
1 Warning light – unlit
2 Fuel gauge segments – illuminated
3 Side stand warning – unlit
4 Oil level warning – unlit
5 Battery warning – unlit

Sensor simulating leads disconnected in turn
1 Warning light – flashes when 2, 3, 4 or 5 is disconnected
2 Fuel gauge segments – flashes when white/yellow lead disconnected
3 Side stand warning – flashes when green/white lead disconnected
4 Oil level warning – flashes when blue/red lead disconnected
5 Battery warning – flashes when pink lead disconnected

Fig. 7.28 LCD fuel gauge and warning unit test – ZX750 A model

Wiring and Connector Test

Wire	Meter Range	Meter Connections	Meter Reading (Criteria)
Side stand warner	x 1 Ω	○One meter lead → Green/white wire ○Other meter lead → Black/yellow wire	○0 Ω when side stand is up. ○∞ Ω when side stand is down.
Oil level warner	x 10 Ω	○One meter lead → Blue/red wire ○Other meter lead → Black/yellow wire	○Less than 0.5 Ω when engine oil level is higher than "lower level line" next to the oil level gauge. ○∞ Ω when engine oil level is much lower than the "lower level line."
Battery electrolyte level warner	10 V DC	○Meter (+) → Pink wire ○Meter (−) → Black/yellow wire	○More than 6 V when electrolyte level is higher than "lower level line." ○0 V when electrolyte level is lower than "lower level line."
Fuel gauge and low fuel warner	x 10 Ω	○One meter lead → White/yellow wire ○Other meter lead → Black/yellow wire	○1 – 117 Ω

Fig. 7.29 LCD fuel gauge and warning unit circuit wiring check – ZX750 A model

37 Electronic tachometer: testing – KZ/Z750 L and KZ750 F1 models

The electronic tachometer circuit used on the L3 and L4 models and on the US KZ750 F1 model is similar to that described in Section 27 of Chapter 6. Note however, that no voltmeter function is included. In consequence, a changeover switch is not fitted, and the checks described in paragraph 4 onwards may be disregarded.

38 Fuel gauge circuit: testing – KZ/Z750 L models

1   The L3 and L4 models are equipped with a conventional fuel gauge system operated from a float type sender unit housed in the tank.
2   If the gauge itself is suspected of being faulty, disconnect the 2-pin connector from the sender unit. Switch on the ignition and check that the gauge reads 'E' (empty). Now connect a length of insulated wire between the black/yellow and the white/yellow wires on the female side of the connector. The gauge needle should now swing across to the 'F' (full) position. If this does not occur, the gauge can be considered faulty and should be renewed.
3   The fuel level sender unit can be checked in the same way as

described in Chapter 6, Section 25, paragraphs 5 and 6. For details of the test connections for the power supply to the gauge circuit, refer to Fig. 6.13.

39 Electrical modifications: general

1   There are a number of detail changes to the electrical system components of the later models. In general these are of a cosmetic nature, such as the fitting of a differently shaped headlamp or revised stop/tail lamp lens. These changes have little effect on working procedures other than repositioning of fixing screws or bolts.
2   In the case of the ZX750 A (GPz) models, the method of headlamp mounting changed in line with the different fairings fitted. Inspection of the model concerned will indicate the headlamp mounting arrangement.
3   Instrument panel arrangements vary considerably, again for styling reasons. In some cases the electrical side of the instrumentation is simplified, this being the case with the L3 and L4 models which are not fitted with the computer warning system. Where procedural alterations occurred, these are described in the preceding Sections and in the accompanying wiring diagrams.

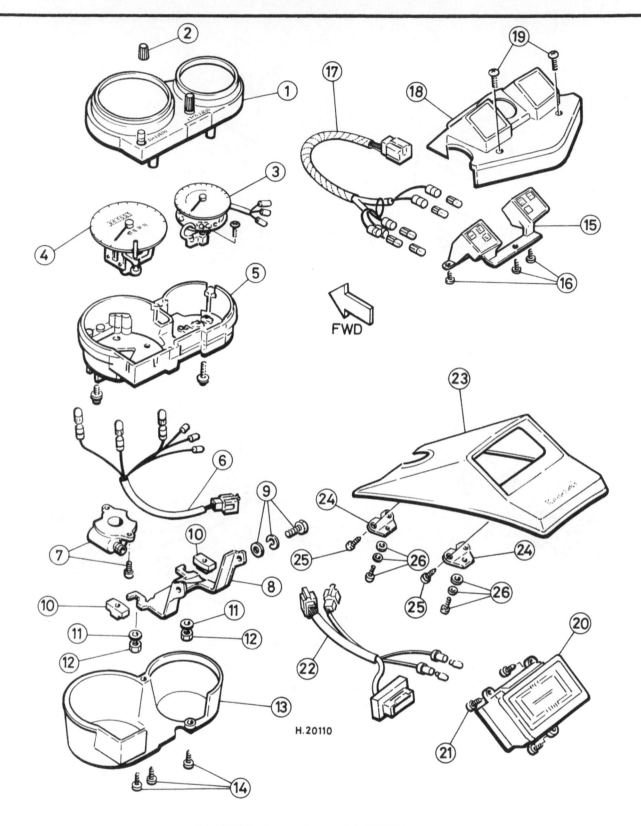

Fig. 7.30 Instrument console – ZX750 A model

1	Top cover	7	Speedometer drive	14	Screw – 3 off	21	Screw – 3 off
2	Reset knob	8	Mounting bracket	15	Warning light display	22	LCD unit wiring
3	Tachometer	9	Bolt and washers – 2 off	16	Screw – 3 off	23	Cover
4	Speedometer	10	Damping rubber – 2 off	17	Warning light wiring	24	Mounting bracket – 2 off
5	Instrument housing	11	Washer – 2 off	18	Cover	25	Screw – 4 off
6	Instrument illumination wiring	12	Nut – 2 off	19	Screw – 2 off	26	Screw and washers – 2 off
		13	Lower cover	20	LCD unit		

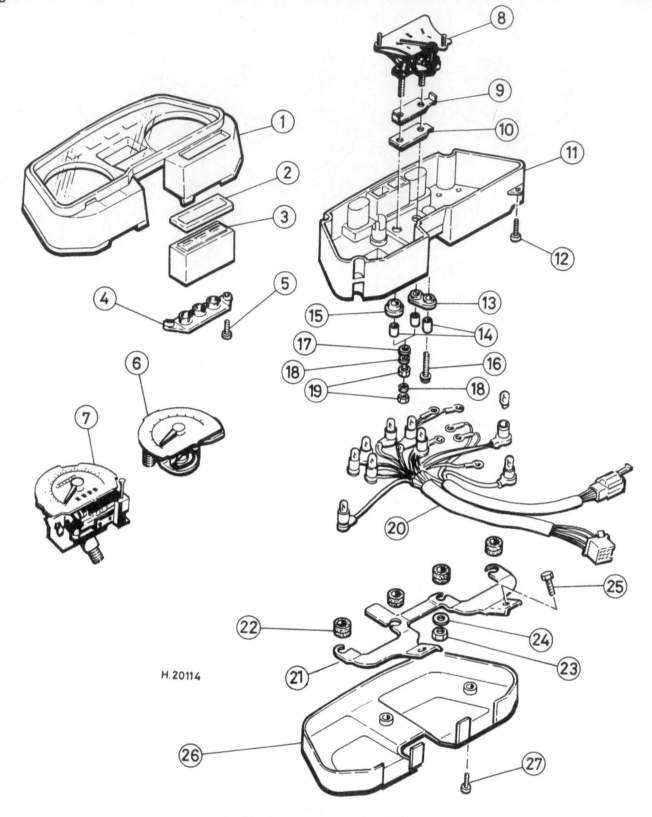

Fig. 7.31 Instrument console – Z750 L model

H.20114

1	Top cover	8	Fuel gauge	15	Damping rubber
2	Plate	9	Plate	16	Screw
3	Warning lamp display	10	Damping rubber	17	Washer – 2 off
4	Bulb holder insert	11	Instrument housing	18	Spring washer – 4 off
5	Screw – 2 off	12	Screw – 4 off	19	Nut – 4 off
6	Tachometer	13	Damping rubber	20	Instrument wiring
7	Speedometer	14	Spacer – 3 off	21	Mounting bracket
				22	Grommet – 4 off
				23	Nut – 2 off
				24	Washer – 2 off
				25	Bolt – 2 off
				26	Lower cover
				27	Screw – 3 off

Wiring diagram – component key

1	Battery	35	Front brake lamp switch
2	Alternator	36	Engine stop switch
3	Regulator/rectifier	37	Starter button
4	Starter motor	38	Fuel level sender
5	Starter solenoid	39	Headlamp switch
6	Fuses	40	Reserve lighting unit
7	Ignition coil	41	Headlamp flash button
8	IC ignitor	42	Horn button
9	Spark plugs	43	Indicator hazard switch
10	Pick-up coil	44	Turn signal switch
11	Headlamp	45	Headlamp dip switch
12	Parking lamp	46	Starter lockout switch
13	Speedometer lamp	47	Combination switch
14	Tachometer lamp	48	Side-stand switch
15	Voltmeter	49	Neutral switch
16	Neutral indicator lamp	50	Oil pressure switch
17	Left-hand turn signal warning lamp	51	Horn
		52	Indicator control unit
18	High beam warning lamp	53	Oil level sender
19	Low fuel level warning lamp	54	Brake lamp failure switch
		55	Tail lamp
20	Right-hand turn signal warning lamp	56	Right-hand rear turn signal
		57	Left-hand rear turn signal
21	Oil pressure warning lamp	58	Rear brake lamp switch
22	Headlamp bulb failure warning lamp	59	Indicator relay
		60	Indicator hazard relay
23	Distance sensor	61	Electrical accessories wiring
24	Brake lamp failure warning lamp	62	Warning lamp self-checker
25	Voltmeter lamp	63	Battery electrolyte level sensor
26	LCD display unit		
27	Reserve lighting warning lamp	64	Fuel gauge
		65	LCD unit warning light
28	Voltmeter switch	66	Junction box
29	Battery electrolyte level warning lamp	67	LCD unit lights
		68	Tachometer
30	Side-stand warning lamp	69	Side-stand relay
31	Oil level warning lamp	70	Side-stand switch release button
32	Right-hand front turn signal		
		71	Diode block
33	Left-hand front turn signal	72	Starter control relay
34	Ignition switch		

200

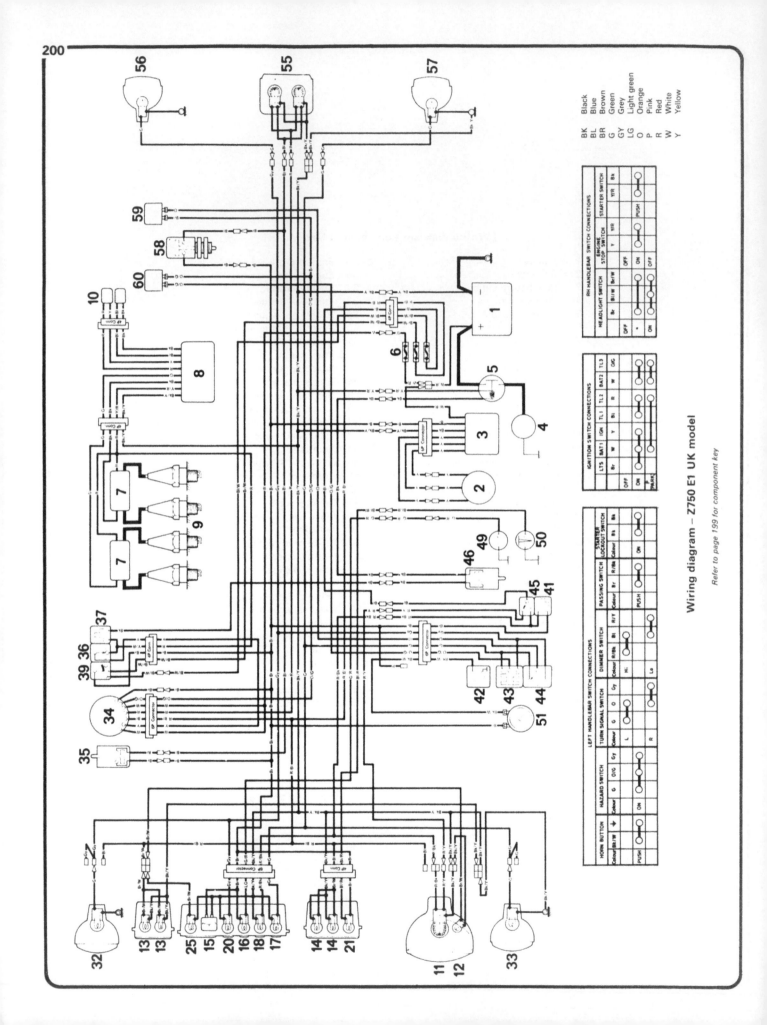

Wiring diagram – Z750 E1 UK model

Refer to page 199 for component key

BK   Black
BL   Blue
BR   Brown
G   Green
GY   Grey
LG   Light green
O   Orange
P   Pink
R   Red
W   White
Y   Yellow

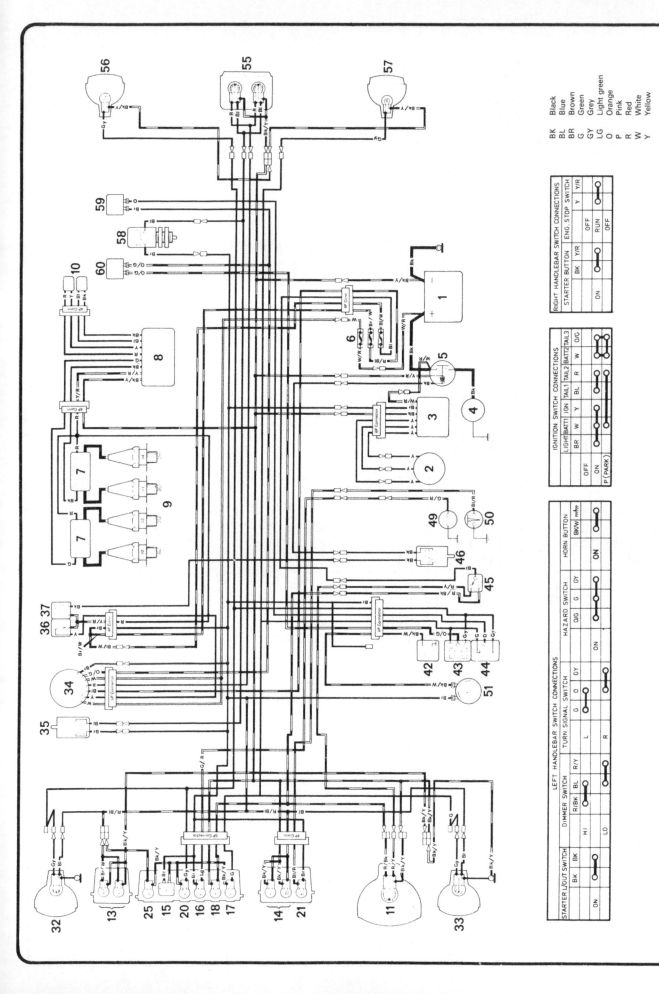

Wiring diagram – KZ750 E1 US model

Refer to page 199 for component key

BK - Black
BL - Blue
BR - Brown
G - Green
GY - Grey
LG - Light green
O - Orange
P - Pink
R - Red
W - White
Y - Yellow

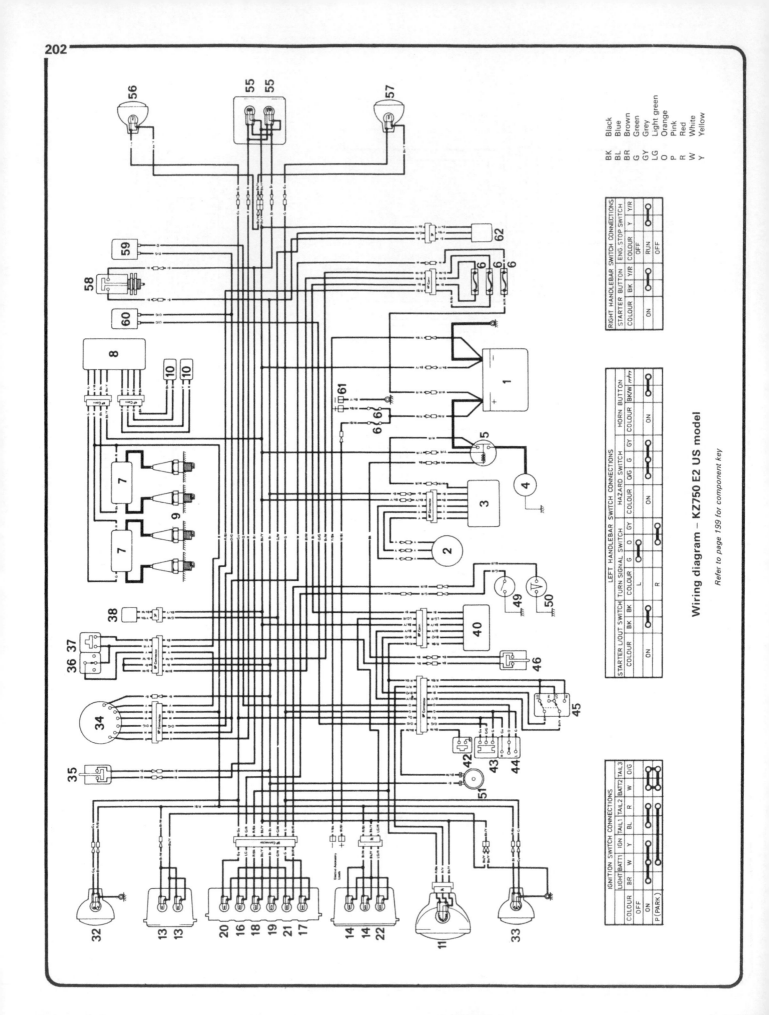

Wiring diagram – KZ750 E2 US model

Refer to page 199 for component key

BK	Black	
BL	Blue	
BR	Brown	
G	Green	
GY	Grey	
LG	Light green	
O	Orange	
P	Pink	
R	Red	
W	White	
Y	Yellow	

RIGHT HANDLEBAR SWITCH CONNECTIONS

STARTER BUTTON		ENG. STOP SWITCH	
COLOUR	BK	Y/R	COLOUR	Y	Y/R
		OFF		OFF
		RUN		
	ON			

LEFT HANDLEBAR SWITCH CONNECTIONS

STARTER L/OUT SWITCH		TURN SIGNAL SWITCH			HAZARD SWITCH			HORN BUTTON	
COLOUR	BK	COLOUR	G	L	GY	COLOUR	O/G	G	GY	COLOUR	BK/W
											ON
	ON	L					ON				
		R									

IGNITION SWITCH CONNECTIONS

	LIGHT	BATT	IGN	TAIL1	TAIL 2	BATT2	TAIL3	
COLOUR	BR	W	Y	BL	R	W	O/G	
OFF								
ON								
P (PARK)								

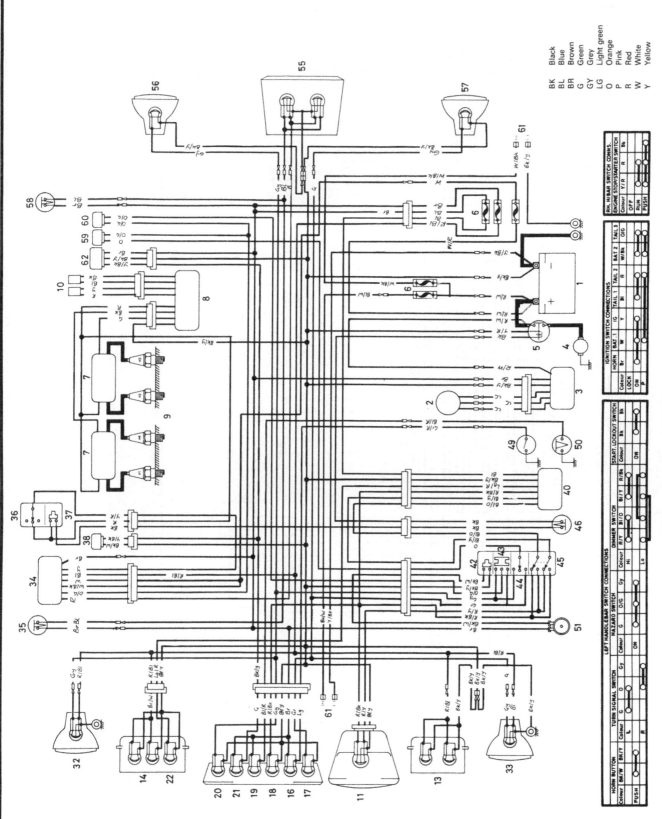

Wiring diagram – KZ750 E3 US model

Refer to page 199 for component key

BK Black
BL Blue
BR Brown
G Green
GY Grey
LG Light green
O Orange
P Pink
R Red
W White
Y Yellow

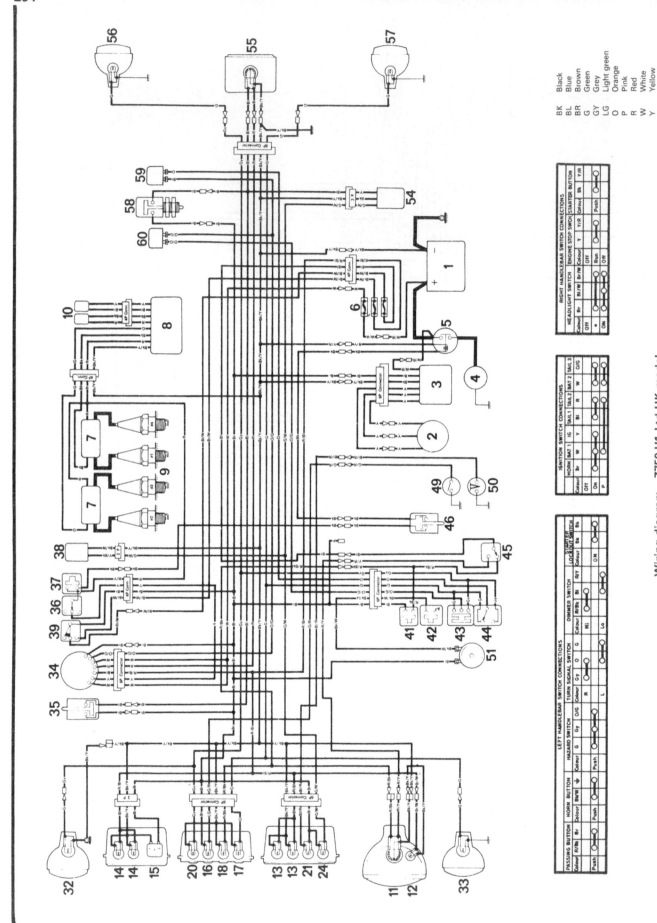

Wiring diagram – Z750 H1 Ltd UK model

Refer to page 199 for component key

BK Black
BL Blue
BR Brown
G   Green
GY Grey
LG Light green
O   Orange
P   Pink
R   Red
W   White
Y   Yellow

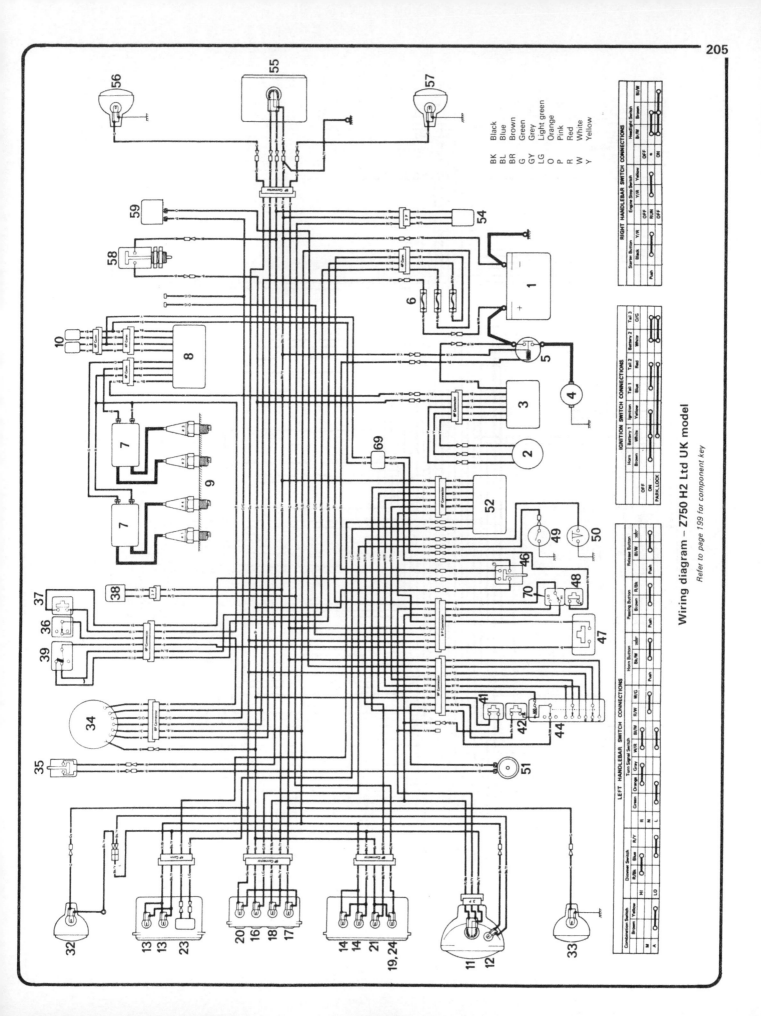

Wiring diagram – Z750 H2 Ltd UK model

Refer to page 199 for component key

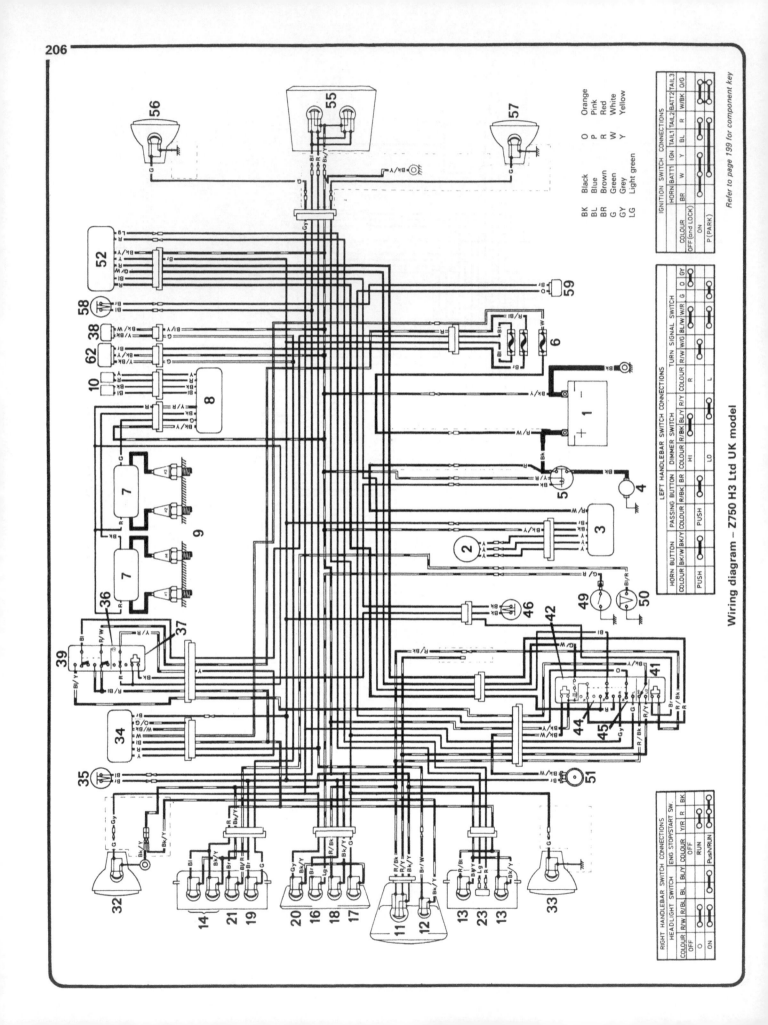

Wiring diagram – Z750 H3 Ltd UK model

Refer to page 199 for component key

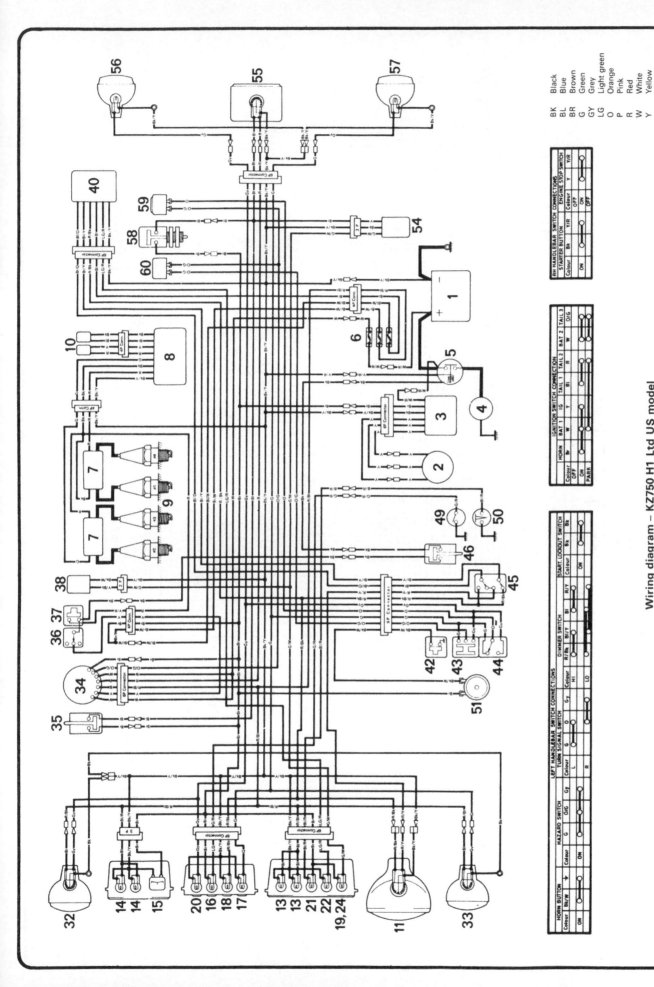

Wiring diagram – KZ750 H1 Ltd US model

Refer to page 199 for component key

BK   Black
BL   Blue
BR   Brown
G   Green
GY   Grey
LG   Light green
O   Orange
P   Pink
R   Red
W   White
Y   Yellow

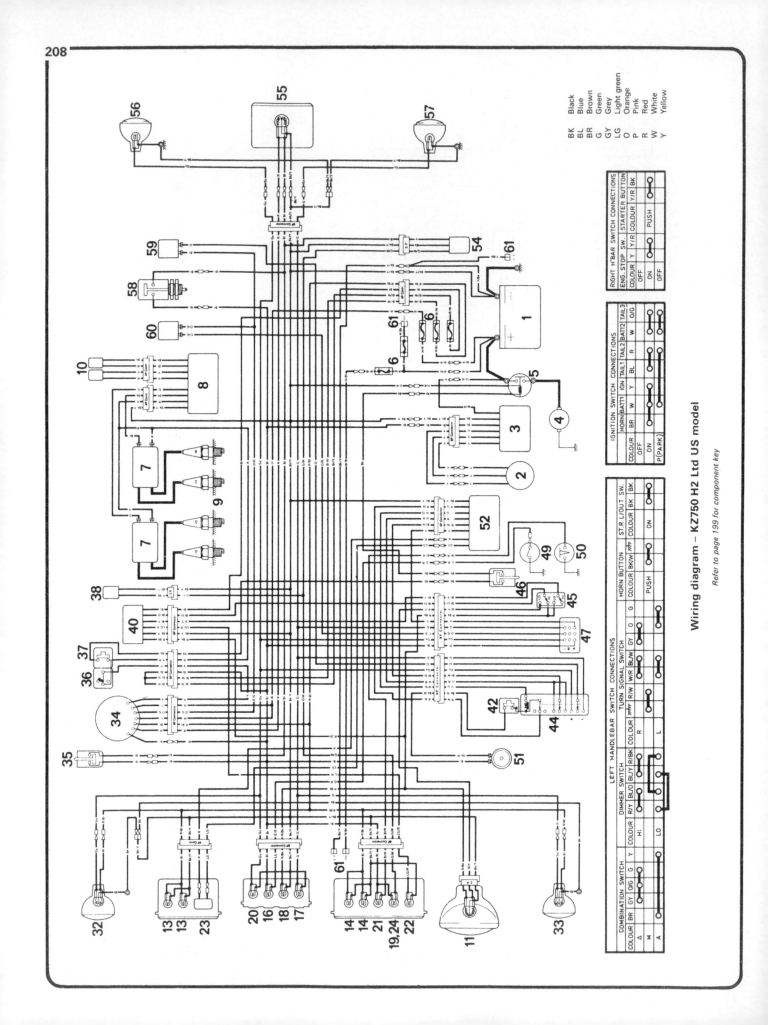

Wiring diagram – KZ750 H2 Ltd US model

Refer to page 199 for component key

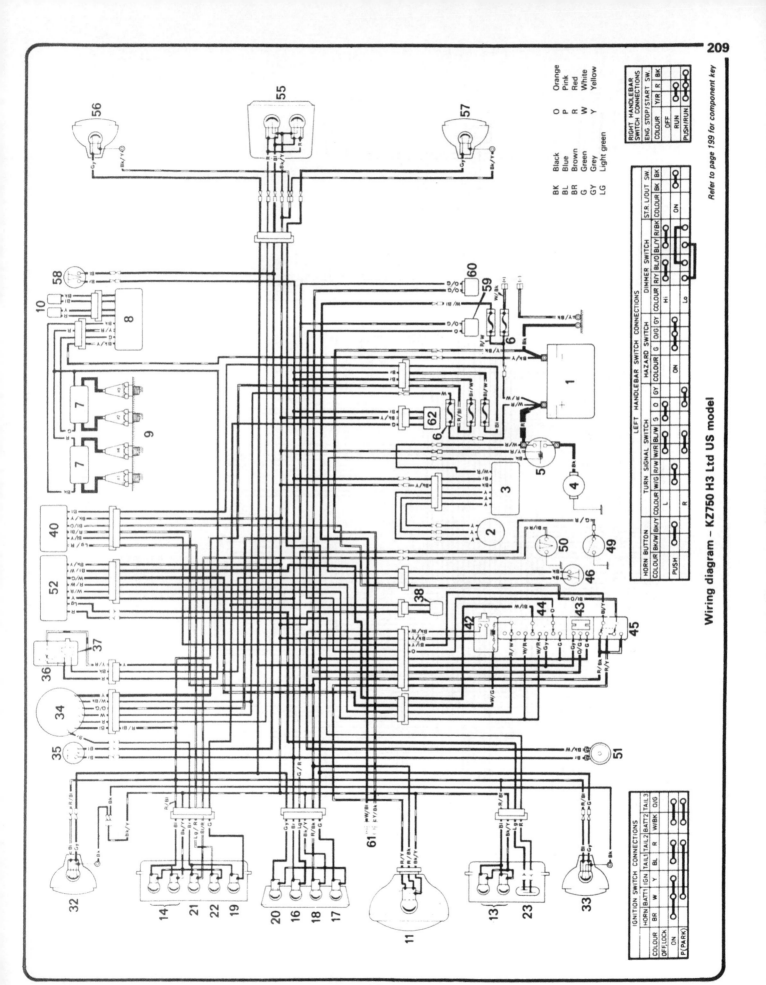

Wiring diagram – KZ750 H3 Ltd US model

Refer to page 199 for component key

210

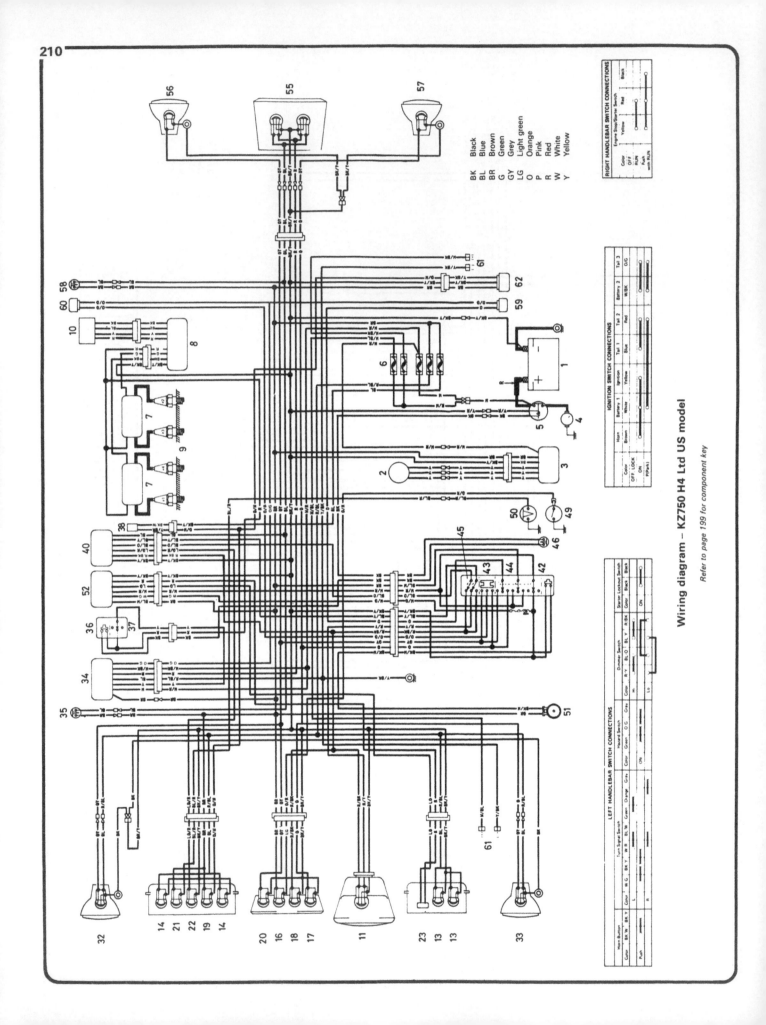

Wiring diagram – KZ750 H4 Ltd US model

Refer to page 199 for component key

Color key:
BK — Black
BL — Blue
BR — Brown
G — Green
GY — Grey
LG — Light green
O — Orange
P — Pink
R — Red
W — White
Y — Yellow

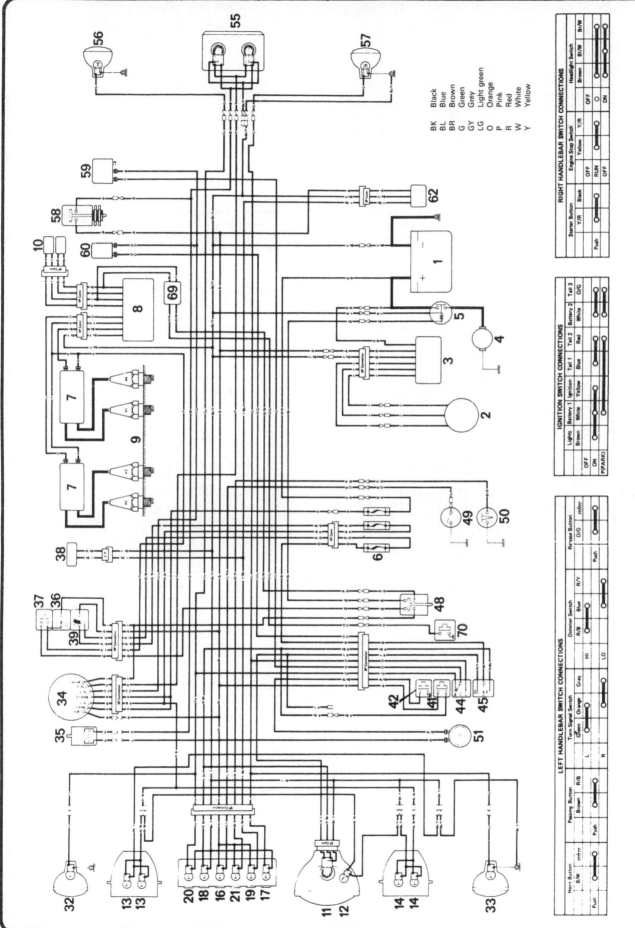

BK Black
BL Blue
BR Brown
G   Green
GY Grey
LG Light green
O   Orange
P   Pink
R   Red
W   White
Y   Yellow

Wiring diagram – Z750 L1 UK model

Refer to page 199 for component key

RIGHT HANDLEBAR SWITCH CONNECTIONS

Starter Button		Engine Stop Switch		Headlight Switch	
	Y/R	Black	Yellow	Y/R	Brown	Bl/W	Br/W
			OFF		OFF		
			RUN		ON		
Push			OFF				

IGNITION SWITCH CONNECTIONS

Lights	Battery 1	Ignition	Tail 1	Tail 2	Battery 2	Tail 3
	White	Yellow	Blue	Red	White	O/G
	Brown					
OFF						
ON						
P(PARK)						

LEFT HANDLEBAR SWITCH CONNECTIONS

Passing Button		Turn Signal Switch			Dimmer Switch			Release Button	
	R/B	Green	Orange	Grey	R/B	Blue	R/Y	O/G	
	Brown	L	R		HI	LO			
Push								Push	

Horn Button	
	B/W
Push	

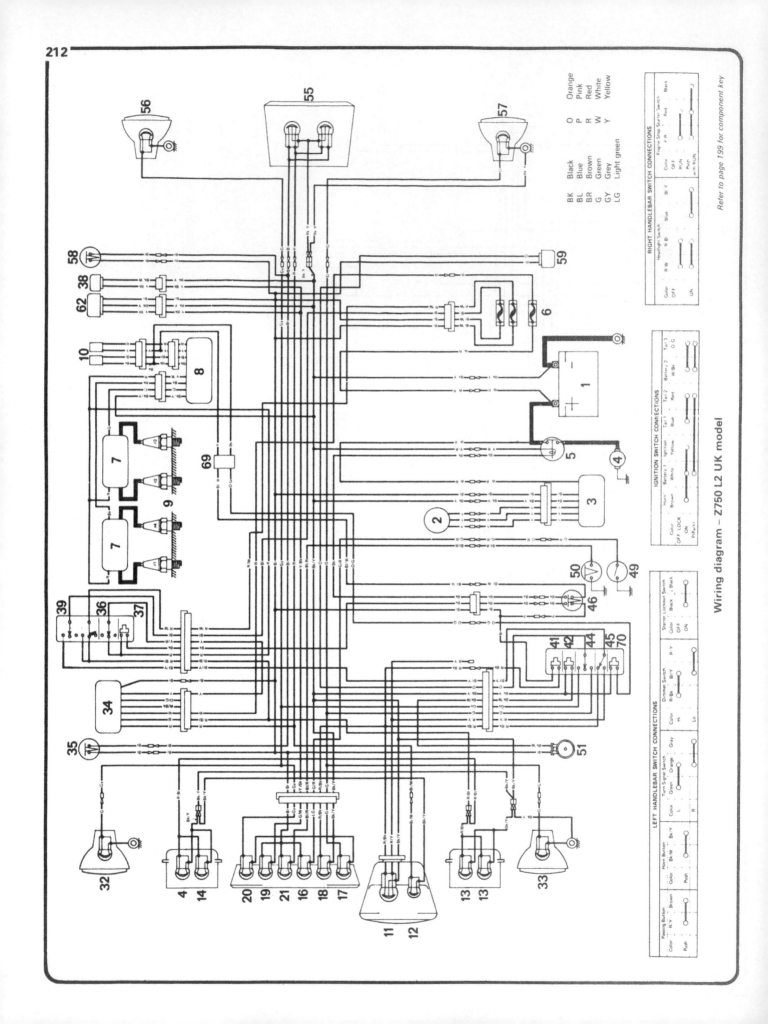

Wiring diagram – Z750 L2 UK model

Refer to page 199 for component key

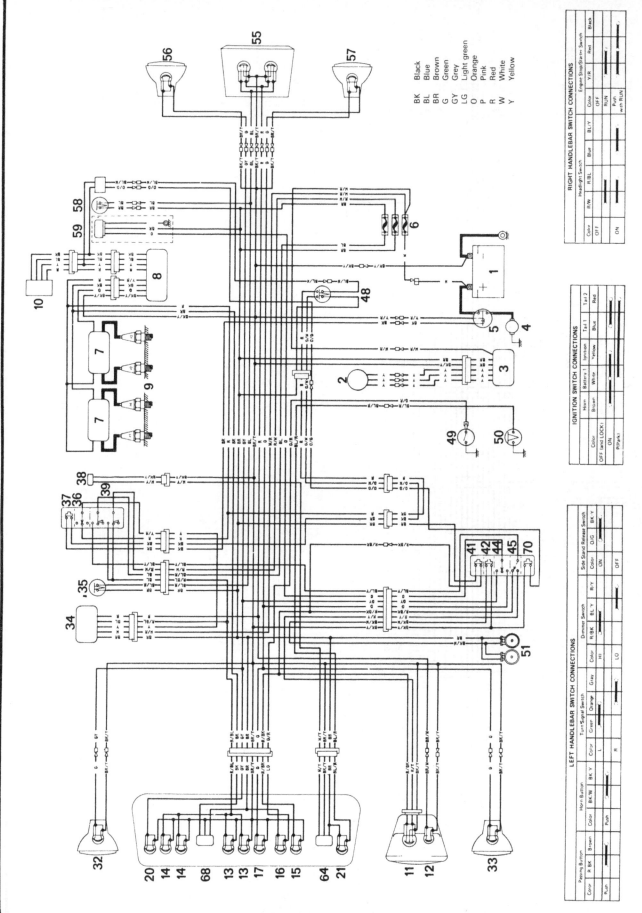

BK Black
BL Blue
BR Brown
G Green
GY Grey
LG Light green
O Orange
P Pink
R Red
W White
Y Yellow

Wiring diagram – Z750 L3 UK model

Refer to page 199 for component key

RIGHT HANDLEBAR SWITCH CONNECTIONS

Engine Stop/Starter Switch

Color	Y/R	BL/Y	Blue	R/BL	Red	Black
OFF						
RUN						
Push with RUN						

Headlight Switch

Color	R/W				
OFF					
ON					

IGNITION SWITCH CONNECTIONS

	Horn	Battery 1	Ignition	Tail 1	Tail 2
Color	Brown	White	Yellow	Blue	Red
OFF (and LOCK)					
ON					
P (Park)					

LEFT HANDLEBAR SWITCH CONNECTIONS

Side Stand Release Switch

Color	R/Y	O/G	BK·Y
ON			
OFF			

Dimmer Switch

Color	R/BK	BL·Y
HI		
LO		

Turn Signal Switch

Color	Orange	Green	Grey
L			
R			

Horn Button

Color	BK·W	BK·Y
Push		

Passing Button

Color	R BK	Brown
Push		

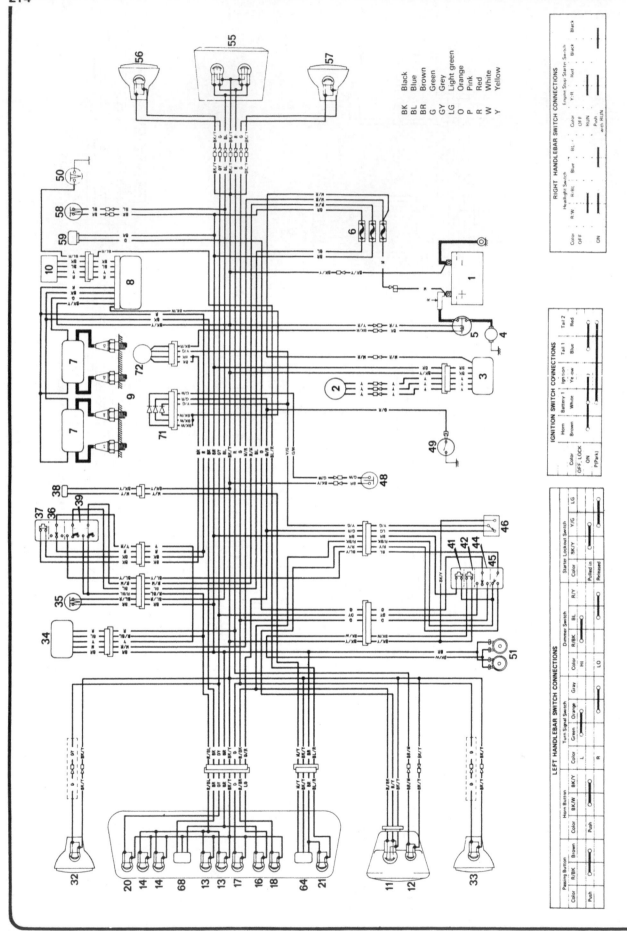

Wiring diagram – Z750 L4 UK model

Refer to page 199 for component key

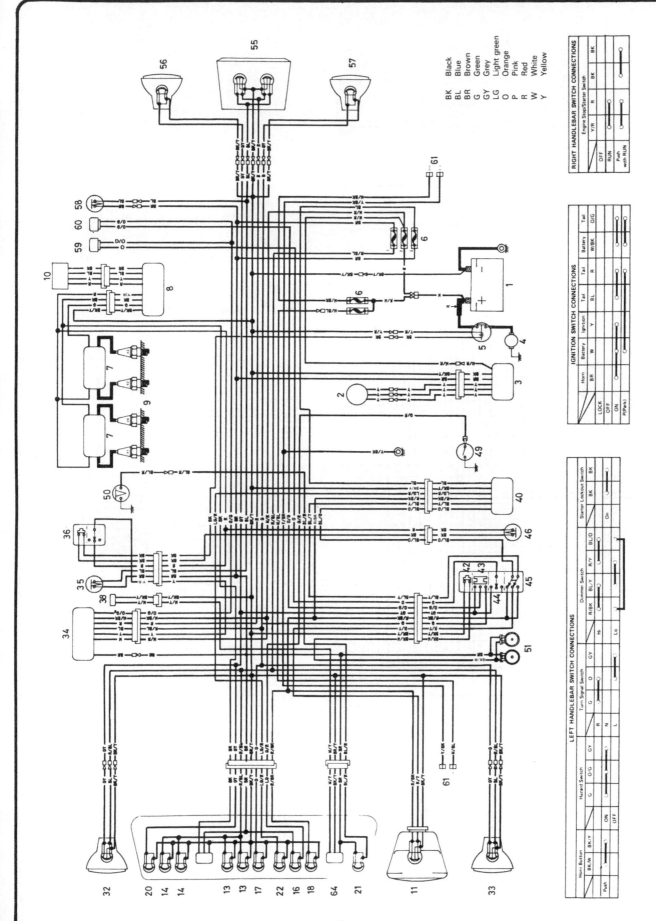

Wiring diagram – KZ750 L3 US model

Refer to page 199 for component key

BK	Black		
BL	Blue		
BR	Brown		
G	Green		
GY	Grey		
LG	Light green		
O	Orange		
P	Pink		
R	Red		
W	White		
Y	Yellow		

216

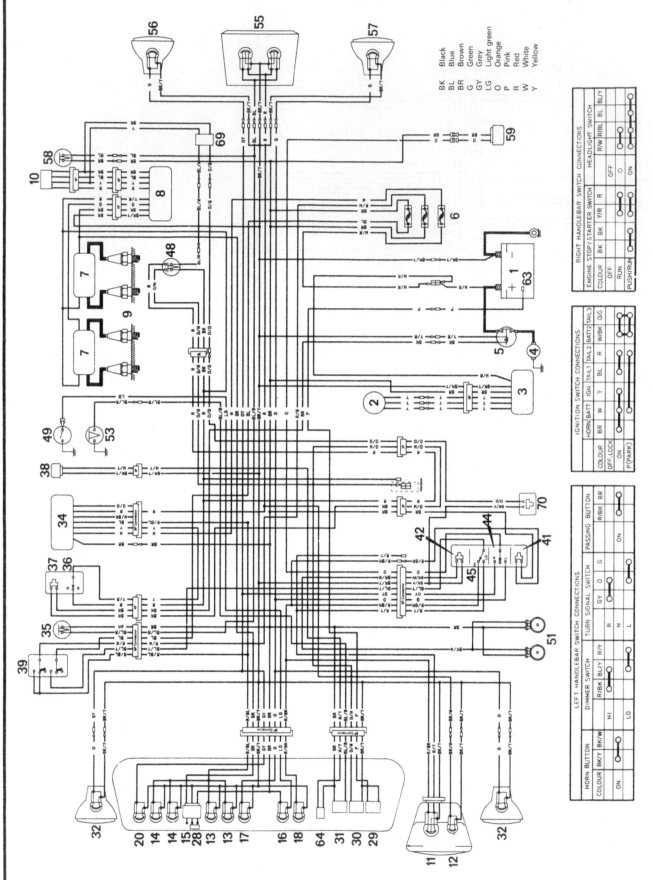

Wiring diagram – Z750 R1 GPz UK model

Refer to page 199 for component key

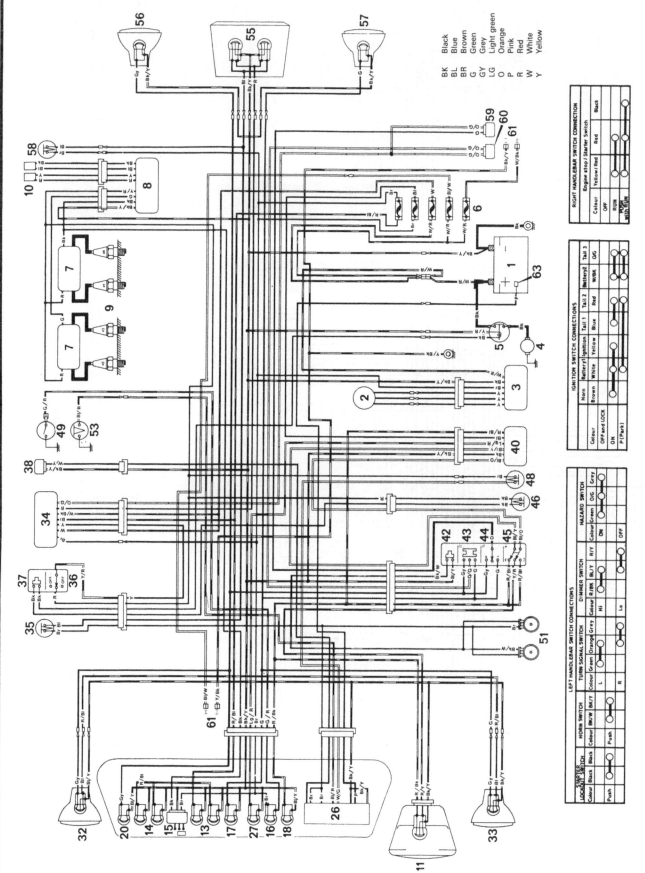

Wiring diagram – KZ750 R1 GPz US model

Refer to page 199 for component key

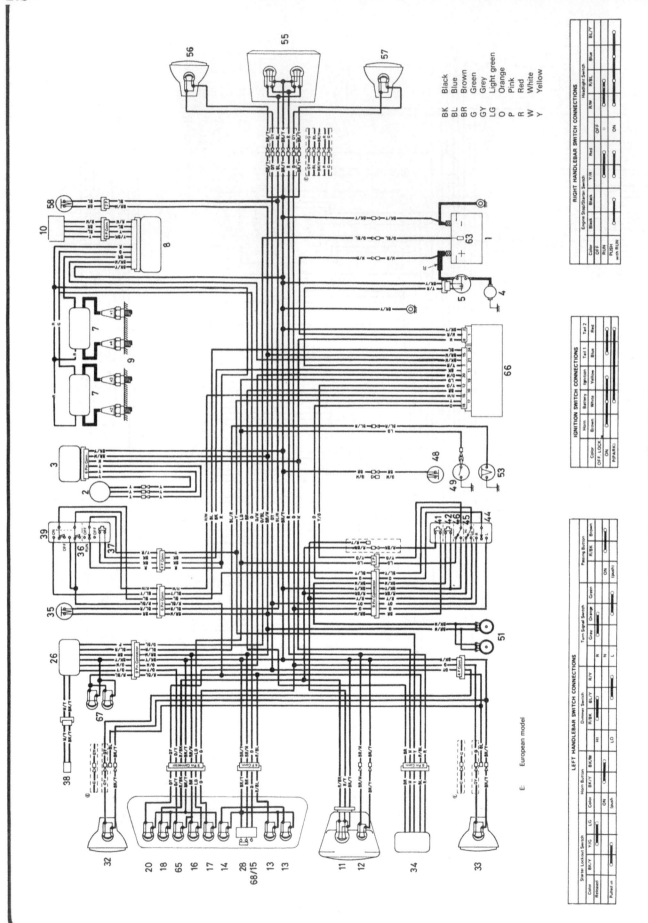

Wiring diagram – ZX750 A1 GPz UK model

Refer to page 199 for component key

	Black
BK	Black
BL	Blue
BR	Brown
G	Green
GY	Grey
LG	Light green
O	Orange
P	Pink
R	Red
W	White
Y	Yellow

E: European model

RIGHT HANDLEBAR SWITCH CONNECTIONS

Headlight Switch

Color	R/W	R/BL	Blue	BL/Y
OFF				
ON				

Engine Stop/Starter Switch

Color	Black	Y/R	Red
OFF			
RUN			
PUSH with RUN			

IGNITION SWITCH CONNECTIONS

Color	Horn	Battery	Ignition	Tail 1	Tail 2
	Brown	White	Yellow	Blue	Red
OFF LOCK					
ON					
P(PARK)					

LEFT HANDLEBAR SWITCH CONNECTIONS

Passing Button

	R/BK	Brown
ON (push)		

Turn Signal Switch

	Gray	Orange	Green
R			
N			
L			

Dimmer Switch

	R/Y	R/BK	BL/Y
HI			
LO			

Horn Button

Color	BK/Y		
ON (push)			

Starter Lockout Switch

Color	BK/Y	Y/G	LG
Released			
Pulled in			

219

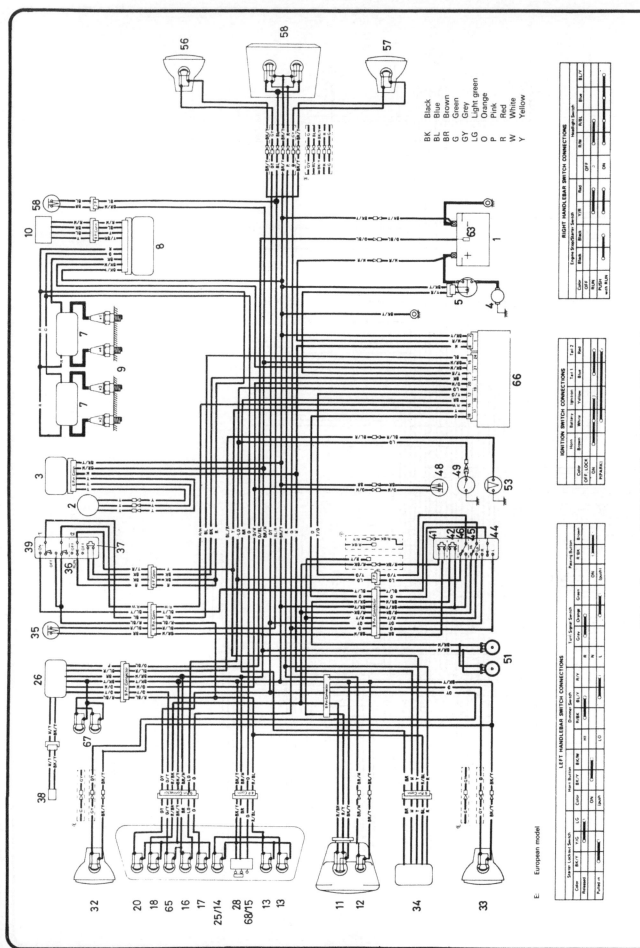

Wiring diagram – ZX750 A2, A3 and A5 GPz UK models

Refer to page 199 for component key

E: European model

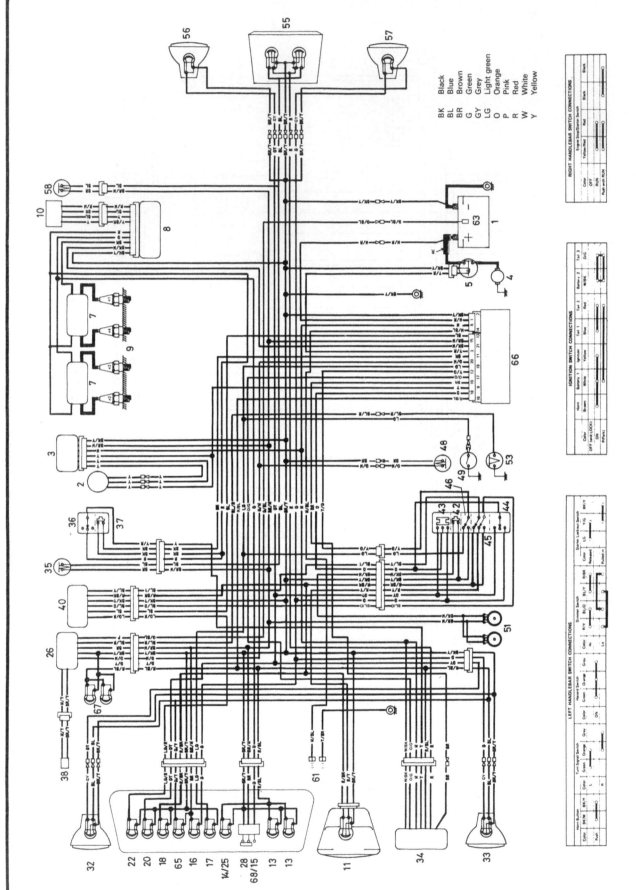

Wiring diagram – ZX750 A1 GPz US model

Refer to page 199 for component key

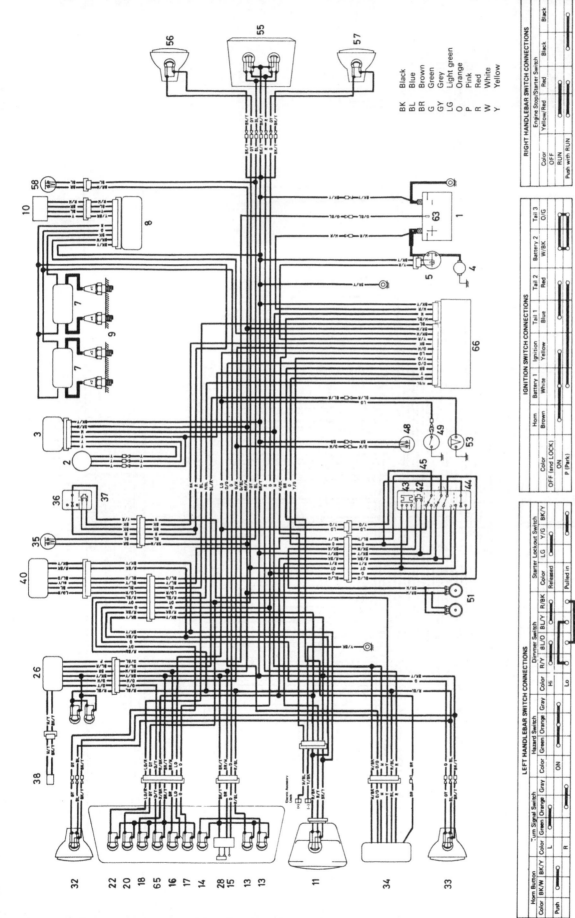

Wiring diagram – ZX750 A2 and A3 GPz US models

Refer to page 199 for component key

BK	Black
BL	Blue
BR	Brown
G	Green
GY	Grey
LG	Light green
O	Orange
P	Pink
R	Red
W	White
Y	Yellow

RIGHT HANDLEBAR SWITCH CONNECTIONS

Engine Stop/Starter Switch

Color	Yellow/Red	Black	Black
OFF			
RUN	Red		
Push with RUN			

IGNITION SWITCH CONNECTIONS

Color	Horn Brown	Battery 1 White	Ignition Yellow	Tail 1 Blue	Tail 2 Red	Battery 2 W/BK	Tail 3 O/G
OFF (and LOCK)							
ON							
P (Park)							

LEFT HANDLEBAR SWITCH CONNECTIONS

Horn Button, Turn Signal Switch, Hazard Switch, Dimmer Switch, Starter Lockout Switch

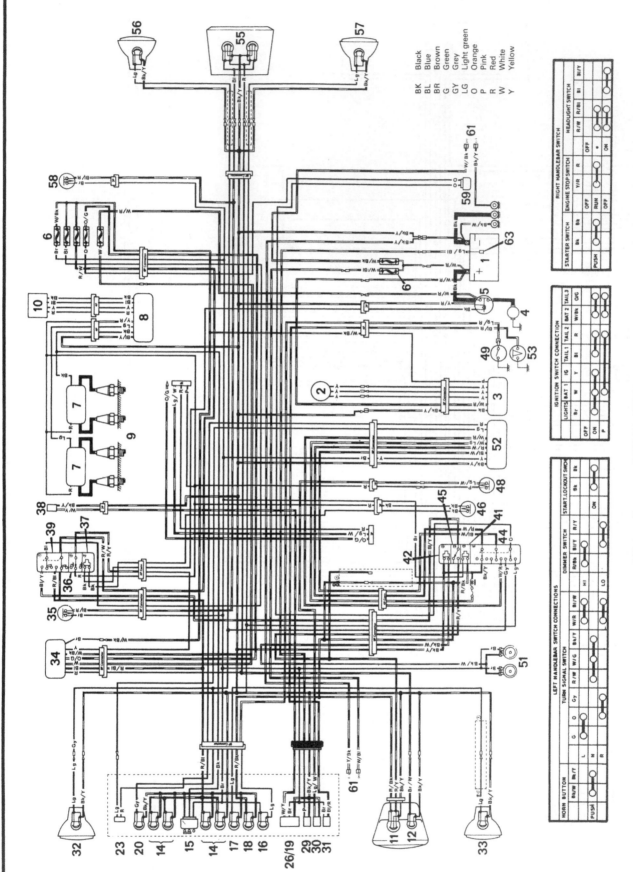

Wiring diagram – Z750 P1 GT750 UK model

Refer to page 199 for component key

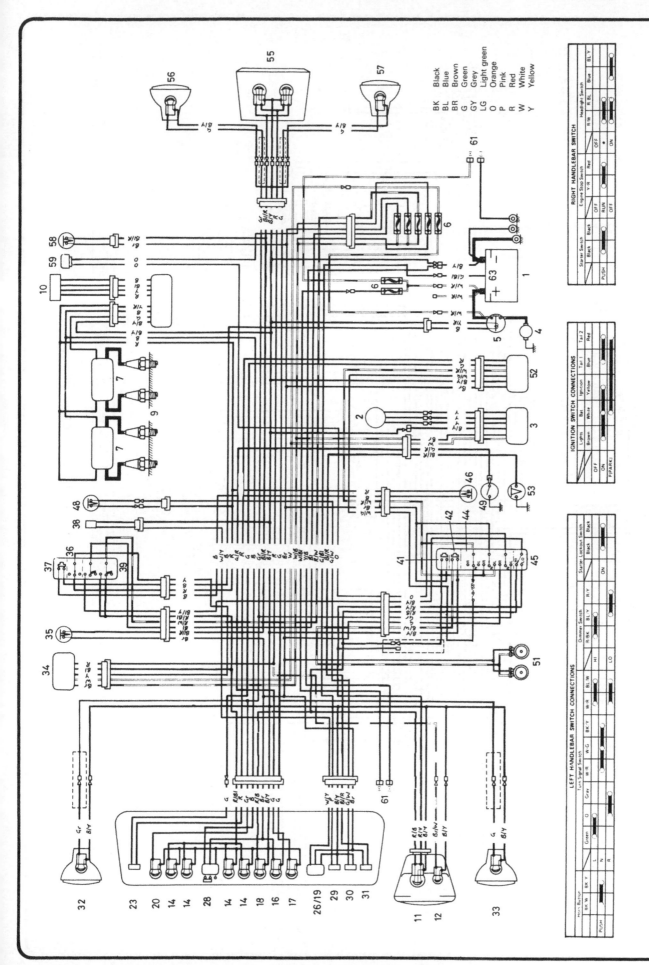

Wiring diagram – Z750 P2 GT750 UK model

Refer to page 199 for component key

BK Black
BL Blue
BR Brown
G   Green
GY Grey
LG Light green
O   Orange
P   Pink
R   Red
W   White
Y   Yellow

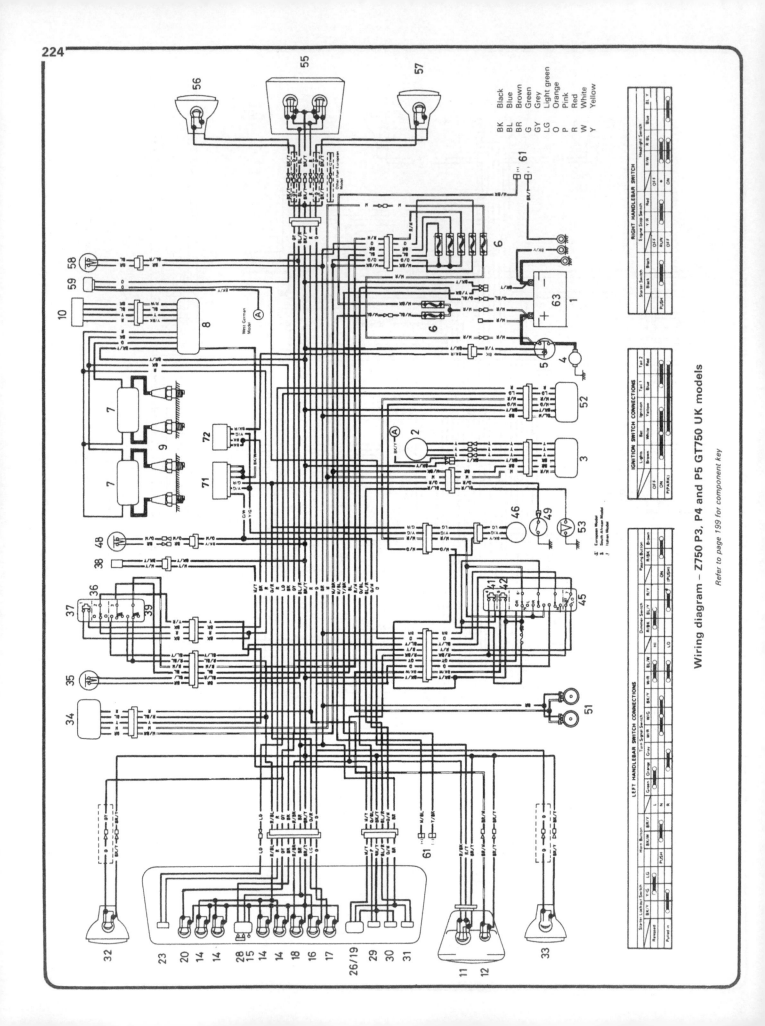

Wiring diagram – Z750 P3, P4 and P5 GT750 UK models

Refer to page 199 for component key

BK Black
BL Blue
BR Brown
G   Green
GY Grey
LG Light green
O   Orange
P   Pink
R   Red
W   White
Y   Yellow

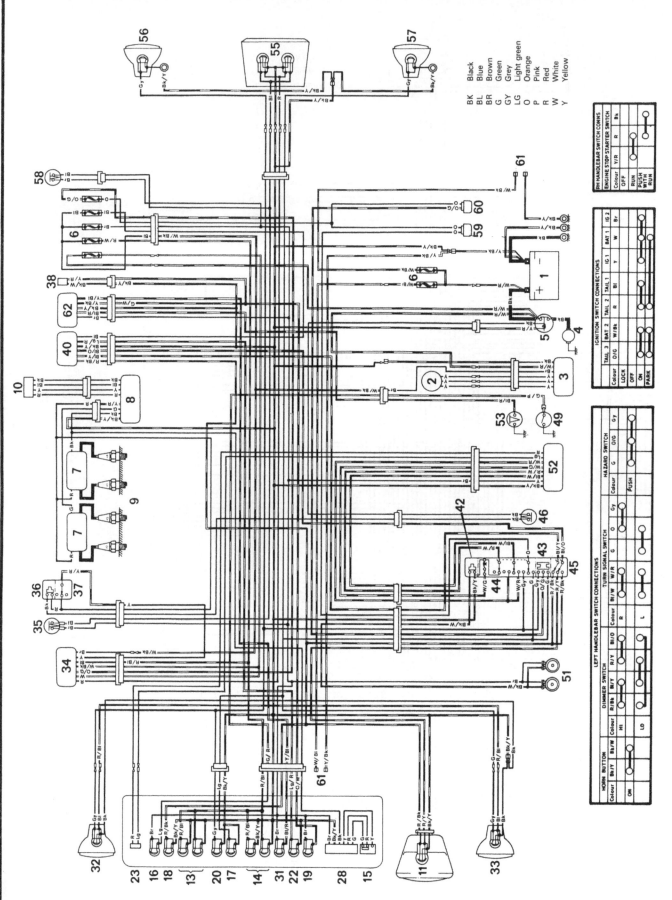

Wiring diagram – KZ750 N1 Spectre US model

Refer to page 199 for component key

225

Wiring diagram – KZ750 N2 Spectre US model

Refer to page 199 for component key

	Color			
BK	Black			
BL	Blue			
BR	Brown			
G	Green			
GY	Grey			
LG	Light green			
O	Orange			
P	Pink			
R	Red			
W	White			
Y	Yellow			

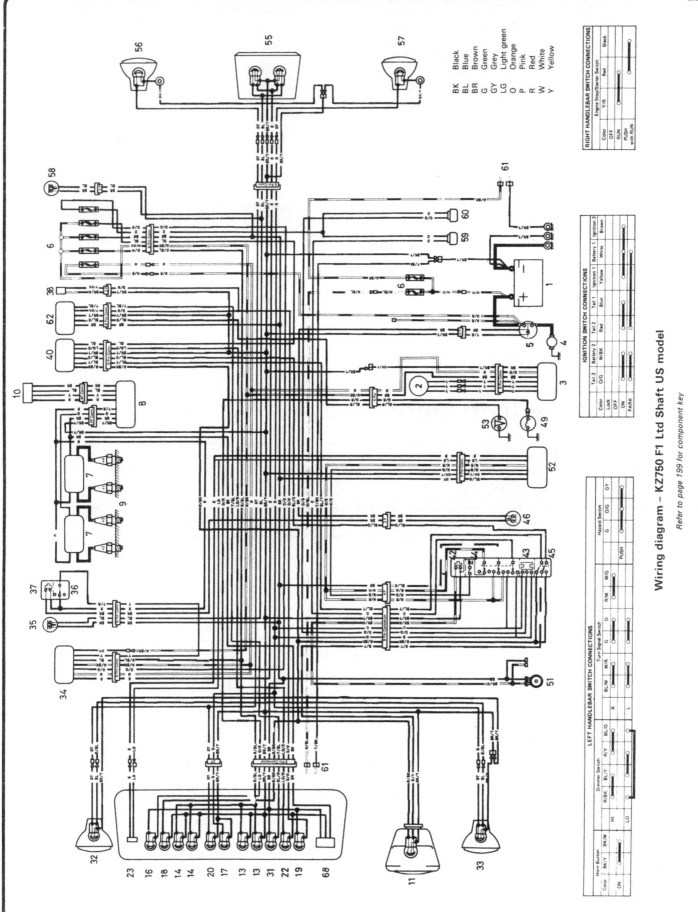

RIGHT HANDLEBAR SWITCH CONNECTIONS

	Engine Stop/Starter Switch		
Color	Y/R	Red	Black
OFF			
RUN			
PUSH with RUN			

IGNITION SWITCH CONNECTIONS

	Tail 3	Battery 2	Ignition 1	Tail 2	Tail 1	Ignition 1	Battery 1	Ignition 2
Color	O/G	W/BK	Red	Blue	Yellow	White	Brown	
Lock								
OFF								
ON								
PARK								

LEFT HANDLEBAR SWITCH CONNECTIONS

		Hazard Switch		
		G	O/G	GY
Color		G	O/G	GY
PUSH				

	Turn Signal Switch			
	BL/W	W/R	G	O
R/W				
R				
L				

	Dimmer Switch		
	R/Y	BL/Y	
R/BK			
HI			
LO			

	Horn Button	
Color	BK/Y	
ON		

Wiring diagram – KZ750 F1 Ltd Shaft US model

Refer to page 199 for component key

BK	Black
BL	Blue
BR	Brown
G	Green
GY	Grey
LG	Light green
O	Orange
P	Pink
R	Red
W	White
Y	Yellow

English/American terminology

Because this book has been written in England, British English component names, phrases and spellings have been used throughout. American English usage is quite often different and whereas normally no confusion should occur, a list of equivalent terminology is given below.

English	American	English	American
Air filter	Air cleaner	Number plate	License plate
Alignment (headlamp)	Aim	Output or layshaft	Countershaft
Allen screw/key	Socket screw/wrench	Panniers	Side cases
Anticlockwise	Counterclockwise	Paraffin	Kerosene
Bottom/top gear	Low/high gear	Petrol	Gasoline
Bottom/top yoke	Bottom/top triple clamp	Petrol/fuel tank	Gas tank
Bush	Bushing	Pinking	Pinging
Carburettor	Carburetor	Rear suspension unit	Rear shock absorber
Catch	Latch	Rocker cover	Valve cover
Circlip	Snap ring	Selector	Shifter
Clutch drum	Clutch housing	Self-locking pliers	Vise-grips
Dip switch	Dimmer switch	Side or parking lamp	Parking or auxiliary light
Disulphide	Disulfide	Side or prop stand	Kick stand
Dynamo	DC generator	Silencer	Muffler
Earth	Ground	Spanner	Wrench
End float	End play	Split pin	Cotter pin
Engineer's blue	Machinist's dye	Stanchion	Tube
Exhaust pipe	Header	Sulphuric	Sulfuric
Fault diagnosis	Trouble shooting	Sump	Oil pan
Float chamber	Float bowl	Swinging arm	Swingarm
Footrest	Footpeg	Tab washer	Lock washer
Fuel/petrol tap	Petcock	Top box	Trunk
Gaiter	Boot	Torch	Flashlight
Gearbox	Transmission	Two/four stroke	Two/four cycle
Gearchange	Shift	Tyre	Tire
Gudgeon pin	Wrist/piston pin	Valve collar	Valve retainer
Indicator	Turn signal	Valve collets	Valve cotters
Inlet	Intake	Vice	Vise
Input shaft or mainshaft	Mainshaft	Wheel spindle	Axle
Kickstart	Kickstarter	White spirit	Stoddard solvent
Lower leg	Slider	Windscreen	Windshield
Mudguard	Fender		

Index